Natural Medicines

An Encyclopaedia of Complementary Healing Arts and Sciences

The Authors

Dr. Debasish Kundu a Ph.D in Spagyric Homeopathy from an American University is a renowned authority on Complementary and Alternative Medicine(CAM), he is a Visiting Professor of Khulna Homeopathic Medical College and Hospital, affiliated to Ministry of Health, Peoples Republic of Bangladesh, Vice President of American Nutritional Medical Association, CA, U.S.A., and International Homeopathic Medical Society, U.S.A., and Editor in Chief, International Journal of Homeopathy and Natural Medicines, Science Publishing Group, USA, Editor in Chief, The Internet Journal of Herbal and Plant Medicine, USA. He has also authored and edited numerous books and journals on CAM, Sexology and Homeotherapeutics.

Dr. Malik A.K. Awan is a leading personality of natural health movement in USA, he is a reknowned scholar in Psychology, Homeopathy/Holistic Medicine and Religious Studies. He is also a Doctor of Common Law and CEO/President of American Nutritional Medical Association, USA, International Holistic Medical Society, International Homeopathic Medical Society. He has edited many journals and authored many books and articles on Natural Medicine. He is a laurate, Albert Schweitzer Award for Medicine.

Natural Medicines

An Encyclopaedia of Complementary Healing Arts and Sciences

— *Third Thoroughly Revised Edition* —

– *Authors* –

Dr. Debasish Kundu

M.D.H., F.F.Hom, F.I.H.M.S., Ph.D. (U.S.A.), D.Sc. (Colombo)

Dr. Malik A.K. Awan

B.S., M.S. (Psychology), M.D. (A.M.), D.C.L., D.D. (U.S.A.), Ph.D. (London)

Kruger Brentt
P u b l i s h e r s

2 0 1 7

Kruger Brentt Publishers UK. LTD.
Company Number 9728962

Regd. Office: 68 St Margarets Road, Edgware, Middlesex HA8 9UU

Library of Congress Cataloging-in-Publication Data

Kundu, Debasish, author.
Natural medicines : an encyclopaedia of complementary healing arts and sciences / authors, Dr. Debasish Kundu, Dr. Malik A.K. Awan. -- Third thoroughly revised edition.
    pages cm
    Includes bibliographical references and index.
    ISBN 978-1-78715-002-7 (hardbound)

    1. Naturopathy. 2. Alternative medicine. I. Awan, Malik A. K., author. II. Title.

RZ440.K86 2017   DDC 615.535   23

For information on all our publications visit our website at http://krugerbrentt.com/

Foreword

From the first existence of human life on the planet that one of the primary concerns focused on health and wellness , and subsequent treatment of the disease .

According to World Health Organization (WHO), disease is defined as " a state of complete physical, mental, and social well-being and not merely the absence of disease or infirmity ."

According to the world organization, health is much more than not having an infection, inflammation , a physical trauma ... is a complete state of physical, mental and social well-being.

Since the appearance of allopathy , the focus of attention was always the disease and not the person who has the disease . The consequent specialization within allopathy with the respective compartmentalization treatment of certain organs / body systems , made conventional doctors look the patient as a disease in an organ or system and not as someone who has his health affected in a whole.

How many are those who search our help in complementary and alternative medicine consultation, and their complaints are "I'm not well"; but allopathic evaluation and diagnostic tests reveals nothing wrong according to pattern wich is considered normal. And these people are sick - do not have a complete state of physical, mental and social well-being!

Since the beginning of humanity, extensive range of sciences and methods of diagnosis and appropriate treatment of absence of physical, mental and social well-being were developed. Many of them obviously with social, cultural and religious influences from the area of the planet they lived. Some of these sciences resorted to fauna, flora, minerals and developed specific techniques for use in the restoration of the affected states of people.

Is there only one correct way to diagnose and treat all altered states of health? Is there any medicine infallible treating people?

There are several ways to treat a human being? We don't need to think hard to answer these questions. Ancient people made their diagnoses and treatments with ancestral knowledge and treatments that nature brought them.

The life expectancy of a human being lies in eighty years .... too little time to apprehend so much knowledge; because perhaps the ideal, complete and full way to approach a state of physical, mental and social imbalance.... is using all forms already known and developed by man. However, all of them have given us concrete evidence (and many with scientific studies to prove it) that several complementary and alternative medicines HEAL!

In Europe, when something new comes along, people usually are skeptical, suspicious. It is said that first it is strange, then guts up.In Portugal for example, after several decades is now accepted by the medical community and the public in general that hospitalized patients with cancer or neurosurgical pathologies, significantly improve after Reiki sessions ...

The Republic of Ireland, for nearly a decade that is already possible that admitted to hospital persons request evaluation and treatment by a homeopath, inside that same hospital ...

In India, homeopathy goes hand in hand with conventional medicine. And in many situations both complement each other in accordance with allopathic and homeopathic practitioners.

In remote regions of Mongolia, there is only the healer who makes the diagnosis by ancient wisdom and use its remedies made from local herbs.

And people not only improve, they HEAL!

Over time people will realize, accepting and believing in alternative therapies to allopathic medicine, not least for the victories that these alternative medicines have demonstrated over time.

Unlike western medicine, alternative and complementary medicines are not intend to suppress the disease, but to restore the balance of that body. And in balanced body, there is no disease. There is only a state of physical, mental and social well-being!!!

In this book the authors have compiled some of the many forms of diagnosis, known forms of treatment, to develop a state of full balance. This work started in the 80s, now sees an update and complementation in this third edition. To the authors, people devoted to study, research, development of natural medicine, my sincere congratulations and a very special thanks to this fabulous legacy.

Wish you all a great and exciting reading.

Dr. Rui Oliveira

DIP HOM, MSc HOM MED, MSc Sensation Method, NHDTP, Editorial Member of International Journal of Homeopathy & Natural Medicines

Preface to the Second Edition

Drugs and Operative methods have come and gone but the Natural healing methods have survived and flourished because they do not offer any side effects or after effects as modern medicines do.

Nature is regarded as the highest authority to relieve the ailing humanity. Since ancient times herbs and natural healing methods are being used by men. The World Health Organisation in their 1978 Alma Ata Declaration also recommended the Natural Medicines because they are not only effective but affordable to the third world countries as they are unusually cheap.

I am indebted to Prof.Dr.Malik A.Kaiyoom Awan, Ph.D., Who's Who in Natural Medicines for giving me the opportunity to revise and rewrite his admirable work, besides addition of some new chapters.

Dr. Debasish Kundu

Preface to the
Third Thoroughly Revised Edition

Hippocrates' golden rule for healers was to "first, do no harm". "Natural" doesn't always mean safe – after all, opium, belladonna and digitalis are all "natural" poisons. We often hear about the dangers of CAM therapies – but are they really so dangerous? What's being done to protect patients?

We hear from experts at the Committee on the Safety of Medicines about how they monitor the safety of herbal remedies, despite poor reporting systems for adverse effects, and the variable quality of products with the same name. What are the main dangers of CAM therapies – and how can people avoid them? The World Health Organisation Department of Drugs and Policy is so alarmed about the dangers for the public that they are now monitoring safety of CAM following several deaths. There have been calls for much clearer labelling of herbal remedies to allow patients to distinguish between licensed and unlicensed products. We hear practical advice from therapists about side effects, interactions with conventional drugs, and about the indirect dangers of homeopathy resulting in delayed diagnoses of, for example, breast cancer. But, as we know, CAM does not have a monopoly on delayed diagnoses, in the NHS in know, the so called advanced countries, they account for nearly a third of all complaints.

There's no shortage of people who'll tell you that CAM works for them. But anecdote is not the same as the sort of evidence that comes from testing new pharmaceuticals in rigorous clinical trials. And in today's cash-strapped NHS, value for money is crucial. How should we establish whether CAM therapies "work", and is the scientific trial the only reliable method?

A new research institute in the UK has been set up to throw some light on the effectiveness of CAM. We follow the progress of some examples of CAM therapy trials, shedding light on the challenges of this sort of "test". Despite a shortage of research funding, there is a growing body of rigorous, published, scientific evidence on CAM. Several new journals are now devoted exclusively to "alternative" medicine, remarkable one is International Journal of Homeopathy and Natural Medicines.

Is the popularity of Holistic Medicine simply a new age fashion for everything "natural"? Does its focus on the prevention of illness feed the health obsessions of the "worried well"? Or could it be a sign of our growing dissatisfaction with medical science which often can't provide a quick fix? What's the attraction?

Why do so many therapists and users across cultures, building up a picture of the range of reasons people have for adopting CAM. We ask why the western world has adopted so many ancient therapies traditionally associated with non-western cultures - from Chinese medicine to Ayurvedic or Shiatsu massage. Some however suggested that CAM's popularity is dictated by fashion, a renewed interest in the paranormal, and the growing numbers of "worried well" who, despite the longer and safer lives we now lead, are obsessed with their health. Whatever its popularity now, will CAM or Holistic Medicine still be around in 100 years time?

Well, an unique feature of this edition that we have included all major complementary healing arts and sciences known to mankind in it.

Dr. Debasish Kundu
Dr. Malik A. Kaiyoom Awan

Contents

1
Introduction

Birth of Diseases

According to the "Wonder Tales of Ancient Greece" in those early days men had lived in perfect happiness. They had never known pain or disease; sorrow or trouble of any kind. Prometheus, a person from earth when came to know that the Gods had held back the secrets of fire from men, which could make their livelihood easier and better, one day stole from heaven and brought to earth a piece of fire. He taught people how to cook, how to work in metals and clay, which they had never dreamed about.

But as a result of what Prometheus had done, Gods became very angry and decided to punish Prometheus, also to prevent men from becoming equal of the Gods. By the order of Zeus, the king of Heaven, Hephaestus, God of fire created a lovely human maiden, she was given the name 'Pandora'. Each of the Gods gave to her some Wonderful Gifts in order to make her attractive to mankind. Aphrodite gave her beauty, Hermes gave her quick wits and Apollo gave her a musical voice.

Hermes brought Pandora to earth, and Epimetheus, the brother of Prometheus got attracted to her beauty and married her. Gods blessed the couple and presented a mysterious box which was closely sealed and Pandora, had been told that she must not open it.

But soon Pandora was eager to know what was inside the box but her husband advised her not to open as told by the Gods. Sometime later when Epimetheus went out in the garden to gather flowers to present his lovely wife, then Pandora examined the beautiful pictures cared on the box and heard some strange voices from inside the box, 'Save us, Pandora? Open, Open set us free.' The voices were tempting and Pandora broke the seal to open the box and then flew out a great number of ugly winged creatures, some of them bit Pandora, she cried out in pain and terror and Epimetheus came to her rescue, the winged creatures also bit Epimetheus and then they flew out of the window and soon other cries of pain and terror were heard. The winged creatures were in fact the evils of pain, disease and sufferings, which Zeus, the king of heaven, had sent to earth.

It is strongly believed that the following 5 books are the real glorious books from Almighty, given by him to the different people at different times which are as under:

1. Vedas to ancient Indian philosophers.

2. Bible to Jesus Christ

3. Torait to Moses

4. Zaboor to David

5. Quraan to Mohammad.

It is recommended that every people has to pay respect in all other books beside his/her own book of religion.

The ALMIGHTY GOD according in Holy Quraan and the Bible old and the new testament created Adam who was made with the dirt and gave him life. After that Eve was created from the left rib of Adam. The both were given strict orders to live in the GARDEN OF HEAVEN, eat whatever they need, drink whatever they want to. Beside this the ALMIGHTY ALLAH (GOD) told ADAM AND EVES that every thing in the Garden of Paradise is allowed in you, except the fruit of one tree. The word of tree in Arabic is 'SHAJJAR.'

SHAJJAR IS CONSISTING OF Roots, Trunk, Branches and the leaves. The God says clear words in Holy Quraan 'DON'T GO NEAR THIS TREE (SHAJJAR).

Different Philosophers have different opinion on the word SHAJJAR. Some have said that it was a tree of apple, some said it was wheat and wine and some unanimously agreed to Grapes. But in our opinion the God asked them in a very respectable manner that you can live, you can eat, you can drink, you can sleep and do whatsover you can, but you are strongly forbidden not to touch the tree of Woman to produce evils of disease, pain or sufferings.

It is general rule or principle that if one commit any crime he is liable for the punishment under the existing normal rules of the society, therefore we should understand that whenever we do any act against the nature, the nature gives us the punishment for that. For example the ALMIGHTY GOD has said in the Holy Bible as well as the Holy Quraan repeatedly that do not eat the meat of Swine and Camel except the emergency of life. Where you think you are going to die and there is no alternative to eat or survive you can eat the Swine (Pig), otherwise no, but the human being instead of valid Instruction of God frequently eat and bingeing with Swine, so Allah has to give them punishment of mal-Nutrition and they become ill of Hypertension, Blood Pressure and many other problems like over weight, *etc.*

According to the Holy hooks, the Devil came and tell the Adam and Eve why they don't eat the fruit of that particular tree, so Adam and Eve became greedy and they had eaten the fruit, alter which they were naked in the Garden and they tried to hide themselves with leaves *etc.* In the mean-while the voice of the God came and said, you know that I strongly prohibited this act for you which you did and that now you be ready for the punishment of this act. I send you on the earth where you will have to live for a certain period of lime.

When Adam and Eves went to Physical intimacy with each other the HOT AND COLD MIXED, in one each other where the health started deteriorating. Religiously it can be said that the deteriorating conditions in man and the woman is outcome of sin which they have committed in the Garden of Heaven against the will of God and the God has given them the punishment of being sick, poor and working hard for their livelihood.

So when the Human being stepped on the earth he brought the basic three diseases with them which according to research and valid proof they can be called the mothers of all the diseases. Those are appended hereunder:

1. Psora
2. Sypllls
3. Sycosis.

According to the Homeopathic system of Treatment, any disease now inforce or before comes under one of each category. And an Homeopathic practitioner can not treat any symptom under any condition until he is able to find out the present stage of any chronic and/or acute diseases affiliated with one of the above mentioned three categories, because the Homoeopathic system fix the categories of medicine in accordance with the above principle for they can be more benefited towards the woes of the sufferers.

Any how very shortly we will express something about the NUTRITION: The food produces the Nutrition.

CLASSES OF FOOD: Proteins, starches and oils (fats).

NUTRITION: Produces energy and warmth to the body.

It comes from animals, vegetables or the air and water we absorb animal produces all meats, eggs, dairy food and proteins in their chemistry preliminary give the vital resistance to the body and in many cases the Nutrition properly intake exterminate the disease from the body.

The essence of the science of nutrition is that the body needs a full and constant supply of vital nutrients in the balance form. The Science of Nutrition is know how to keep the cells of the body properly fed, so the span of cell life can be at its peak.

2

Acupressure

Acupressure is an ancient art of healing believed by some people to be even older than acupuncture. It involves the use of the fingers (and in some cases, the toes) to press key points on the surface of the skin to stimulate the body's natural ability to heal itself. Pressing on these points relieves muscle tension, which promotes the circulation of blood and qi to aid in the healing process.

What's the Difference between Acupressure and Acupuncture?

Acupressure and acupuncture are actually quite alike. In fact, acupressure is sometimes referred to as "needleless acupuncture," because both forms of healing use the same points to achieve the desired results. The main difference between the systems is that an acupuncturist stimulates points by inserting needles, whereas an acupressurist stimulates the same points using finger pressure.

How Does Acupressure Work?

Like acupuncture, acupressure involves the stimulation of certain points on the body. Stimulating these points can trigger the release of endorphins, chemicals produced by the body that relieve pain. When endorphins are released, the pain is blocked, and the flow of blood and oxygen to the affected area is increased. This causes the muscles to relax and promotes healing.

In acupressure, as with most traditional Chinese medicine concepts, local symptoms are considered an expression of the whole body's condition. A person with a tension headache, for instance, may actually be suffering from pain or stress in the shoulder and neck. An acupressurist would focus not only on relieving pain and discomfort, but removing the source of that pain and discomfort, before it develops into a more serious condition.

Besides relieving pain, acupressure can help rebalance the body by lowering stress and tension levels and strengthening the immune system. Certain acupressure points can also relieve tension in the chest, enabling people to breathe more deeply and sleep better, and there has been anecdotal evidence that acupressure can help pregnant women have a quicker, less painful delivery.

Advantages and Limits of Acupressure

Acupressure has several advantages over most other forms of healing:

It is extremely cost-effective. No special equipment is required; the only items a person needs to perform acupressure effectively are their own fingers and/or toes.

It can be performed anywhere, at any time. A person doesn't need to practice in a particular setting or a particular time of the day to experience its effects.

It can be performed alone. Self-acupressure has been performed in Asia for thousands of years. In the United States, many people can learn acupressure from an experienced practitioner or health care provider. In addition, there are a wide range of books and videotapes on the subject so that people can learn how to self-perform acupressure in the comfort of their own homes.

It is very safe, as long as a person follows the instructions provided by a licensed health care professional. No drugs are involved with acupressure; hence, there is no opportunity for drug-related side-effects to occur.

It offers a great deal of benefits to the patient. Correctly performed, acupressure increases circulation, reduces tension and enables the body to relax. Reducing tension, in turn, strengthens the immune system and promotes wellness.

However, acupressure is not without its limits. Applying acupressure too abruptly, or using too much force during treatment, can lead to bruising and discomfort. Great care should be used when applying pressure to points on or near the abdomen, groin, armpits, or throat. Special care should also be taken by pregnant women or those with recently-formed scars, burns, infections or skin lesions. Patients should always consult with a licensed health care professional before using acupressure or any other form of health care.

3

Acupuncture

The Chinese healing art of acupuncture is one that can be dated back at least two thousand years. Some authorities maintain that acupuncture has been practiced in China for even four thousand years. Though its exact age is vague, what is certain is that up until the recent twentieth century, much of the population of the world was uninformed about acupuncture, its origins, and its capacity to promote and maintain good health. Even today in relatively "advanced" nations such as the United States there are many who hold acupuncture under the stereotype of a new or radical medicine, one which would almost always be a second choice after more familiar Western approaches to handling illness. Following a brief synopsis of the theory of acupuncture, the following text will, to a limited extent, elucidate the vast history of this ancient medicine and assert that it is neither new nor radical.

According to Chinese Embryology when two people come together for sex, there is an exchange and blending of sexual energies. When a man and a woman have sex and conception results, the sexual energy blends to form "Pre-Heaven Jing" in the newly conceived individual. Both the father and the mother supply Jing.

The developing embryo and fetus has no independent Jing of its own. It's totally dependent on the Pre-Heaven Jing supplied by the mother and father and on nourishment from the mother's Kidneys. One of the functions of Jing is it acts like a blueprint and master control. It turns things on an off during development.

1st Month

The newly conceived being is like a drop of essence (jing), like a pearl of dew on a leaf. The mother experiences morning sickness, dizziness, and a lack of appetite. Her pulse is floating and tight. She should eat nourishing and easily digestible cooked foods. She is allowed to eat some sour foods if cravings arise. Barley, which makes the fetus grow normally, is beneficial at this time. Avoid acupuncture on the Liver meridian during this time as the fetus is nourished through this meridian. Blood circulation of the mother is impaired so she should be careful not to over exert herself physically. She should avoid becoming anxious and fearful. The fetus feels pain when the mother is exposed to cold and fear when the mother is exposed to heat.

2ⁿᵈ Month

The fetus is now like a peach flower and is situated at Zhong ji – Ren 3 (just above the pubic hairline). The placenta is being formed at this time and the mother may experience dizziness, blurred vision, nausea, vomiting, lack of appetite and slight fatigue. The mother should not eat pungent/spicy foods and should avoid excessive sexual activity and physical work. Joint pain is quite common at this time. The fetus is now nourished through the Gallbladder channel and therefore should not be needled. This is also the time that the Original Qi of the fetus comes into being so the mother should stay relaxed, breath properly and meditate. Exposure to excessive cold may bring about miscarriage and exposure to excessive heat may make the fetus weak.

3ʳᵈ Month

The fetus is now like a cocoon. It begins to look like a silkworm, large at one end and small at the other. The mother experiences nausea and vomiting and should avoid extreme cold or heat. Included now with any treatment of the mothers disease should be that of 'calming the fetus'. The fetus's body shape and sex are still changing and are under the influence of external stimuli of the mother. If she wants a son she should practice shooting arrows, if she wants a daughter she should spend time with jewellery and pretty things. For a nice looking child she should spend time feasting her eyes on jade and for a good hearted child she should spend as much time as possible sitting quietly. During this month the fetus is nourished through the Heart channel and therefore should not be needled.

4ᵗʰ Month

The fetus and the placenta this month collect the Field of Elixir (Dan Tian). The mother should not eat Rabbit and should not be exposed to cold. During this time the fetus begins to absorb essence from the mother's Kidneys to form blood vessels. The mother should now eat rice, fish or wild goose, this makes the fetus's Qi and blood strong, its ears and eyes sensitive and bright, and its channels free from obstructions. In this month the fetus is nourished from the San Jiao meridian and should not be needled. The Yang organs are formed this month. The mother should keep herself calm and free from any emotional disturbance, and be moderate in her eating.

5ᵗʰ Month

It is in this month that the gender differentiation occurs. A strange test; if one stands behind the pregnant woman and calls her, if she turns to the left it is male; if she turns to the right it is a female. Also if she is carrying a boy she will prefer sour foods. At this time the fetus is firmly embedded in the uterus and is steady. It is the 5ᵗʰ month that the fetus begins to receive Essential Qi of the Heart from the mother. This is to establish temperament. It is therefore advisable to sleep long hours, bathe and change clothes often, stay away from strangers, wear enough clothes and get exposure to sunshine. The mother should eat wheat, beef and lamb. She should also balance sweet with sour foods and bitter with salty. The fetus is nourished this month through the Spleen meridian and should not be needled there. The limbs are formed. The mother should not eat too much or go hungry. She should not eat drying foods or have Moxibustion or hot compresses or water bottles applied.

6ᵗʰ Month

In this month the muscles and hair grow. If it is a boy his movements will be felt more on the left and if it is a girl the feeling will be more on the right. The fetus is now described as being like a fish in the mother's abdomen. During this month the fetus receives the mother's Lung Qi, which forms sinews and gives the child its Corporeal (physical) soul, great care should be taken with this in mind. The mother should take light exercise and not stay indoors all the time. She should go to the countryside to look at horses or dogs running around and should eat meat of wild animals. Doing this will make the fetus grow strong sinews, muscles, skin, and spine. The fetus is nourished through the Stomach meridian and should not be needled there. The mouth and eyes are formed and the mother should eat sweet foods but not in excess. The fetus will now move more which may lead to restlessness.

7ᵗʰ Month

In the seventh month the fetus grows big, the eyes are bright, the nose has Qi, the ears are open and the mouth can taste. At this time the mother is usually very healthy, but if there is any problem one should implement 'calming the fetus'. The fetus begins to receive Essential Qi from the mother's Liver to form bones. The mother should take enough exercise to promote the circulation of Qi and blood by flexing and extending her joints. Her bedroom should be dry. She should avoid cold foods and eat rice, which will nourish the fetus's bones and teeth. The Lung meridian nourishes the fetus this month and therefore should not be needled. The skin and hair develop, the mother should not shout or speak in loud tones or wear too thin clothes. She should also bathe less often.

8ᵗʰ Month

This month the Qi of the fetus becomes active and this may cause some irritability with a feeling of oppression in the chest, lack of appetite and all food may taste sweet or stale. Problems at this time should be checked by regulating the Stomach and 'calming the fetus'. The Essential Qi from the mother's Spleen now promotes the growth of the skin. At this time the mother should avoid emotional upsets and practice quiet breathing to maintain her Qi, which will promote a moist lustrous skin on the fetus. The Large Intestine meridian nourishes the fetus this month and therefore should not be needled. The mother should avoid over eating and getting angry.

9ᵗʰ Month

Now the fetus is very large, reaching into the hypochonrial regions. The mother still may feel oppression in the chest and a lack of appetite. This is the time when the fetus receives Essence from the mother. All the organs are formed. The mother should eat sweet foods, wear loose clothes and not live in dampness. The fetus is now nourished through the Kidney meridian and therefore should not be needled.

10ᵗʰ Month

The fetus is now ready to emerge. The limbs stretch as it naturally turns and starts its downward descent. All Yin organs are developed and the Yang organs are free from obstructions. The mother should concentrate her Qi on the lower Dan Tian below the umbilicus to promote the growth of the fetus's joints and mental faculties. The fetus is now nourished through the Bladder meridian and therefore should not be needled. 'When the melon is ripe it falls off the plant'. The mother goes into labor.

Basic to traditional Chinese medicine is the dualistic cosmic theory of the yin and the yang. The yang, the male principle, is active and light and is represented by the heavens; the yin, the female principle, is passive and dark and is represented by the earth. The human body, like matter in general, is madeup of five elements: wood, fire, earth, metal, and water. With these are associated other groups of five, such as the five planets, the five conditions of the atmosphere, the five colours, and the five tones. Health, character, and the success of all political and private ventures are determined by the preponderance, at the time, of the yin or the yang; and the great aim of ancient Chinese medicine is to control their proportions in the body.

One of the most important concepts of Chinese medicine is that of natural balance. From this idea of balance arises the fundamental theory of yin and yang. According to this theory, life takes place in the alternating rhythm of yin and yang.

Day gives way to night, night to day; a time of light and activity (Yang) is followed by darkness and rest (Yin). Flowers open and close, the moon waxes and wanes, the tides come in and go out; we wake and sleep, breathe in, breathe out. Yin/Yang is a constant, continual flow through which everything is expressed on the one hand and recharged on the other. They are an inseparable couple. Their proper relationship is health; a disturbance in this relationship is disease.

The paradoxical nature of yin and yang is further illustrated in an excerpt from the Huang Di Nei Jing, or "The Yellow Emperor's Canon of Internal Medicine," which is considered to be the best known and earliest of Chinese medical texts:

Yang has it's root in Yin

Yin has it's root in Yang

without Yin, Yang cannot arise

Without Yang, Yin cannot be born

Yin alone cannot arise; Yang alone cannot grow

Yin and Yang are divisible but inseparable.

The well-known symbol of the yin-yang further demonstrates that nothing is pure Yin or pure Yang; black and white embrace and intertwine in perfect symmetry, each side containing a small seed of its opposite. The conclusion drawn from this theory is that good health entails the balance and harmony of all that is yin and all that is yang within the body.

When such a proper balance of forces exists, the body has achieved a healthy circulation of the life force qi (roughly pronounced "chee"). In Chinese medicine it is theorized that the human body, as well as every other living thing, has a natural flow of qi throughout it. Qi is said to travel the body along channels called "meridians," of which there are mainly fourteen. Qi flows constantly up and down these pathways, and when the flow of qi is insufficient, unbalanced, or interrupted, yin and yang become unbalanced, and illness may occur. An understanding of the relationship between the body, yin and yang, and qi is necessary to understand the utility of acupuncture.

Acupuncture can be described as the insertion of very fine needles (sometimes in combination with electrical stimulus or with heat produced by burning specific herbs, called Moxibustion) into the skin at specific acupuncture points in order to influence the functioning of the body. Traditionally, there are 365 acupoints on the body, most of which have a specific energetic function. Some are the meeting of meridian pathways while others are junctions with an internal pathway of the meridian. Some points tend to move qi towards the interior of the body while others bring energy to the surface. The choice of acupuncture points varies from patient to patient and from treatment to treatment and relies on very careful diagnoses of different kinds. Diagnosis entails the observation of the body through looking, touching, smelling and listening. In diagnosis, detailed questions are asked about the history of the illness and about such things as the patient's taste, smell, and dreams. Conclusions are drawn from the quality of the voice, and note is made of the colour of the face and of the tongue. The most important part of the investigation, however, is the examination of the pulse. Wang Shu-ho, who wrote the "Pulse Classic," lived in the 3rd century BC, and innumerable commentaries were written on his work. The pulse is examined in several places, at different times, and with varying degrees of pressure. The operation may take as long as three hours. It is often the only examination made, and it is used both for diagnosis and for prognosis. Not only are the diseased organs ascertained but the time of death or recovery may be foretold.

One of the primary and fundamen tal diagnostic methods of traditional Chinese medicine is pulse taking, which is far more intricate than pulse taking in the West.

It has been said to take upwards of fifteen years to master this diagnostic art.

Today Acupuncture enjoys what may be its greatest popularity to date. It is important to note that this popularity, however, is a fairly recent achievement of the medicine. In the early 20th century, China, as the rest of Asia, experienced a flood of Europeans and American influence. As early as the late 1890's the European germ theories of Koch, List, and Pasteur were starting to arrive in China, marking the beginning of Western medicine in the Far East. By 1912, acupuncture was in precipitous decline, barely able to counter this growth of biomedicine. At the same time, traditional Chinese medicine had gained a small hold in Europe and North America but was far from accepted, and by the beginning of the First World War, the art of acupuncture was close to cultural extinction in China.

Only a handful of background information is required to understand this sharp decline of acupuncture. What Europeans first introduced to China was not medicine or culture; it was a narcotic drug called opium. Designed to profit the Westerners who organized it, the opium trade grew exponentially throughout the 19th century, having appalling effects on Chinese society. It was a social horror, and it very aptly illustrated the subhuman status in which many Westerners held the Chinese as well as the creations of Chinese culture, including acupuncture. To add to the trauma, a famine in 1878-1879 left 9 million people in unimaginably horrifying conditions. There was no public sanitation, and open sewers and garbage-laden streets were the standard urban scene. It was in this atmosphere of decline, loss of self-determination, and inconceivable human suffering that the Republic of China was formed in 1911. China quickly developed a desire to modernize, and its people began to turn to Western medicine. Given this background information, it is not difficult to understand why the beginning of the 20th century was a devastating time for acupuncture, theoretically as well as practically. In the fall of 1915, an order was issued demanding that medical, pharmacy, and veterinary students meet the qualifications established by Western nations. By 1929, registration of traditional doctors was ceased, and an announcement was made by the Ministry of Health proposing the abolishment of traditional medicine altogether.

Though in rapid decline, not everyone stood for the disappearance of traditional Chinese medicine. In the 1920's a group of traditional practitioners formed the zhong yi ("Chinese medicine") movement seeking to salvage their medicine. Through this movement, the group attempted to create a force that could resist the enroachment of xi yi, "Western medicine." Though unsuccessful at the time, this movement laid the foundation for the reappearance and modernization of traditional medicine following the Second World War.

The resurgence of acupuncture in China after the Second World War had a relatively simply understood cause, namely "the greatest and most intractable public health problem of any nation in the world," as stated by the United Nations Relief Organization. Among the first tasks of the People's Republic was coming up with a way to provide health care for a huge society. China found itself with less than 40,000 Western physicians and approximately 500,000 disorganized and crudely trained traditional practitioners attempting to serve a population of a half billion people. Thus, the clear and pragmatic reason for the post-war promotion of traditional medicine: there was no other choice.

A compromise had to be reached. Conservative politicians saw traditional medicine as an aspect of Chinese culture to preserve against the onslaught of Western enroachment. Meanwhile, political modernists saw it as, at best, "a collection of empirical tricks, some of which might be useful". With the help of a general feeling of preserving the essence of Chinese culture, traditional medicine was saved from abandonment or possible prohibition, but not without being scientifically modernized. And so, starting in the 1950's, traditional medical schools with standardized curricula were established to teach basic biomedical sciences, traditional pharmacotherapy, and acupuncture. It was then, too, that the zhong yi movement triumphantly re-emerged. For the first time in the 20th century, Chinese medicine had "a firm cultural position, a clear economic role, and a powerful political patron"

Efforts to integrate Western and Chinese medicines continued through the 1950's with limited success. Having neither money nor prestige, traditional practitioners, who were not themselves united, feared eventual elimination. Furthermore, "popularization and expansion had further diluted what had never been a homogenous system". Despite these drawbacks, the late 1950's did experience one of the most important developments of acupuncture, at least from the viewpoint of Western physicians. It is during this time that aggressive acupuncture techniques were developed and that acupuncture was first utilized as an anesthetic and analgesic during surgery.

When Westerners started flooding China after Nixon's visit, it was these techniques they were shown. They brought these back to the West, where they were popularized and became a primary focus for research. Another factor that greatly contributed to the westward migration of acupuncture appeared in the 1970's in the form of acupuncture programs sponsored by the World Health Organization. Physicians from many countries came to China to take advantage of these programs, and since the early 1980's many similar programs have been established for non-physician acupuncturists as well. Today, affiliations exist between many Western acupuncture schools and traditional medical schools in China, allowing students to travel abroad to partake in pre- and post-graduation programs. In the 1980's by the so-called "three roads" policy allowed for the individual development of traditional medicine, biomedicine,

Important Acupuncture Points with Functions

Point #	Location	Functions
Liver 3	2 cun up from the base of the big toe and the next toe	Source point for the liver; Part of Four Gates
Liver 5	5 cun above inside ankle bone	Breaks through blockages; Emotional breakthrough
Liver 8	2 cun up from the knee bend (inside)	Liver Go point; Treats all knee problems
Gallbladder 8	1.5 cun above the top of the ear	Extraordinary for migraines, headaches w/vomiting and cluster headaches
Gallbladder 9	1.5 cun above and 0.5 cun back from top of the ear	Extraordinary for migraines, headaches w/vomiting (strong for the vomiting) and cluster headaches
Gallbladder 12	Behind the bone that is behind the earlobe	
Gallbladder 20	Just below the base of the skull on meridian line	Pool of wind for GB; Flu point
Gallbladder 21	On the top of the Trapezius	Emotional release point; moves chi in a downward direction
Gallbladder 25	Tip of the last rib on the back	Main steroid release point; Prolongs effectiveness of treatment
Gallbladder 33	3 cun above the knee (outside)	Treats/Great for all knee problems
Gallbladder 34	Between the Tibia and Fibula at the knee (outside)	GB Go point;
Gallbladder 39	3 cun above ankle (outside)	Treats any pain that goes in a downward direction
Spleen 6	3 cun above the middle of the ankle bone (inside)	Taken with Sp 7 and Lv 5 to treat any inflammation or stagnation in the body (dual meridian vortex)
Spleen 7	6 cun above the middle of the ankle bone (inside)	The Leaky Valley; Taken with Sp 6 and Lv 5 to treat any inflammation or stagnation in the body (dual meridian vortex)
Spleen 9	Bend in the top of the Tibia (hook); inside hook inside knee, or notch at head of Tibia	
Spleen 15	4-4.5 cun outside navel, on nipple line (where nipple and navel lines intersect)	
Stomach 9	On Carotid pulse	Stomach meridian landmark (on neck)
Stomach 17	Center of nipple	Stomach meridian landmark (on chest)
Stomach 25	2 cun lateral to navel	

Contd...

Contd...

Point #	Location	Functions
Stomach 30	On Femoral pulse	Stomach meridian landmark (in groin area)
Stomach 35	4 corners of the kneecap	Tim takes with Gb 33
Stomach 36	1 cun down from bottom outside kneecap	Stomach Go point; Treats all pain from the waist up; Cardinal point for the Stomach; Tim takes with Gb 34
Stomach 38	5 cun down from bottom outside kneecap	Master point for all pain below the waist; taken with St 39 & 40 to create vortex
Stomach 40	1 cun behind 38 (does a jog)	Cardinal point for shoulder pain; taken with St 38 & 39 to create vortex
Stomach 41	Center of the top of the ankle (inside)	Important for ankles
Lung 5	0.5 cun inside elbow crease	Go point for the Lung; Cardinal point for lungs
Lung 9	On bend of wrist (outside)	Cardinal point for skin
Large Intestine 4	In webbing between thumb and pointer finger	"Joining of Valleys" infinite uses
Large Intestine 11	End of skin fold at elbow (outside)	Go point for the LI
Large Intestine 13	3 cun above end of skin fold at elbow (outside)	Main hormone balancing point; used with LI 14 to balance hormones
Large Intestine 15	Head of Humerus, outside shoulder	Important for shoulder and hip pain
Heart 8	Bend ring and pinky finger down on palm; point is between these two fingers	Sedative point; Very important for wrist pain
Small Intestine 3	Outside end of crease at base of the pinky finger	Important for pneumonia; Flu point; Very good for neck pain
Small Intestine 10	2 cun above the end of the armpit crease	Prolongs effectiveness of treatment
Conception Vessel 2	Top center of pubic bone	Landmark for CV (bottom of meridian)
Conception Vessel 5	3 cun up from the center of the pubic bone	
Conception Vessel 8	Center of navel	Landmark for CV;
Conception Vessel 12	4 cun up from center of the navel (half way between navel and sternum)	
Conception Vessel 16	Tip of the Xiphoid Process	
Conception Vessel 17	Center of Sternum between nipples	

and their combination, integrated medicine. Since then, more than 2000 Chinese physicians have graduated with a thorough training in both traditional and Western medicine.

"Siravedhana(Acupuncture) and Marma Chikitsa were prevalent and highly accepted therapies during RigVeda and AtharvaVeda and flourished during the vedic period.The details of treatment which Sushruta described in the Sushruta Samhita, the concept and method of Marma and Siravedhanam [acupuncture] are amazing, Acupuncture, is one of the components of age-old Shalya Chikitsa(Surgery) and is the Indian study of the Dhanwantari (Dhanwantari was the father of Ayurveda) School. Chinese literature of Acupuncture when decoded answers to it. In fact 24 channels (meridians) of Chinese Acupuncture are nothing else than Sushruta's 24 Dhamanis while points on channels are 700 Siras of Sushruta. In fact, this knowledge has been spread to the nearby countries around India mainly during the regime of 'Buddha' and the countries developed it in their own way.

In his writings Sushruta called the acupuncture points as marmas, although marmas also consist of one or more 'heads' called siras. Marmas are the overall area where trauma can be inflicted whereas siras are the pin points of these marmas.

Acupuncture is thus a medical science of Indian origin and that it was originated from China is a myth only".

It is evident that this medicinal art is indeed one of the oldest and most complex that exists, based on ideas and theories formulated over hundreds of years. Acupuncture has come a long way since its origins and has won an uphill battle against time, misunderstanding, and criticism to gain the great popularity that it enjoys today. Despite its successes, however, there are still many legislative and public opinion battles to be fought in terms of its acceptance and utilization as a modern medicine. It is hoped that within the next several years, acupuncture will break through the remaining political barriers and that the world will see practitioners of traditional Chinese medicine as primary care providers with licensing in all states, much the way chiropractors are today. Much of the friction arises from the western need to see hard scientific evidence of how acupuncture works. Advocates of the practice are optimistic that western science will eventually catch up to the complexities of acupuncture and provide a logical explanation for its consistent success. Traditional Chinese medicine offers a comprehensive, integrated and well-tested theory for its application in acupuncture, which is inexpensive, non-invasive, safe and effective, is surely a key form of treatment for the future.

4

Alexander Technique

Frederick Matthias Alexander (1869-1955) was a Tasmanian reciter who gradually developed a condition of voice hoarseness which impaired his ability to perform to the point where he felt that he no longer could practice his profession. He sought medical help, and his doctor suggested he should rest his voice completely for two weeks before giving an important recital. Alexander followed his advice, and there was an improvement in his condition. When he again started to recite his voice deteriorated a new, and in the end, he could not complete his performance.

Alexander thus postulated that the loss of voice was caused by something he did when attempting to recite, something his doctor agreed seemed plausible. What that something was, unfortunately, the doctor could not tell him. Based on this, Alexander set out to discover the causes for his hoarseness by himself.

After long observation of himself using several mirrors, Alexander noticed that whenever he attempted to recite, several habits were brought into play that he had not been aware of: he was pulling his head back, depressing his larynx and sucking in breath through his mouth.

Upon deciding not to do these things which appeared counter-productive for his reciting, Alexander made his first important discovery: he was not able to stop himself from doing them at the critical point when he started to recite. He also discovered that his misuse of the neck and head when reciting was no isolated behaviour – indeed his total pattern was uncoordinated, each part contributing to the malfunctioning of the others.

Through continued observation and experimentation, Alexander eventually realised that this inability to stop the undesirable reactions was due to the fact that, when wanting to recite, he was relying on his feeling as to what was "the right way" of doing this. Through prolonged habitual misuse, the kinaesthetic sense becomes untrustworthy and can therefore not be relied on.

Through consciously being aware of reactions, the possibility exists to say "no" to respond in a habitual way, and thus opening possibilities for new, conscious behaviour to take place that is not dependent on the habitual feeling of what is "right".

By this discovery Alexander, through work on himself, rid himself of his hoarseness, and also started to teach this technique to other people, since it by then had become apparent to Alexander that this mis-use that had been causing him so much trouble was also affecting everyone else. Within years, he was very well-renowned and eventually moved to London to teach his Technique.

The Alexander Technique teaches how to become aware of one's habitual reactions, and how to change them.

The Alexander Technique is not a cure or a treatment, it is a teaching through which the student can learn to change and improve his functioning. This has curative properties as a by-product.

The Alexander technique can be described as a technique for developing constructive awareness. Awareness of what is going on inside and outside ourselves - while it is going on, and helping us to improve the way we handle ourselves in any situation.

As awareness grows, we begin to notice many destructive habits embedded in the way we normally react to different stimuli.

These habits feel right and normal to us just because they are habitual. Like Alexander discovered, we are often doing something completely different than we think we are: our kinaesthetic sense- the sense of where we are, our balance, where our limbs are positioned in space - is faulty. The Alexander Technique is a method of self-discovery, which explores how we "use ourselves" in ordinary activity - how we walk, sit, stand, breathe, speak, react to stress. Learning about our own habits allows us to gradually free ourselves of our unconscious patterns that interfere with our natural coordination and spontaneity. The actor will move towards a state of psycho-physical well-being so important for imaginative growth and the development of his means of expression.

Mis-use is the Alexandrian word used to describe actions that are irrelevant or counter-productive to the aim one is trying to attain, this can be, for example, an unnecessary lifting of the shoulder when shaving, or a constant holding of breath when trying to perform a difficult task. All these habits draw energy from us and have a cumulative negative influence on the physical parts involved. Economy is a characteristic of good use.

The Alexander Technique aims to encourage all the body's processes to work more efficiently as an integrated, dynamic whole.

Start by sitting in a chair. Get comfortable. Do you find yourself slouching or sitting upright? Now stand up. How did you do it? Did you place your hands on your thighs or on the arms of the chair and force yourself up?

Now sit back down and try it this way. Sit upright in the chair and relax your head upward. Concentrate on relaxing your head and thinking about a line travelling through the top of your head. To stand up, without pushing off your thighs or knees, relax upward and forward. Simply lean forward, tilting that imaginary line so it continues to direct your body up and forward.

How did that feel? Was it easier than using the pushing off method? Did you feel less tension in your neck and back when relaxing upward?

Alexander found that the most useful change he could make was to mentally direct his neck to be free so that his head, followed by his body, could release in an upward direction – delicately, without any stiffening or undue effort. This is the guiding principle of all movements in Alexander technique, including walking and running. Next time you go for a walk or jog, try concentrating on letting that line from the top of your head direct you. Relax upward and tilt that line somewhat forward. Let your body follow and see if you feel more comfortable.

In the Alexander technique, there is no separation of "mental" from "physical", and thus it is said that it deals with the organism as a whole. This was formulated by Alexander long before it was fashionable to say such things. In order to improve one's functioning, one needs to do things differently, and in order to do differently it is necessary to be aware and to learn how to think differently.

Normally it is recommended to have between 25 and 30 lessons with as close interval as possible when new to the Alexander Technique. Since one is working against the habits of a lifetime it is sometimes said "The more, the merrier."

Many students also come to experience the Alexander Technique as a continuous learning experience and thus go on having occasional or regular lessons long after this initial period.

The Technique

Good use of the body means moving the body with maximum balance and coordination to all parts so that only the effort absolutely needed is expended. It's not about getting to the correct position and holding it there, it's an ever rejuvenating thought process which becomes the basis for your movement through the rest of your life.

Use this technique constantly and especially before and even during performing. It will help relax you and also release excess tension in the voice and body to allow for better vocal production and emotional flexibility.

The Basic Movement

Check in with your body.

Where do you feel tension?

Are you holding your breath?

Keeping your eyes open, look around the room.

Look to the right slowly, what do you notice?

What muscles are engaged to do this?

Are you holding your breath anywhere?

Look back to centre and then to the left.

What muscles are engaged to do this?

Are you holding your breath anywhere?

Slowly tilt your head back to look at the ceiling

What muscles are engaged to do this?

Are you holding your breath anywhere?

Slowly tilt your head forward and look down to the floor.

What muscles are engaged to do this?

Are you holding your breath anywhere?

The Alexander Technique

1. Never hold your breath.
2. Breathe in slowly and easily through the nose, and out naturally using the nose and mouth.
3. While maintaining this breathing pattern, imagine the following thoughts.
4. A string is coming out of the centre of my head and is being stretched straight upward.

Please Note: Upward is away from your torso and hips wherever you are in space, so if you're bending at the waist to the left, upward may mean imagining the stretch and release toward the left wall.

The use of breath is like a two way stretch or a rubber band being stretched.

As you slowly breathe in through the nose and the breath stretches and drops into your lower abs, imagine the above movement happening as you say the phrase above to yourself.

Repeat Section 2 using this breath and imaging technique. Notice if you have less tension and more efficient use of movement within your body.

According to Glenna Batson of Winston-Salem State University, North Carolina, USA, the exact mechanism by which Alexander Technique achieves its effects is unknown because Alexander taught in relative isolation from the scientific community, he did not substantiate his work with scientific theory. Sherrington, Magnus, Coghill, and others offered limited support for Alexander's top-down concept of motor control. Alexander's frequent dictum was, "the head leads, and the body follows," implying that upright balance largely is governed by a "head-neck motor system." Alexander believed that the head-neck relationship was particularly important for organizing efficient movement because of its sensory acuity. Researchers still are unravelling the complex role of head and neck postural reflexes and the regulation of multiple degrees of freedom for voluntary action. Substantial, though inconclusive, research has been conducted on the morphology and specialization of head-neck afferents and reflexes for head-neck-eye control during standing and walking. The muscle spindle afferents of the cervical longus colli and multifidi are densely distributed, although research on the role of these spinal afferents in postural support has been restricted to lumbar (low back) pain.

Alexander intuited that the primary control helps the human body suspend dynamically in gravity. Rather than a columnar model of the spine, (*i.e.*, vertebral bodies characterized as superincumbent weight of stacked bones), the spine is flexible, floating, and continually reorganizing within the dynamic constraints of the moment through neuromuscular and fascial mechanisms. Contemporary proponents have explained this concept of primary control using an engineering model of structural "tensegrity," or an energy medicine model of fascial plasticity. Engineer Buckminster Fuller popularized the concept of Tensegrity through his geodesic dome, a large structure in which the rigid compression members of the structure "float" within the soft tensile members, much like a tent. Applied to the human body, bones (compression members) "float" in an ever-changing web of tension (myofascial and neuromuscular tension).

5

Aromatherapy

Aromatherapy is an ancient healing art, which has been used for centuries.

It's also a rewarding and simple therapy, which you can experiment with yourself. It uses natural essences extracted from plants, trees, fruits and herbs to treat a wide range of conditions, from stress to sexual appetite.

If you've never tried aromatherapy before just start with a few oils and a carrier oil (almond or olive) for mixing. As you discover more oils, some will probably become your favourites. One of the easiest ways to experiment is to use oils that help you relax. Among the most effective are lavender, geranium and rose.

Lavender is one of nature's greatest relaxants. There are many varieties of lavender available; it's inexpensive and one of the most versatile oils around. For an instant calm fix, simply pop a few drops onto a tissue and inhale. Or, to create a calm atmosphere, add a few drops to a bowl of boiling water and place in a room. The lavender oil will slowly evaporate and fill the room.

If you have more time, immerse yourself in the magical mellowness of an aromatherapy bath. Run the water and then add about 8-10 drops of lavender oil, swish around with your hand and then jump in. Try and spend a least ten minutes lying there. You will emerge renewed.

When you need chilling big time, reach for the rose. It's the rarest and most expensive essential oil - two tons of fresh petals in full bloom are needed to make a kilogramme of rose oil but if you can afford some, it's a real investment. Rose has extraordinary de-stress properties. Use it sparingly (even a couple of drops are effective) in the bath or diluted for a massage.

Aromatherapy means "treatment using scents". It is a holistic treatment of caring for the body with pleasant smelling botanical oils such as rose, lemon, lavender and peppermint. The essential oils are added to the bath or massaged into the skin, inhaled directly or diffused to scent an entire room. Aromatherapy is used for the relief of pain, care for the skin, alleviate tension and fatigue and invigorate the entire body. Essential oils can affect the mood, alleviate fatigue, reduce anxiety and promote relaxation. When inhaled, they work on the brain and nervous system through stimulation of the olfactory nerves.

The essential oils are aromatic essences extracted from plants, flowers, trees, fruits, bark, grasses and seeds with distinctive therapeutic, psychological, and physiological properties, which improve and prevent illness. There are about 150 essential oils. Most of these oils have antiseptic properties; some are antiviral, anti-inflammatory, pain-relieving, antidepressant and expectorant. Other properties of the essential oils which are taken advantage of in aromatherapy are their stimulation, relaxation, digestion improvement, and diuretic properties. To get the maximum benefit from essential oils, it should be made from natural, pure raw materials. Synthetically made oils should be avoided.

Aromatherapy is one of the fastest growing fields in alternative medicine. It is widely used at home, clinics and hospitals for a variety of applications such as pain relief for women in labour pain, relieving pain caused by the side effects of the chemotherapy undergone by the cancer patients, and rehabilitation of cardiac patients.

Aromatherapy is already slowly getting into the mainstream. In Japan, engineers are incorporating aroma systems into new buildings. In one such application, the scent of lavender and rosemary is pumped into the customer area to calm down the waiting customers, while the perfumes from lemon and eucalyptus are used in the bank teller counters to keep the staff alert.

Essential oils are very effective in the treatment of stress and in the encouragement of relaxation.

Scents of vanilla, orange blossom, rose, chamomile and lavender {and other floral fragrances) have a noticeable calming effect on the way you feel.

Scents of lavender, sandalwood and nutmeg help you shrug off the ill-effects of stress.

Patchouli oil helps eliminate anxiety and lifts the mood (it is also said to be an aphrodisiac.).

Clary sage is a sedative and tonic.

Ylang-ylang is euphoric, regulator, sedative, and tonic. Use it in moderation. It can cause headaches in some people.

Lavender is probably the most useful of them all for relaxation. It is a sedative and a tonic. It helps you to relax, at the same time, it also eases aches and pains, such as headache. Sprinkle four drops on a tissue and inhale deeply for sudden stress. Our ancestors carried with them handkerchiefs perfumed with lavender water or rose water.

The most effective methods for using essential oils to help calm the mind and relax the body include massage, baths, and vaporization. You can use them singly or in combination

Aromatherapy had been around for 6000 years or more. The Greeks, Romans, and ancient Egyptians all used aromatherapy oils. The Egyptian physician Imhotep recommended fragrant oils for bathing, massage, and for embalming their dead nearly 6000 years ago. Imhotep is the Egyptian god of medicine and healing. Hippocrates, the father of modern medicine, used aromatherapy baths and scented massage. He used aromatic fumigations to rid Athens of the plague.

The modern era of aromatherapy is dawned in 1930 when the French chemist Rene Maurice Gattefosse coined the term aromatherapy for the therapeutic use of essential oils. He was fascinated by the benefits of lavender oil in healing his burned hand without leaving any scars. He started investigating the effect of other essential oils for healing and for their psychotherapeutic benefits.

During world war II, the French army surgeon Dr. Jean Valnet used essential oils as antiseptics. Later, Madame Marguerite Maury elevated aromatherapy as a holistic therapy. She started prescribing essential oils as remedy for her patients. She is also credited with the modern use of essential oils in massage.

Aromatherapy works the best when it works on the mind and body simultaneously.

Essential oils stimulates the powerful sense of smell. It is known that odours we smell have a significant impact on how we feel. In dealing with patients who have lost the sense of smell, doctors have found that a life without fragrance can lead to high incidence of psychiatric problems such as anxiety and depression. We have the capability to distinguish 10,000 different smells. It is believed that smells enter through cilia (the fine hairs lining the nose) to the limbic system, the part of the brain that controls our moods, emotions, memory and learning.

Studies with brain wave frequency have shown that smelling lavender increases alpha waves in the back of the head, which are associated with relaxation. Fragrance of Jasmine increases beta waves in the front of the head, which are associated with a more alert state.

Scientific studies have also shown that essential oils contain chemical components that can exert specific effects on the mind and body. Their chemistry is complex, but generally includes alcohols, esters, ketones, aldehydes, and terpenes. The effect of these chemical components are summarized in the accompanying table given on next page.

Each essential oil contains as much as 100 chemical components, which together exert a strong effect on the whole person. Depending on which component is predominating in an oil, the oils act differently. For example, some oils are relaxing, some soothes you down, some relieves your pain, *etc.* Then there are oils such as lemon and lavender, which adapt to what your body needs, and adapt to that situation. (These are called "adaptogenic"). The mechanism in which these essential oils act on us is not very well understood. What is understood is that they affect our mind and emotions. They leave no harmful residues. They enter into the body either by absorption or inhalation.

A fragrance company in Japan conducted studies to determine the effects of smell on people. They have pumped different fragrances in an area where a number of keyboard entry operators were stationed and monitored the number of mistakes made as a function of the smell in the air. The results were as follows: When exposed to lavender oil fragrance (a relaxant), the keyboard typing errors dropped 20 percent. When exposed to jasmine (an uplifting fragrance), the errors dropped 33 percent.

Flower Essences and their Corresponding Chakras

7th Energy Center-Chakra

Flower: Saguaro (*Cereus giganteus*): Guardian.

Table

Name of the Chemical Component	Properties of the Chemical Component	Essential Oils that Contain the Chemical Component
Aldehydes	Anti-inflammatory,calming, sedative and anti-viral.	Characteristic lemon-like smell, such as lemon grass, lemon balm, citronella, eucalyptus
Alcohols	Bactericidal (kills bacteria), stimulant, energizing, vitalizing, antiviral, diuretic. Our pancreas produce 32 kinds of alcohol for use in human metabolism.	Rose, petitgrain, rosewood, peppermint, myrtle, tea tree, sandalwood, patchouli, and ginger
Phenols	Strongly bactericidal, tonic, stimulates immune system, invigorating, warming. Can produce slight liver toxicity if taken high doses for extended periods of time. Used in lip balms and cough drops.	Clove, cinnamon, thyme, oregano, savory, cumin.
Cetone (Ketones)	Wound healing, mucolytic (eases the secretion of mucous), stimulates new cell growth. used as a nail polish.	Camphor, rosemary, sage, eucalyptus globulus and hyssop
Terpenes	Very stimulating, potential skin irritants, anti-viral properties.	Lemon, orange, bergamot, black pepper, pine oils, nut meg and angelica.
Sesquiterpenes	Anti-phlogistic (moves fluids), anti-inflammatory, sedative, anti-viral, anti-carcinogenic, bacteriostatic and immune stimulant	Blue chamomiles, immortelle, tansy, yarrow and tagetes.
Esters	Anti-fungal, sedative, calming, spasmolytic, fungicidal, anti-inflammatory.	Roman chamomile, lavender, clary sage, petitgrain, bergamot.
Lactones (part of ester group)	Anti-inflammatory, mucolitic	Arnica, elecampane
Ethers	Harmonizing to the nervous system. antiseptic, stimulant, expectorant (increases secretions), spasmolytic, and diuretic.	Cinnamon, clove, anise, basil, tarragon, parsley, and sassafras.

Saguaro promotes a cleansing effect and releases stagnation and tension at all levels throughout our body. Helps us to restore, stretch, and expand beyond any self-imposed limitations, Guides us toward expanding awareness and embracing new ways of being.

Location: Crown Experience:Understanding/Blissfulness Element: Magnetum.

Quality: Awareness: Colour: White/Gold.

6th Energy Center-Chakra

Flower: Aster (*Machaeranthera tephrodes*)/Illumination.

Aster offers spiritual balance and upliftment through vision and wisdom, opening our ability to "see deeper."like a shining star,Aster gives us illumination and insight, helping us to accept our paths of spiritual evolution and whole life living, Location: Brow Experience : Intuition/Insight Element: Radium.

Quality: Inner Guidance Colour: Purple/Violet.

5th Energy Center-Chakra

Flower: Desert Larkspur (*Delphinium scaposuam*)/Communication.

Desert Larkspur gives us guidance toward higher spiritual truths and ways of living.

Offers gracefulness in communication, self-expression, and ease in communicating our needs, especially during transitional life changes and shifting life cycles.

Location: Throat Exprience: Expression/Imagination Element: Ether.

Quality: Communication Colour: Blue.

4th Energy Center-Chakra

Flower: Wild Rose (*Rosa arizonaca*)/Love.

Wild Rose is especially beneficial for those who show indifference, lack of interest,unconcern, unresponsiveness, have little or no emotion, and seem "spiritless". Wild Rose essence helps us to enliven and restore a vital sense of living and being, compassion and love.

Location: Heart Experience: Compassion/Forgiveness Element: Air, Quality:Love Colour: Pink/Green.

3rd Energy Centre-Chakra

Flower: Sunflower (*Hehanthus annus*)/Empowerment.

Sunflower restores youth and innocence, fun and play, liveliness and pleasure, bringing life to the child within us and directing us toward purpose and positive thought, joy and humor. Gives us the ability to think and reason, to gather strength and power from deep within.

Location: Solar Plexus Experience: Mental/Purpose Element: Flower.

Quality: Empowerment, Colour: Yellow.

2ⁿᵈ Energy Centre-Chakra

Flower: Pomegranate (*Punica granatum*)/Abundance.

Pomegranate stirs passion and acceptance of creative life force energy, Empowers intuition and receptivity, bringing an abundance of positive inner and outer resources, Helps us to understand how our emotional/mental states affect who we are, the way we pro-create, and how we relate to and experience our sexuality.

Location: Spleen/Pelvic Region Experience : Sexuality/Procreation Element :Water.

Quality : Emotional Health Colour: Orange.

1ˢᵗ Energy Centre-Chakra

Flower: Indian Paintbrush (*Castilleja chromosa*)/Creativity.

Indian Paintbrush encourages us to take time to be still and relax into silence. Stirs creative, inspirational, passionate, visionary and artistic self-expression, while feeling grounded and connected to our roots, Encourages us to find the needed resources that will meet our survival needs so we can live a fuller more creative life.

Location: Root/Base of the Tailbone Experience: Survival/Security Element:

Earth Quality: Physical Well being, Colour: Red.

Aromatherapy is particularly effective for stress, anxiety, and psychosomatic induced problems, muscular and rheumatic pains, digestive disorders and women's problems, such as PMS, menopausal complaints and postnatal depression. The human sensory steams when exposed to lemon fragrance (a sharp, refreshing stimulant), the mistakes fell by a whopping 54 per cent! Below is a summary of the results from clinical studies:

Behavior

Considerable evidence exists that fragrant compounds have a profound effect on our mind and behavior. Animal studies have found that hyper excited mice (as a result of consuming a large quantity of caffeine) was calmed by the aroma of lavender, sandalwood, and other oils sprayed into their cages. The same mice were found to become very irritable when exposed to the aroma of orange terpines, thymol, and some other substances. These oils were all detected in their bloodstream after about an hour.

Sleep

In a study reported in the British Medical Journal Lancet, elderly patients slept "like babies" when a lavender aroma was wafted into their bedrooms at night. These patients had complained of difficulty falling asleep and had to take sleeping pills to get sleep prior to the aromatherapy.

Postpartum Discomfort

In a double blind study, 635 women applied lavender oil to their perineal area (part of the body between the vagina and the rectum) after child birth. The women reported a distinct improvement between the third and fifth day. (The discomfort is the worst during this time for patients in the control group).

Colds

It has been well established that chicken soup is good for cold (both historically and scientifically). Studies were conducted to find out whether the effect was due to the action of the hot steam on the lining of the nostrils or whether the aroma of the chicken soup has anything to do with it. The results indicated that chicken soup was more effective than the steam indicating the effectiveness of the aroma.

Stress

In a study conducted at the Memorial Sloan-Kettering Hospital in New York, patients undergoing magnetic resonance imaging (MRI) reported 63 per cent less claustrophobic after getting exposed to the aroma of vanilla. There was no change in their heart rate. Obviously, the aroma reduced their anxiety probably by the pleasant memories evoked by the vanilla aroma or by some other physiological response.

In another study, 122 patients who were in an intensive care unit, reported feeling much better when aromatherapy was administered with the oil of lavender (compared to when they were simply given a massage or allowed to rest.) No changes in the patients who were given aromatherapy was observed in the blood pressure, respiration, or heart rate. As we mentioned before, Japanese have reported less mistakes by keypunch operators when exposed to fragrance.

Male Sexual Response

Circulation to the male sexual organ was found to have improved substantially by treatment with licorice or lavender with pumpkin pie. Doughnut with black licorice was also very effective. Men who are considered the most sexually active responded well to lavender, cola, and oriental spice; older males preferred the fragrance of vanilla. These studies prove that aromatherapy is an effective therapy for the treatment of male impotence from the blood vessel disease or from psychogenic factors.

Aromatherapy

Clary Sage

Properties

Warming, soothing, aphrodisiac, anticonvulsive, antidepressant, antiphlogistic, antiseptic, antispasmodic, aphrodisiac, astringent, bactericidal, carminative, cicatrisant, deodorant, digestive, emmenagogue, hypotensive, nervine, regulator (or seborrhoea), tonic, uterine.

Uses:

Menstrual problems, depression, anxiety, high blood pressure, muscular aches and pains, asthma, throat infections, whooping cough, colic, cramp, dyspepsia, flatulence, acne, boils, dandruff, hair loss, inflamed conditions in skin, oily skin and hair, ulcers, wrinkles, the genito-urinary system disorders such as menorrhoea, labor pain, dysmenorrhoea, leucorrhoea and nervous system is orders such as frigidity, impotence, migraine, nervous tension and stress related disorders.

Eucalyptus

Properties

Antiseptic, analgesic, anitneuralgic, antirheumatic, antispasmodic, antiviral, balsamic, cicatrisant, decongestant, deodorant, depurative, diuretic, expectorant, febrifuge, hypoglycaemic, parasiticide, prophylactic, rubefacient, stimulant, vermifuge, vulnery.

Uses

Muscular aches and pains, poor circulation, rheumatoid arthritis, sprains, asthma, bronchitis, catarrh, coughs, sinusitis, throat infections, cystitis, leucorrhoea, chicken pox, colds, epidemics, flu, measles, nervous system disorders such as debility, headaches, neuralgia and skin disorders such as burns, blisters, cuts, herpes, insect bites, insect repellent, lice, skin infections, wounds.

Geranium

Properties

Soothing, refreshing, relaxing, antidepressant, astringent, antihaemorrhagic, anti-inflammatory, antiseptic, cicatrisant, deodorant, diuretic, fungicidal, haemostatic, stimulant, tonic, vulnerary.

Uses

PMT, adrenocortical glands and menopausal problems, nervous tension, neuralgia, apathy, anxiety, stress related conditions, sore throat, tonsillitis, cellulitis, engorgement of breasts, poor circulation and skin complaints such as acne, bruises, broken capillaries, burns, congested skin, cuts, dermatitis, eczema, hemorrhoids, lice, oily complexion, mature skin, mosquito repellant, ringworm, ulcers and wounds.

Jasmine

Properties

Analgesic (mild), antidepressant, anti-inflammatory, antiseptic, antispasmodic, aphrodisiac, carminative, cicatrisant, expectorant, galactagogue, parturient, sedative, tonic (uterine).

Uses

Depression, nervous exhaustion and stress related conditions. Jasmine is said to produce a feeling of optimism, confidence and euphoria. It is especially good in cases of apathy, indifference, or listlessness. Jasmine is also used for catarrh, coughs, hoarseness, laryngitis, dysmenorrhoea, frigidity, labor pains, uterine disorders and other skin problems such as dry, greasy, irritated, sensitive skin and for muscular spasms and sprains.

Lavender

Properties

Analgesic, anticonvulsive, antidepressant, antimicrobial, antirheumatic, antiseptic, antispasmodic, antitoxic, carminative, cholagogue, choleretic, cicatrisant, cordial, cytophylactic, deodorant, diuretic, hypotensive, insecticide, nervine, rubefacient, sedative, stimulant, tonic, vulnerary. An excellent first aid oil. It soothes cuts, bruises and insect bites.

Uses

One of the most versatile therapeutic essence. It is used for nervous system disorders such as depression, headache, hypertension, insomnia, migraine, nervous tension, stress related conditions, PMT, sciatica, shock and vertigo. Lavender is useful in treating skin conditions such as abscesses, acne, allergies, athlete's foot, boils, bruises, burns, dandruff, dermatitis, earache, eczema, inflammations, insect bites and stings, insect repellant, lice, psoriasis, ringworm, scabies, sores, spots, all skin types, sunburn and wounds. Other applications of lavender is for the treatment of disorders such as lumbago, muscular aches and pains, rheumatism, sprains, asthma, bronchitis, halitosis, laryngitis, throat infections, whooping cough, abdominal cramps, colic, dyspepsia, flatulence, nausea, cystitis, dysmenorrhoea, leucorrhoea and for flu.

Lemon

Properties

Refreshing, antiseptic, stimulating, anti-anaemic, antimicrobial, antirheumatic, antisclerotic, antiscorbutic, antispasmodic, antitoxic, astringent, bactericidal, carminative, cicatrisant, depurative, diaphoretic, diuretic, febrifuge, haemostatic, hypotensive, insecticidal, rubefacient, stimulates white corpuscles, tonic.

Uses

Warts, depression, acne and indigestion. arthritis, cellulitis, high blood pressure, nosebleeds, obesity (congestion), poor circulation, rheumatism, asthma, throat infections, bronchitis, catarrh, dyspepsia, colds, flue, fever and infections. Other applications of lemon are in the treatment of anemia, brittle nails, boils, chilblains, corns, cuts, greasy skin, herpes, insect bites, mouth ulcers, spots, and varicose veins.

Peppermint

Properties

Digestive, cooling, refreshing, mentally stimulating, analgesic, anti-inflammatory, antimicrobial, antiseptic, antiviral, astringent, carminative, cephalic, cholagogue, cordial, expectorant, hepatic, nervine, stomachic, sudorific, vermifuge and antispasmodic.

Uses

Muscle fatigue, bad breath, toothache, bronchitis, indigestion and travel sickness, neuralgia, muscular pain, palpitations, asthma, sinusitis, spasmodic cough, for digestive system disorders such as colic, cramp, dyspepsia, flatulence, nausea and skin problems such as acne, dermatitis, ringworm, scabies and for toothache.

Petitgrain

Properties

Soothing, calming, antidepressant. Also antiseptic, antispasmodic, deodorant, digestive, nervine, stimulant (digestive, nervous), stomachic, tonic.

Uses

Skin problems, apathy, irritability and depression, convalescence, insomnia, nervous exhaustion, stress related conditions, dyspepsia, flatulence, acne, excessive perspiration, greasy skin and hair and for toning.

Rosemary

Properties

Analgesic, antimicrobial, antioxidant, antirheumatic, antiseptic, antispasmodic, aphrodisiac, astringent, carminative, choleretic, cordial, diaphoretic, digestive, diuretic, fungicidal, hepatic, hypertensive, nervine, parasiticide, restorative, rubefacient, stimulant of circulatory, adrenal cortex and hepatobiliary systems, stomachic, tonic.

Uses

Muscle fatigue, colds, poor circulation, aches and pains and mental fatigue. Debility, headaches, hypotension, neuralgia, mental fatigue, nervous exhaustion, stress related disorders, dysmenorrhoea, leucorrhoea, asthma, bronchitis, whooping cough, colitis, dyspepsia, flatulence, hepatic disorders, hypercholesterolaemia. Rosemary is also used for treatment of disorders such as arteriosclerosis, fluid retention, gout, muscular pains, palpitations, poor circulation, rheumatism and for the treatment of skin conditions such as acne, dandruff, dermatitis, eczema, and greasy hair. Other uses are as an insect repellent, for promoting hair growth, and for the treatment of scabies, scalp stimulation, lice and varicose veins.

Sandalwood

Properties

antidepressant, antiphlogistic, antiseptic, aphrodisiac, astringent, antispasmodic, bactericidal, carminative, diuretic, expectorant, fungicidal, insecticidal, sedative and tonic.

Uses

Depression, insomnia, nervous tension, stress related complaints, cystitis, diarrhoea, nausea, bronchitis, catarrh, coughs, laryngitis, sore throat, acne, dry, cracked and chapped skin, aftershave, greasy skin, moisturiser.

Tea Tree

Properties

Antifungal, antiseptic, anti-infectuous, anit-inflammatory, antiviral, bactericidal, balsamic, cicatrisant, diaphoretic, expectorant, fungicidal, immono-stimulant, parasiticide, vulnerary.

Uses

Dandruff, mouthwash, cuts, insect bites, pimples, abscess, acne, athlete's foot, blisters, burns, cold sores, herpes, oily skin, rashes, spots, veruccae, warts, wounds, asthma, bronchitis, catarrh, coughs, sinusitis, tuberculosis, whooping cough, vaginitis, cystitis, colds, fever, flu, infectious illnesses such as chicken pox.

Ylang Ylang

Properties

Antidepressant, anti-infectious, euphoric, aphrodisiac, relaxant, antiseborrhoeic, antiseptic, hypotensive, nervine, regulator, sedative (nervous), stimulant (circulatory), tonic.

Uses

Depression, nervous tension, high blood pressure, hyperpnoea (abnormally fast breathing), tachycardia, palpitations and digestive upsets. For skin care such as for acne, hair growth, hair rinse, insect bites, irritated and oily skin. For other nervous system disorders such as frigidity, impotence, insomnia, and stress related disorders

The following oils and spices are useful for relaxation.

- ☆ Basil
- ☆ Benzoin Tincture
- ☆ Bergamot
- ☆ Black Pepper
- ☆ Cajeput
- ☆ Camomile, Roman
- ☆ Camphor
- ☆ Cardamom Seed
- ☆ Cedarwood
- ☆ Cinnamon
- ☆ Clary Sage
- ☆ Clove
- ☆ Coriander
- ☆ Cypress
- ☆ Eucalyptus
- ☆ Fennel
- ☆ Frankincense
- ☆ Geranium
- ☆ Ginger
- ☆ Grapefruit
- ☆ Hyssop
- ☆ Jasmine
- ☆ Juniper
- ☆ Lavender
- ☆ Lemon
- ☆ Lemongrass
- ☆ Marjoram

- ☆ Melissa
- ☆ Myrrh
- ☆ Neroli
- ☆ Orange
- ☆ Patchouli
- ☆ Palma Rosa
- ☆ Peppermint
- ☆ Pine
- ☆ Rose Otto
- ☆ Rosemary
- ☆ Rosewood
- ☆ Sage
- ☆ Sandalwood
- ☆ Tea Tree
- ☆ Thyme
- ☆ Ylang Ylang

The Remedy for Stress

The following should be mixed into 5 tablespoonful of the carrier oil 2 drops neroli, 2 drops tangerine,3 drops bergamot the mix should be massaged on the face, temples, sinus points, forehead, base of skull.

Remedy for Depression

This mix is best when dabbled on a pillow, or placed on a handkerchief underneath a pillow.

Remedy for Headache

The following should be mixed with the carrier oil.

3 drops lavender, 3 drops peppermint, 2 drops rosemary and massaged on the face, temples, sinus points, forehead, base of skull.

Remedy for Depression

This mixture does not need to be mixed with carrier oil because it is going directly into the water in the spray bottle:

2 drops of lemon, 2 drops clary sage, 2 drops sandalwood,2 drops ylang-ylang (pronounced elaing, elaing)

This mix is best when used in a spray bottle.

Remedy for Muscle Pain

The following should be mixed with the carrier oil:

3 drops birch,2 drops cyprus,1 drop lemon, 1 drop lavender and massaged on the tender/sore area.

This mixture does not need to be mixed with carrier oil because it is going directly into the water in the spray bottle:

- ☆ 2 drops of lemon
- ☆ 2 drops clary sage
- ☆ 2 drops sandalwood
- ☆ 2 drops ylang-ylang (pronounced elaing, elaing)

This mix is best when used in a spray bottle.

6

Astromedicine

The main topic of astromedicine is the question: why it comes to an illness. Causes for diseases are always looked for in the life situation of the person concerned. From the astrological point of view it comes to an illness or an accident without exception only if the development of a person is blocked.

The goal of the astromedical horoscope interpretation is to find out which prenatal or early childhood experience is underlying the block of development and can so be considered as part of the cause for disease.

What are and what Causes Blocks of Development?

Blocks of development are karmic. We bring them along with us into life. During childhood we learn to have the same blocks that we used to have in former lives in order to go through a life long process of solving them.

In periods when we do no steps to take over the responsibility for our own happiness and the development of it in life, the flow of energy in our bodies can get stuck and we unconsciously install muscular blocks of it.

As development blocking it is usually considered that humans do not live themselves, but the dependence on or overlay of their true identities by the external world. Therefore in an astromedical consultation we try to put attention to the fact that everyone carries the self healing forces in him or herself.

For the purpose of development and the solution of our karma we consider self treatment to be more important than the medical treatment by the external world.

We can solve our Karma only ourselves. According to german law for the healing professions as astrologers we are not allowed to give medical treatment or diagnosis of diseases.

But what we can do and what we actually do is : We give you a list of those homeopathic remedies that refer to the astrological constellations in your birth chart. We so define your homeopathic constitutional remedies as those remedies, that are suitable for you in helping you to learn to solve the karmic and psychological blocks of development in your personal case.

The Vedas are the oldest scriptures known to exist, they are well over 5,000 years old. The vast ocean of Vedic knowledge encompasses both material and spiritual knowledge; it is perfect knowledge that was revealed by the sages of Vedic culture. Astromedicine is part of the Vedic literature, and has been preserved and handed down in the guru sishya parampara, the chain of apostolic succession, since time immemorial. Astromedicine is a form of astrology that existed thousands of years ago in what is now known as the Indian sub-continent. It gradually spread by diffusion into other cultures such as the Babylonian, Persian, Greek, Chinese, *etc.*

Astromedicine being an annexure to the Vedas, is thoroughly soaked in Vedic thought and philosophy. This necessitates that the practioner of the science must have earnestly assimilated the Vedic philosophy and lived the life with all its attendant spiritual practices.

The birth-map or horoscope is a symbolic representation of the Earth, planets, and stars at the time of your birth. It is a divine language, a mystic cryptogram by which God has communicated what a person's destiny will be, knowing which a wise person will exercise his freewill to optimize the result.

The view point of Astromedicine is that life is an interplay of both "Fate" and "Freewill." Our fate being the reaction to our previous exercise of freewill. In our lives we are often faced with choices presented to us by our environment. We have the freedom to chose between "a" and "b," but once that decision has been made we must accept the reaction to that decision. The reaction may be immediate or it may be delayed by thousands of lifetimes. In any case, the reaction, pleasant or unpleasant, will come at the appointed time determined by higher authorities. As the famous politician and economist Chanakya (371 BC to 283 BC) says, "Just as a calf can find its mother in a herd of 10,000 cows; in the same way your actions (reactions) will find you." The reactions to life's actions are inescapable by everyone except those who are devoted to God in Bhakti Yoga. When a karmic reaction ripens and fructifies it creates a new situation, an environment that presents you with more choices on which to exercise your freewill. Thus life is an interplay of both fate and freewill as previously stated.

Let us give an example of how this works. Suppose someone is born into a poor family because of his bad karma from a previous lifetime. This person is raised in a surrroundings with crime and poverty. Let us say, hypothetically, that he has two choices. He can blame society for the evils he faces and thus try to victimize society by taking to a life of crime and thus acquire wealth. Or he can take responsibility for his situation as he finds it and try to better himself by education and honest endeavor. Suppose he chooses the first option of crime and robs several institutions. Eventually he will get caught and will face consequences, which will be a reaction for his criminal behavior. But even while in captivity he still has options, though more limited than those of a free man. He can choose to be a hardened criminal or to be a model prisoner and go straight. Each choice will have a reaction. Thus we can see that life is an intricate interplay of Fate (Karma) and Freewill.

Getting back to the horoscope. By looking at the birth chart, the Astromedicine practitioner studies the interaction of the planets, signs, houses, and the aspects between them. This is the language of the stars. Every thing known is in the portfolio of one or more planets, signs, or houses. And it is by studying the interactive trends in the chart that the practitioner can ascertain so much information about a person.

The various elements in the chart: planets, signs, nakshatras (lunar mansions), houses, elements, modalities, polarities, aspects; are the components of this divine celestial language and play an important role for the health of the person concerned.

The planets represent the following:

☆ Sun : Atma (soul), self, self-realization, influence, prestige, power, valor, health, eye, general well being, heat, splendor, father, king, royalty, royal favor.

☆ Moon: The heart, understanding, inclinations, emotions, sleep, happiness, good name and fame, facial luster, mother, royal favor, affluence, travel, water reservoirs.

☆ Mars: Stamina, courage, desire, anger, scandal, diseases, enemies, opposition, controversies, weapons, commander of an army, land, immovable properties, younger brother, relations such as cousins.

☆ Mercury : Intelligence, discrimination, speech, expression, education, learning, mathematics, logic, astrology, medical knowledge and profession, writing, publishing, acting as a middle man in trade or politics (diplomacy), dancing, mixture of things, leafy trees, testing of precious stones, charms (amulets), maternal uncles, friends.

☆ Jupiter : Wisdom, learning, corpulence, acts of religious merit, devotion to God, ancestors and superior beings, holy places, scriptures, proficiency in learning, philosophy, giving alms or donations, benevolence, wealth, respect, sons, religious, preceptors, fruit, fruit trees.

☆ Venus : Spouse, marriage, sexual matters, pleasures of the senses, singing, poetry, scents, ornaments, jewellery, all articles of luxury, cooperation with and from others, flowers, flowering trees, beauty, buying and selling, cows, watery places.

☆ Saturn :Longevity, life, death, adversity, calamities, disrespect, diseases, poverty, livelihood, servility, unrighteous conduct, learning of sciences and foreign languages, agricultural pursuits, minerals, oils, things buried deep in the ground and coming out there from, servants (male and female), service, theft, cruel deeds, malice, lameness, very old persons.

☆ Rahu: (North Node of the Moon):Paternal grandfather, fallacious argument, harsh speech, gambling, movement, travelling, outcastes, foreigners, snakes, snake bite, theft, wickedness, widow(er), intrigue with a widow(er), skin diseases, itches, eczema, acute or sharp pain in the body, hiccoughs, swelling in the body.

☆ Ketu: (South Node of the Moon): Maternal grandfather, consumption, pain, fever, wound, witchcraft, causing trouble to enemies, horned animals, dog, cock, birds with spots or of variegated colors, philosophy, salvation.

The meanings of the signs of the Zodiac are well known to everyone or can easily be found out so we will not explain them here.

The following are brief meanings for the twelve houses of the horoscope suitable for our basic educational purposes.

- ☆ First House: self, head, body, personality, mental temperament, health.
- ☆ Second house: eyes, face, upper part of the throat, speech, wealth, family.
- ☆ Third House: arms, lower part of the throat, shoulders, brothers, sisters, valor.
- ☆ Fourth House: chest, heart, mother, landed property, friends, conveyances, happiness.
- ☆ Fifth House: stomach, education, intelligence, sons, daughters.
- ☆ Sixth House: the region of the navel, illness, enemies, debts, distress.
- ☆ Seventh House: partnership, sexual relations, spouse, the part of the body below the waist and down to the genitals.
- ☆ Eighth House: genitals, anus, death, legacies.
- ☆ Ninth House: hips, thighs, religion (dharma), devotion to God, prosperity, religious preceptor (guru), father.
- ☆ Tenth House: knees, back, status and position in life, activity, honor from government.
- ☆ Eleventh House: calves of legs, friends, income.
- ☆ Twelfth House: expenditure, loss, sexual enjoyment, left eye, teeth, confinement.

The above information is grossly simplified but gives you a bit of an idea of what is being dealt with. Considering that there are nine planets, twelve signs, and twelve houses, you can imagine what the possible combinations and permutations can be. The energies of the planets, signs, and houses, interact with each other in myriad ways. By studying the combinations, a well trained astrologer can know an amazing amount of information about the owner of the chart. A remarkable feat considering that the original input data consisted of only the date, time, and place of birth.

Natal Reading

The Natal chart is the birth chart. The reading of this chart is what most people consider to be the sum total of astrology. In a general reading of the Natal chart one should expect to be told who you are, where you have been, and where you are going.

In truth, most people do not understand themselves, and in fact are quite confused about who they are, how their mind works, why it works that way, *etc.* The vast majority of people don't really know who or what they are on either the psychological or spiritual plane. Therefore, the first thing that is done is an analysis of: psychology, character, and life potential. The astrologer discusses the person's qualities, intellectual abilities and aptitude, people skills and relationships with family, friends, and loved ones, financial acumen and status, education, travel, health, *etc.*

Life potential indicates a person's predisposition and predilection in certain areas. For example, a certain combination of planets may indicate a person with acute intelligence and love of knowledge, this would give a strong potential for academic achievement or a career in research, teaching, scholarship, *etc.*

This initial analysis serves several functions: 1) To bring the seeker a big step closer to achieving the Socratic ideal of "Know thyself." 2) To verify the accuracy of the birth time and the horoscope. A good astrologer will be able to paint a picture that should be instantly recognizable by the seeker. The seeker should feel illuminated by the knowledge as well as secure and confidant that he is dealing with an astrologer who is qualified. This is especially important since in most cases this will be the first time that the seeker will have met the astrologer. 3) The astrologer will also understand the psychic nature of the seeker and be better able to guide him.

At some point the discussion will turn to your past in this lifetime. Astrologers deal with this in different ways. Some go deep into the past attempting to give a few details about the life. The idea is that by correctly identifying certain life events they will gain the confidence of the seeker. This is often done at the expense of giving a good psychological analysis which is more beneficial to the seeker. Why is a psychological analysis more beneficial? Because most of the time the psychological analysis will reveal items that the person is unaware of, with the added benefit that often many things about the past are brought up and detailed anyway. Whereas an astrologer who just mentions a few incidents from the past life of the seeker has not really told them something they didn't know before. And by neglecting to analyze the personality in depth and give satisfactory explanations he has deprived the seeker of much valuable knowledge that could not be gained even after a score of psychological examinations.

Depending on the situation and the needs of the seeker, the astrologer will usually give an outline of the seeker's recent past and the present situation that caused him to consult the astrologer in the first place.

Then the astrologer will proceed to discuss the up-coming future trends and planetary periods that the seeker can expect to experience both externally in the world and internally in terms of psychological attitudes. The internal and external realities are both very intimately related, for a negative mental attitude can attract and often is the precursor to equally negative experiences in external reality. The reverse is also seen, that a few reverses in life can instill a negative and depressed world view in the seeker creating a vicious cycle of gloom and doom.

This is where the real skill of the astrologer is tested. A real Vedic astrologer is not just someone who makes some predictions and then stands aside. Rather he is a Vedic Brahmin who has been trained in yoga, philosophy, counseling, *etc.* The astrologer, if he foresees difficulty ahead, should be able to provide the seeker with strategies for meeting the difficulties in a mature and wise way. This after all is the benefit of Vedic Astrology and the proper use of your freewill. To "Act with Wisdom" as opposed to "Reacting in Ignorance." Simply knowing that difficulty lies ahead may be sufficient to avert real danger. You may still be exposed to the problem but you will know what to do because you were not taken by surprise. In cliche terms:"To be fore-warned is to be fore-armed."

In 1992, Hurricane Andrew destroyed a large area of Florida. There was much physical damage to property, but considering the horrific nature of the storm the loss of human life was practically nil. Why? Because the National Weather Service had warned residents of the approach of the storm and they in turn had time to prepare for the inevitable. They evacuated dangerous areas, boarded up windows, cached food and water, sought high ground, and took shelter. They acted with wisdom. We can only imagine what the death toll would have been if Andrew had pounced on Florida with no warning at all. The carnage would have been astronomical and the people would have been simply reeling from the death blow not knowing what to do next. This is reacting in ignorance. That is, reacting to a situation after it has already struck without any fore knowledge that it would happen. This is why, generally, there are few casualties from hurricanes compared to earthquakes. At present, hurricanes can be predicted by the weather service, but no government agency can predict earthquakes.

Of course, the previous discussion should not unnerve you into thinking that the astrologer is only going to give out dire predictions of impending ruination and existential calamity. This is hardly the case; we just dealt with a worst case scenario first. Just as often, and much easier and pleasant for the astrologer, are predictions of up-coming times of good fortune and happy times free of difficulty and full of opportunity. People love to hear about good fortune, and, frankly, it is much easier on me when I have to give predominantly good news. You know what happened to the bearer of bad news in ancient times.

People are funny though. Most people have heard of the saying: "Make the best use of a bad bargain." Often hardship can bring the best out of a person. It has been observed that the only difference between the carbon in a diamond and that in a graphite lead pencil, is that the former has been subjected to tremendous heat and pressure for a long time, then chiselled and polished to perfection. So while giving people bad news is not so pleasant, they often take it more seriously than when I give good news.

The opposite of "Make the best use of a bad bargain;" is "To squander a good opportunity." Or: "To rest on your future laurels." Surprisingly, people often think that if good times are ahead they don't have to do anything. Wrong. When an astrologer indicates that there is smooth sailing ahead it is not time to take it easy, but rather to strive harder because now that resistance is lower you are more likely to succeed. But you must still endeavor with determination, just that now it will be easier. And who knows? The cycle may change. So it is the sign of wise a astrologer who advises the seeker to maintain and even build up his momentum in good times. That way if things should take a turn for the worst, the seeker will be able easily overcome the problem. If you are driving on the flats and you see a hill up ahead you speed up; the momentum of the car will carry you over the top with less strain.

The usual situation with the vast majority of people is that their future karma will be a mixture of good and bad. So it is up to the astrologer to sort out as much as possible which is which and advise how to minimize the negative and maximize the positive with suitable strategies.

The process that I have just described is how the astrologer tells the seeker "Who he is, where he has been, and where he is going."

How Long will it Take?

You should be prepared to spend from 1 to 3 hours, or longer in some cases, with your astrologer, for this initial reading. The length of time will be determined by the subjects covered, questions asked and the intricacy of the case. I have described a general first reading. Sometimes a person has a specific problem they want to deal with such as career, marriage, health, *etc.* The astrologer should adjust to fit the needs of the seeker. In so doing, however, several of the items I described may be left out.

If the birth time is not accurate or unknown, the astrologer may have to rectify it by asking certain questions and adjusting the time according to your answers. This all takes time.

Your Part

Up to now we have only discussed what procedure the astrologer is likely to employ, and what subjects he will speak on. But what about you? What should you say or do? Should you say anything or just listen? The following recommendations will help you to get the most out of your reading:

Before going to the astrologer find out if he is going to tape record the session. If not, get permission to bring your own tape recorder and record it yourself. Bring enough tapes to adequately cover the meeting, more is better than less.

Before starting the session pre-arrange with the astrologer that if you make a particular signal he should turn off the tape recorder. This indicates you want to discuss something that you don't want to be recorded. You may not want others who may listen to your tape to know some personal details.

It is a good idea to note down any questions or areas in your life that you want covered in the reading. It is generally best to let the astrologer know this in advance so that he will be able to focus extra attention on those areas. Some topics may require extra research.

While the astrologer is speaking we recommend that you avoid breaking his train of thought. Bring a note pad, if you have a question write it down, if it hasn't been resolved before the reading is over then you should ask it and resolve it then. Breaking the astrologers train of thought means going off on tangents, or asking a question and before he can answer it fully you ask another question. Of course there should be feedback, both positive and negative, so that the astrologer knows that he is on the right track. Astrology is, after all, a difficult science. If the astrologer makes a mistake, let him know, he will adjust his calculations. If he is right, confirm it by a nod or whatever so that the astrologer will continue in confidence. Whatever the case, don't just sit there like a piece of wood. Interact. But do so in such a way as is conducive to the reading. This may mean that you have to go into a long explanation of your situation, or it may mean that you listen while he explains. As long as you stay on track.

Be honest with your astrologer. You wouldn't lie to your physician would you? If you don't feel comfortable telling the astrologer the truth about your situation then find an astrologer who you can trust. Misleading your astrologer could have serious consequences. Another thing, don't try to trick or test your astrologer. If you intend to have a long standing relationship with anyone, what to speak your astrologer, it is not good by starting out on such a footing. The astrologer will be able to tell that there is something afoot and this just hampers the process.

What has been said above about your part in a natal reading also applies, with necessary adjustments, to other kinds of readings.

Chart Updates

A person's life cannot be encapsulated in a few hours. Because it is impossible to do everything in one sitting, people often have follow up readings that cover certain time frames. People usually ask about what the next six months or one year will hold in store for them. Many people have their charts updated around the beginning of the New Year, or their birthday and have a report for the upcoming year. You may want a chart updated because you are contemplating a new business, moving your residence, having marital, legal or health problems, *etc.* There are no hard and fast rules regarding when or why to have your chart done, or for what span of time. Let your needs be the guideline for this.

Follow up readings are almost exclusively predictive in nature. (Unless for some reason the first reading didn't cover all the main points. This sometimes happens when the seeker asks the astrologer to focus on just one area, thus excluding others which the seeker may want covered in a subsequent reading.) They will reveal, in more detail, the types of situations, both physically and mentally, that you will find yourself in. Times of strength or weakness, happiness and sadness, romance, love, health, illness, travel, education, *etc.* These types of readings are quite popular with business people, executives, and anyone (and that means you) who needs to have the best information available for decision making and planning. These type of people know the value of a good advisor who can suggest strategy for progressing in material or spiritual life. We shall momentarily diverge and take a glimpse at what Vedic thinkers have said about the need of good counsel, then we shall return back to the topic.

The Need for Counsel

Since ancient times good astrologers have been recognized as among the best of advisors. Chanakya Pandit (circa 1550 BC) has said that every king should have ministers to guide him:

"One without an adviser has no certainty of counsel."

"All things begin with counsel."

"The country prospers by proper ministerial counsel."

"Through ministerial eyes other's weaknesses are seen."

"Counsellors (ministers) are the ones who see the true implications of what ought to be done and what ought not to be done."

"Governance is possible only with assistance. A single wheel does not move. Hence ministers should be appointed and their counsel listened to."

Of the many ministers to the king Chanakya explains that:

"The royal astrologer should know reading, writing and arithmetic, explain well and interpret the secrets of the planets." And, "The Brahmin who knows mathematics, interprets scripts, is interested in the ancient scriptures and is able to interpret the secrets of planets is worthy of worship like a god."

And finally Varaha Mihira said: "There is no better boat than a horoscope to help a man cross over the troubled sea of life."

Though you may not be a King you can still avail yourself of good astrological counsel to help you plan your life.

Chart Updates Continued

Closely related to update charts but of generally longer duration are charts that focus in on a particular planetary period or subperiod. These will normally cover time frames several years in duration. I would still advise anyone who had such a longer forecast done to have his chart updated at least annually if not every six months. The reason is simple, the shorter the span of time investigated, the more detail revealed.

Compatability Charts

The potential of the natal chart has not been exhausted yet. It has often been said that "Marriages are made in heaven." And the cynics have added: "But often end in hell."

The situation is common, you meet someone, you are attracted, get attached, jump into the relationship, but after sometime find out that it is not right for you. Usually after much pain. There must be a better way. There is!

It is a time honored tradition in India to compare and match horoscopes of prospective marriage partners. This usually is done by the parents who were detached from the effects of hormones. Not surprisingly, in the hands of a good astrologer there is a high success rate and happy marriages.

The problem that exists among astrologers around today is that few are actually trained in how to do this properly. The vast majority of Vedic astrologers (at least in this country) simply rely on a detailed comparison of the Moon in the respective charts. This is called the Kuta method. This is valuable but somewhat simplistic, though way ahead of the Western method of comparing Sun signs.

An experienced astrologer employs a system that thoroughly examines the two horoscopes on three tiers:1) The general strength of each chart individually with especial attention paid to capacity for relationship and marriage; 2) A detailed examination of compatibility between the two charts based on all the planets not just the Moon; 3) Calculations to determine the durability of the relationship (many people are attracted, seem compatible, but afterwards end in divorce court, giving rise to the saying "Marry in haste; lament at leisure").

For persons who are capable of maintaining a relationship (obviously it will be impossible to match up someone if they don't have the karma to marry) this Vedic method of horoscopic matching is quite useful by saving you a lot of time, energy, money, and protecting your sensitive emotions from a lot of pain. All that you need to know is the date, time, and place of birth of the people involved.

Just be sure that the astrologer you consult to do a compatibility reading is experienced in doing them and uses a sophisticated technique as outlined above.

Though compatibility studies are generally done for romantic purposes it is not limited to this. The same principles can be adjusted and used in conjunction with other astrological techniques to develop an excellent method for selecting personnel. If you were going to pay someone $50,000, $100,000 or more per year it would make sense to hire someone who not only has a good resume (which could be forged) but someone who fits the job, fits in with the other people on the team, and most of all works well with you. An astrologer could easily determine if the candidate was honest or a cheater and if he had the capacity to fulfill the position. And by comparing his chart to the others he would have to interact with, it could be quickly determined who he would get on with and who he would rub the wrong way. If he fit in with most of the people, you could take him on the team and just make sure he had little interaction with the people whose charts showed a negative reaction. Thus by nicely harmonizing workers into an effective team, esprit de corps is instilled, and competitiveness is greatly enhanced.

Parent-Child Relationships

Parents often consult astrologers to find out how they can best guide their beloved children. Astrology can point out what the compatibilities and incompatibilities are between parent and child. An astrologer can suggest strategies for dealing with problem areas in the parent/child relationship, and guide the parent in understanding the nature of their child, what the child's creative potential is, what areas should be encouraged, and also identify the problem areas.

What most people, including some astrologers, don't realize is that your own personal chart contains in it the information about your signs success or failure. All the good signs in the world are not going to help you if it is not your karma to be successful in life.

Medical Astrology

Medical Astrology is the full blown investigation and diagnosis of medical problems using astrology. It was once very much in vogue. Even today in India, Ayurvedic doctors are urged to study astrology. But the difficulty is that to properly practice Medical Astrology you must be a qualified physician. It is a valuable tool of the astrologer/physician but few astrologers fall into this category.

Unless the astrologer is also trained highly in medicine he should limit his medical advice and refer you to a health professional. The same can be said in regard to other spheres such as law, financial planning and marriage counselling. The astrologer does his part then refers you to the appropriate specialist to handle your problem. It is not expected that the astrologer should be an expert in all fields. He should be very knowledgeable, but he should know his own limitations and for the sake of the benefit of the seeker he should refer him to proper sources of help if need be rather than go outside his field of expertise.

Medical astrology is a conglomeration of many branches of astrological knowledge and we shall mention the medical uses of other branches when appropriate. As has already been mentioned from the Natal chart the general health of the individual can be determined, periods of stress and disease can be seen, *etc.*

Understanding of human body and diseases controlled or influenced by the grahas influencing them:

There is a method of measuring the 12 bhāva* to the human bodies, which is based on dividing the human body in three parts on the basis of dreshkāna occupied by the ascendant. If the ascendant falls in the 1st dreshkāna the body part to be divided over the 12 bhāvas is head of the -ātaka. If in 2nd dreshkāna, the body parts from throat to navel are to be divided over 12 bhāvas and if in 3rd dreshkāna then Basti to feet get divided over the 12 bhāvas. For ease of comprehension these divisions are given in a chart form, given on next page.

This is the basic structure that we can use to understand what organ of a person, could be afflicted.

In vedic astrology, Bhāva (Sanskrit: bhāva, 'state, condition') is a term denoting a fixed zodiacal division of the sky from the perspective of an observer. It corresponds to the concept of "house" in Western astrology. Each house has associated kāraka (Sanskrit: 'significator') planets that can alter the interpretation of a particular house. Each Bhāva spans an arc of 30

Bhāva	1st Dreshkāna	2nd Dreshkāna	3rd Dreshkāna
1st bhāva/lagna	Risen portion of lagna is left side of head and the yet to rise portion is the right side of the head	Risen portion of lagna is left side of throat and the yet to rise portion is the right side of the throat, where it joins trunk	Risen portion of lagna is left side of basti and the yet to rise portion is the right side of the basti
2nd bhāva	Right eye	Right shoulder	Right side of anus and sexual organ
3rd bhāva	Right ear	Right arm	Right testiclel ovary
4th bhāva	Right side of nose	Right Parshva (portion between stomach and back *i.e.* side)	Right thigh
5th bhāva	Right cheek	Right side of heart	Right knee
6th bhāva	Right side of chin	Right side of stomach	Right leg
7th bhāva	Risen portion is left side and yet to rise is the right side of mouth	Risen portion is left of navel and yet to rise represents the right side of navel	Risen portion is left foot and yet rise portion is right foot
8th bhāva	Left side of chin	Left portion of stomach	Left leg
9th bhāva	Left cheek	Left part of heart	Left knee
10th bhāva	Left side of nose	Left parshva	Left thigh
11th bhāva	Left ear	Left arm	Left testicle
12th bhāva	Left eye	Left shoulder	Left side of anus and sexual organ

degrees and therefore there are twelve Bhāvas in any chart of the horoscope. These are a crucial part of any horoscopic study since the Bhāvas, understood as 'state of being' personalize the Rāśis/Rashis to the native and each Rāśi/Rashi apart from indicating its true nature reveals its impact on the person based on the Bhāva occupied. The best way to study the various facets of Jyotica is to see their role in chart evaluation of actual persons and how these are construed.

Having understood the basics, now let us understand what are the diseases connected with different planets. There are many diseases allotted to different grahas by the learned. However here we shall try to understand the basics of individual planet's area of influence on body and their nature to be able to point to the disease that could be troubling the person. If we approach a chart in this fashion, we may be better able to come to near right conclusions as in this time and age there are numerous names to diseases and even if we want to it might be difficult to categorize them in to only nine categories, which is the number of planets available to us and doing so is more likely to confuse than help, as diseases are usually caused by a combination of influence of various planets on the person's relevant bhāvas that we know are connected with his disease. Therefore let us refresh our memory as to the area of influence of planets, as told to us by the sages.

Crystals and Gemstones for Healing

AGATE - Good for transmutation; helps with the emotion of acceptance; gives a mellow, blended aspect; beneficial in stomach area.

AGATE -Eases stress and pain of a loss. Reduces anxiety.

AGATE/BOTSWANA - Use with high-pressure oxygen therapy; smoke inhalation.

AGATE/FIRE - Master healer with color therapy; enhances all essences; grounds and balances; sexual and heart chakra binder; burns energy.

AGATE/MOSS - Emotional priorities; mental priorities; colon, circulatory, pancreas and pulses; blood sugar balance; agriculture.

AGATE/PICTURE - L and R brain imbalances; *i.e.* epilepsy, autism, dyslexia; visual problems; blood circulation to the brain; apathy is eased.

ALEXANDRITE - Low self-esteem and difficulty centering imply need; central nervous system disorders; spleen and pancreas.

AMBER - Fossils from pine tree resin, they add strength to magick and spells, attract love and increase beauty.

AMBER - Memory loss; eccentric behavior; anxiety; inability to make decisions; thyroid, inner ear and neural-tissue strengthener; activates altruistic nature; realization of the spiritual intellect.

AMETHYST - Headaches; blood sugar imbalance; L brain imbalances; edginess; facilitates healing; inner peace; psychic insight; stimulates third eye; aid for meditation, spiritual opening and internal surrender.

AMETHYST - Increases spiritual awareness, used in all workings for peace, love, happiness and protection, used in divination and psychic work.

ANHYDRITE - Heavy metal maism is alleviated.

APACHE TEARS - Apache Tears are "transparent" black obsidian. It is said that these are the tears that Apache women cried when warriors did not return from battle. Simply hold them up to a light or the sun and you can see through them. Good for Protection, grounding, good luck, and they may used for divination.

AQUAMARINE - Fluid retention; coughs; fear; thymus gland; calms nerves; problems with eyes, ears, jaw, neck, stomach, teeth; Mental clarity; meditation.

ATACAMITE - Genitals; VD; thyroid; parasympathetic ganglia.

AVENTURINE - Eliminates psychosomatic ills, fear; skin diseases; nearsightedness; positive attitude; creative insight.

AVENTURINE - Increases mental powers, perception and creative insight, gambler's stone.

AZURITE - Arthritis and joints; surfaces psychic blocks that form physical blocks; helps one let go of old belief systems; dissolves fear and helps transform it into understanding.

AZURITE-MALACHITE - Skin diseases; anorexia; calms anxiety; lack of discipline; powerful healing force to physical body; emotional release.

BERYL - Attracts Deep Romance and Love.

BERYL - Laziness; hiccoughs; swollen glands; eye diseases; bowel cancer.

BLOODSTONE - Increases physical strength and courage, brings victory in the courtroom, used to banish, exorcise.

BLOODSTONE/HELIOTROPE - Circulation; all purpose healer and cleanser; stomach and bowel pain; purifies bloodstream; bladder; strengthens blood purifying_ mysticmagicspells.organs.

CARNELAIN - Promotes self-confidence and peace, eases depression.

CARNELIAN - Grounding; stimulates curiosity and initiative; focuses attention to the present moment; use with citrine on lower 3 chakras; digestion.

CHALCEDONY - Touchiness; melancholy; fever; gallstones; leukemia; eye problems; stimulates maternal feelings and creativity. release.

CHRYSOCOLLA - Emotional balance and comforter; alleviates fear, guilt and nervous tension; facilitates clairvoyance; arthritis; feminine disorders; eases labor and birth; thought amplifier.

CHRYSOLITE - Inspiration; prophecy; toxemia; viruses; appendicitis.

CHRYSOPGRASE - Gout; eye problems; alleviates greed, hysteria and selfishness; VD; depression; promotes sexual organ strength.

CITRINE QUARTZ - Heart, kidney, liver and muscle healer; appendicitis; gangrene; red and white corpuscles; digestive tract; cleanses vibrations in the atmosphere; creativity; helps personal clarity; will bring out problems in the solar plexus and the heart; eliminates self-destructive tendencies.

CLEAR QUARTZ - Holding or increasing energy.

CLEAR QUARTZ - Transmitter and amplifier of healing energy and clarity; balancer, channeler of universal energy and unconditional love; all purpose healer; programmable.

CORAL - Protects against the evil eye, all spells intended to harm the wearer, natural disasters and unfortunate occurrences.

DIAMOND - All brain diseases; pituitary and pineal glands; draws out toxicity, poison remedy.

DIAMOND - Reconciling differences between those who have quarreled. This stone is said to symbolizes peace, fidelity and opulence.

DIOPSITE and ENSTATIATE - rejection; heart, lung and kidney stimulation; self-esteem.

ELIAT STONE - Tissue and skeletal regeneration; detoxification; antidepressant; karmic life acceptance.

EMERALD - Precognition and the ability to see into the future. It is said to nurtures love and beauty, and turns all negative spells back to the sender.

EMERALD - Radiation toxicity; all mental illness; circulatory and neurological disorders; transmits balance, healing and patience; increases psychic and clairvoyant abilities; meditation; keener insight into dreams.

FLOURITE - Bone disorders; anesthetic; hyperkenesis; ability to concentrate; balances polarities; 3rd eye center; mental capacity and intellect.

FLUORITE - Reduces emotional involvement, used to gain perspective, increase mental power.

GARNET - Enhances strength and endurance, protection, healing, enhances self-esteem and encourages success in business.

GARNET/RHODOLITE - Capillaries; skin elasticity; protection from pre- cancerous conditions.

GARNET/SPESSARTINE - Bad dreams; depression; anger; self esteem; hemorrhages; hormone imbalances; inflammations; sexual disease.

HEMATITE - Blood cleanser and purifier; self esteem; augments meridian flows; aids in astral projection.

HEMATITE - Legal Matter, Victory, Healing.

HERKIMER DIAMOND - Enhances dream state; helps alleviate stress; draws toxicity from physical form; balances polarities; increases healing ability; develops ability to "give".

JADE - Good Luck, Protection from disease or Evil spirits. It is said to enhance one's occult powers. It presents serenity and immortality.

JADE - Kidney, heart, larynx, liver, parathyroid, spleen, thymus, thyroid and parasympathetic ganglia healer; strengthens body; longevity.

JASPER - Can protect from pain and guard one's independence. It is said to bring the wearer good fortune and protection from the controlling influences of others.

JASPER/GREEN - Constipation; ulcers; intestinal spasms; bladder, gallbladder and general healer; clairvoyance; balances healer's auric field.

JASPER/PICTURE - Skin, kidneys, thymus and their neurological tissues; betters the immune system; past life recall; overactivity in dream state and hallucinations show a need for it.

JASPER/RED - Liver; stomach troubles and infections.

JASPER/YELLOW - Endocrine system tissue; thymus; pancreas; sympathetic ganglia stimulation; etheric body alignment.

JET - Feminine disorders; teeth; stomach pain; glandular swelling; fevers; hair loss; alignment of lower spine.

KUNZITE - Alcoholism; anorexia; arthritis; epilepsy; gout; headaches; colitis; retardation; memory loss; schizophrenia and manic-depression; phobias; emotional equilibrium; thyroid malignancy; gums; pain; self-esteem.

LAPIS - Neuralgia; melancholy; fevers; inflammations; penetrates subconscious blockages; throat chakra; sore throat; energy focuser for teachers, lecturers and speakers, mental and spiritual cleanser; used on 3rd eye for meditation; eliminates old and negative emotions; use with other healing stones; thought form amplification; helps in creating mantras.

LAPIS LAZULI - Stimulates wisdom, truthfulness, psychic awareness, healing.

LAZULITE - Frontal lobe stimulation; hypertension; liver diseases; immune system.

LEPIDOLITE - A calming stone, this is carried for peace, spirituality and to drive off negativity.

LODESTONE, GREEN - Draws money, employment, strengthens loyalty, used in bindings.

LODESTONE, RED - Draws love, friendship, strengthens loyalty, used in bindings

MALACHITE - Draws money and protection, guards wearer from danger

MALACHITE - Draws out impurities on all levels; balances L and R brain functions; mental illness; co-ordination and vision; radiation eliminator; evil eye protector; all purpose healer, especially in solar plexus and good for healers.

MOLDAVITE- Form of Tektite. Made when meteorite hit the earth thousands of years ago,. it's getting very rare, as it is only found in a small region of Russia. Said to have links to extraterrestrial energies and will greatly enhance the power of other stones. Brings you into communication with your higher self.

MOONSTONE - Draws love, hope and protection, promotes unselfishness and helps settle disputes, induces prophetic dreams.

MOONSTONE - Soothes and balances the emotions; helps eliminate fear of "feeling"; encourages inner growth and strength; aids peace and harmony and psychic abilities; aligns vertebrae; digestive aid.

NANITE - Larynx; lungs; thyroid; parasympathetic nervous system; major muscle tissues.

NATROLITE - Colour; lower intestines; thyroid; sciatic nerve; parasympathetic nervous system.

OBSIDIAN - Protects the gentle from being abused; stabilizer; stomach, intestine and I general muscle tissue healer; bacterial and viral inflammations.

OBSIDIAN - Used for grounding and centering, divination, and increasing spirituality

ONYX - Objective thinking; spiritual inspiration; control of emotions and passions, help eliminate negative thinking, apathy, stress and neurological disorders; also used as a heart, kidney, nerve, skin, capillary, hair, eye and nail strengthener.

OPAL - Good luck, and extra mental powers

OPAL/CHERRY - Red corpuscle and blood disorders; depression; apathy; lethargy; intuition and joy.

OPAL/DARK - Reproductive organs; spleen and pancreas; filters red corpuscles and aids white corpuscles; bone marrow; depression, esp. of sexual origin; balances; amplifies creative and intuitive thought; grounds radical emotional body.

OPAL/JELLY - Spleen and abdominal diseases; cellular reproductive problems; helps absorb nutrients; minimizes wide mood swings; mystical thought amplifier.

OPAL/LIGHT - Balances L and R brain hemispheres for neural disorders; stimulates white corpuscles; helps bring the emotions to mystical experiences; aids abdomen, pituitary and thymus problems.

PEARL - Eliminates emotional imbalances; helps one master the heart chakra; aids stomach, spleen, intestinal tract and ulcer problems.

PERIDOT - Protects against nervousness; helps alleviate spiritual fear; aids in healing hurt feelings and bruised egos; incurs strength and physical vitality; aligns subtle bodies; amplifies other vibrational energies and positive emotional outlook; helps liver and adrenal function.

PETRIFIED WOOD - Good for past life work, nervous conditions, and grounding.

PYRITE - Helps purify the bloodstream and upper respiratory tract; upper intestines; digestive aid; nervous exhaustion; grounding.

QUARTZ/SOLUTION - Lymphatic cancer and circulatory problems; helps the psychologically inflexible.

RHODOCHROSITE - Narcolepsy and narcophobia; poor eyesight; extreme emotional trauma; mental breakdown; nightmares and hallucinations; astral body; kidneys; clears solar plexus of blocked energy; unconditional love forgiveness; evil eye protection; helps one utilize the creative power of the higher energy centers.

RHODONITE - Inner ear; alleviates anxiety; confusion and mental unrest; promotes calm, self worth, confidence and enhanced sensitivity.

RHYOLITE - Balances emotions; self worth; enhances capacity to love; aligns emotional and spiritual bodies; stimulates clarity of self statement.

ROSE QUARTZ - Draws love and happiness. Helps heal broken heart.

ROSE QUARTZ - Heart chakra opener; love and self-acceptance healer for emotional wounds; dissipates anger and tension.

ROYAL AZEL (SUGALITE or LUVALITE) - L and R hemisphere balance; opens crown chakra; heart statement; increases altruism, visions and general understanding; protects against negative vibrations; helps one gain power to balance the physical body.

RUBY - Heart chakra; balances love and all spiritual endeavors; self-esteem; strengthens neurological tissues around the heart; prevents miscarriages.

RUBY - Qualities of Power, Loyalty, and Courage.

RUTILE - Alleviates blockages within the psyche from childhood pressures.

SAPPHIRE - Radiate Gentleness and Peace. It is said that this stone helps bring Justice and truth to light

SAPPHIRE - Spiritual enlightenment; inner peace; colic; rheumatism; mental illness; pituitary; metabolic rate of glandular functions; anti- depressant; aids psychokinesis, telepathy, clairvoyance and astral projection; personal statement; also for pain.

SARDONYX - Mental self control; depression; anxiety and especially for grief.

SMITHSONITE - Eases fear of interpersonal relationships; merges astral and emotional bodies; balances perspective.

SMOKY QUARTZ - Stimulates Kundalini energy; cleanses and protects the astral field; draws out distortion on all levels; good for hyperactivity and excess energy; grounding.

SODALITE - Used for meditation, brings wisdom and calms inner conflicts

SODOLITE - Oversensitivity; helps intellectual understanding of a situation; awakens 3rd eye; cleanses the mind.

SPINEL - Leg conditions, when worn on solar plexus; powerful general healer; detoxification aid.

TIGER'S EYE - Draws wealth and money, protects travelers, increases clarity of thought.

TIGER'S EYE - Mind focuser; helps purify the blood system of pollution and toxins; psychic vision; grounding.

TOPAZ - Balances emotions; calms passions; gout; blood disorders; hemorrhages; increases poor appetite; general tissue regeneration; VD; tuberculosis; reverses aging; spiritual rejuvenation; endocrine system stimulation; releases tension; feelings of joy.

TOPAZ - Give protection to warriors, has the qualities to put demons to flight and vanquishes the evil spells of sorcerers. Can be used for Divining purposes.

TOURMELINE - Dispels fear and negativity and grief; calms nerves; concentration and eloquence improve; genetic disorders, cancer and hormones regulated; raises vibrations; charisma; universal law; tranquil sleep.

TOURMELINE/BLACK - Arthritis; dyslexia; syphilis; heart diseases; anxiety; disorientation; raises altruism; deflects negativity; neutralizes distorted energies, *i.e.* resentment and insecurity.

TOURMELINE/GREEN - Creativity; opens heart chakra; immune system; psychological problems with the father; blood pressure; asthma; balancer; eliminates conflict within.

TOURMELINE/RUBELLITE - Creativity; fertility; balances passive or aggressive nature.

TOURMELINE/WATERMELON - Heart chakra healer; imparts sense of humor to those who need it; balancer; eliminates guilt; nervous system; integration, security and self-containment.

TOURMELINNE/BLUE - Lungs, larynx; thyroid; parasympathetic nerves.

TURQUOISE - Believed to bring love and courage. It is said to be a protector against violence in thought and deed. Also good for reducing bodily and mental tensions.

TURQUOISE - Master healer; protects against environmental pollutants; strengthens anatomy and guards against all disease; improved absorption of nutrients; tissue regeneration; subtle body alignment and strengthening; eye disorders.

ZIRCON - Attracts fame and fortune. It is considered to be a wishing stone. It is also a protector against accidents or natural disasters.

7

Ayurveda

Ayurvedic medicine is an ancient system of medicine which dates back to the vedic era. The word "Ayurveda" is a tatpurusha compound of the Sanskrit word āyus meaning "life" or "life principle", and the word veda, which refers to a system of "knowledge". This contains a comprehensive and complete natural herbo-mineral health care therapy. Ayurveda combines holistic assessment and diagnosis with diet, exercise and herbo-mineral medication.

It is native to India, and used by millions from Nepal, Sri Lanka, China, Tibet and Pakistan. It is now in practice for health care in American and European countries Ayurveda is a 'vibrant living system' which continues to flourish despite the advent of newer systems of medicine. Today you can reap its benefits, acclaimed all over the world, because it is not only about medicines and treatments but a complete guide to a healthy and happy way of life. Ayurveda is the alternative solution for today's ailments which have no cure in modern medical science.

In Hindu mythology many stories, incidents reveal that the sainya chikitsa (treatment of the army) or chikitsa (treatment) was in a developed state in the Vedic era. As described in Rig- Veda many examples unveil that the doctors of Gods, Aswini kumars performed many breathtaking surgeries and they were experts in body implants.

The examples of implantation of the steel legs in place of broken legs of Vishakha, the daughter of King Ravel, implantation of a horse's mouth in place of Dadhichi's head and again replace it with original makes it evident that Sainya Chikitsa was very progressed.

Atharva Veda, Kaushiksutra, Ramayana, Mahabharata and Harshacharit *etc.* novels have the description about the well- equipped doctor in the army quarters.

In Arthashastra (Economy) by Kautilya, there is a discussion about the doctors who possessed Yantra (equipments), Shastra (Tools), Agada (poison), Aushadha (Medicine), Sneha (love), Vastra (clothes), Parichaarak/Parichaarika (Nurses), to cure and heal.

Bharadwaj

He was the first person who learnt Ayurveda from Indra and educated the mankind. Shakatayam has described him in this sense:

The divine son of Brahma, Devarshi Angina had two sons: Utathya and Brihaspati. At the end of KrutaYuga or at beginning of TretaYug, Utathya's wife Mamta and Bruhaspati gave birth to Bharadwaj. The details can be found out in Matsya, Agni Bhahma or Hariwarsha Puranas as well as in Shree Madbhagwat Gita.

It has been described in Charaksanitha that in the midst of Tretayug the disease like temperature came into existence. It wa saftre the introduction of new diseases that After the development of disease Rishis send Bharadwaj to study Ayurveda from Indra. His Ashrama was in Prayag. (Even today this place is quite popular in the priests).

Saints had long life due to Tapobala (Power got from Yagna) and Rasayana (Divyaushadhi *i.e.* divine medicine) while Bharadwaj had the longest life in all these, as pointed in Autareya Aoranyaki.

He was good friend of the King Prushat of Panchal Desh and the father of Dronacharya, (Guru of Kaurav, Pandav). When Drupad, the son of Prushat became King after death of Prushat, Bharadwaj also passed away. Draupadi, the daughter of King Drupad, was the wife of Pandavs. In Mahabharata war, Drupad and Dronacharya both died. In this way it is sure that Bharadwaj was alive from KrutaYuga till some time before the end of Dwapar Yuga.

Maharshi Atreya, the Guru of Agnivesha learnt Ayurveda from Bharadwaj. Many of his successors also had knowledge of Ayurveda.

Aatreya

He was the son of Devarshi Atri, divine son of Bramha. The word Atreya can have different meanings like son of Atri or successor of Atri or Shishya (student) of Atri *etc*. But as in Charaksanhita at different places it is clearly understood that the relation is father- son only.

Atri Rishi was himself the Acharya (teacher) of Ayurveda as stated in, KashyapSanhita. He was the Guru (teacher) of Agnivesha who created CharakSanhita's original Novel 'Agniveshatantra'. He was also recognised by the name Punarvasu and Chandrabhag. The other two names mentioned in CharakSanhita are Bhikshu Atreya and Krishnatreya.

As Atreya and Bhikshu Atreya are mentioned in one instance (together), these two are definitely two separate identities. But names Krishnatreya and Atreya are not mentioned together at any place so it may be said that these two are names of one person.

Meaning which can be taken from this:

Soma: - Chandra, Chandrbhag in Punarvasu Atreya

Krishnatreya: - Durvasa

Dattatreya: - Bhikshu Atreya

But from paragraph of CharakSanhita Sootrasthan, it seems that being a krishnayajurved: Punarvasu can be called as Krishnatreya. He was a teacher (Acharya) of three Ayurvedas.

Agnivesh

Agnivesh was foremost among the disciples of Atreya and the author of the Agnivesh - tantra. Agnivesha is mentioned in Sarngaravadi, Aswadi, Gargadi and Tikakitavadi Ganas of Panini's Astadhyayi. Researchers have opined that 7th cent B.C. as the date of Panini. As Agnivesh is mentioned in more than one ganas it is evident that he existed long back and became a historical figure by the time of Panini.

Charaka

On the second stratum stands Charaka who was the first man to refine the treatise of Agnivesha thoroughly and enlarge it with his interpretations and annotations. His contributions in this respect were so spectacular that the original treatise in its new form began to be known on the name of Charaka himself instead of the original author.

Charaka enlarged the original Agnivesha-tantra in brief (Sutra) style with his annotations (Bhashya). Thus Charaka was the Bhasyakara of Agnivesa's work as was Patanjali for the Astadhyayi of Panini. That is why no wonder that Charaka has been identified as Patanjali, the author of Yogasutra and Mahabhasya.

Some scholars opined that Charaka was one of the branches of black Yajurveda and the persons following this branch formed a sect known as Charaka. Thus perhaps Charaka, the annotator of the Charaka Sanita, was a person belonging to that sect. There was also a branch of Atharvaveda known as 'Vaidyakarana', now extinct, which was perhaps more intimately connected with the tradition of vaidyas who served the masses while moving from village to village. This very mobile character (Carana) might have been responsible for the nomenclature 'Charaka'. This is supported by the theme of the Charak Sanhita, which is based on movement of the scene of activities from one place to another.

Bhaav mishra said that Charaka was the incarnation of Shesha (Naga). This on one hand proves the identity of Charaka and Patanjali and on the other hand, gave rise to speculations that Charaka belonged to the sect of Nagas who at one time were very powerful and established their footholds in several parts of the country.

Sylvan Levi, on the basis of the Chineses version of the Buddhist Tripitaka established that Charaka was the royal physician of the Emperor Kaniska who belonged to 1st or 2nd Cent. AD. The evidence however doesn't support the fact because Kaniska was a staunch Buddhist and the scholars attached to him like Aswaghosa, Nagarjuna *etc.* were all Buddhists while Charaka shows his fifth in vedas, brahmanism and positivism. Moreover, it looks improbable that a freely moving mendicant like Charaka might have accepted the bondage of a royal court. Hence Charaka, the annotator of the Charaka Sanhita, can't be the same person as the royal physician of Kaniska. It is possible that the name 'Charaka' being popular at one time was given to more than one person.

Asvaghosha, the poet laureate attached to Kaniska has not mentioned the name of Charaka though he has said Atreya as the pro founder of the School of Medicine. Had Charaka been in his colleagues he must have mentioned him.

The probable connection of Charaka with Kanishka leads to some more important but hidden points. Kanishka belonged to the Kushana dynasty, which was an offshoot of Shakas who came to India roaming about from Central Asia. As the Kunhan Raja view 'Charaka' not as a Sanskrit word but a Paali word 'Chareka'. Then the word 'Charaka' began to be used for inferior type of people.

In Navanitaka, Charaka is not mentioned as author of the text though Agnivesha is there and the followers of Charaka instead of having been assimilated in the general mass of vaidyas formed a separate group patronized by the Shakya kings. All these facts indicate that Charaka was either himself a Shakya or very close to them so that he had to struggle hard for putting his foot down. Perhaps during the same process, the work of Charaka (the Charaka Sanhita) was mutilated which was redacted and reconstructed by Drudhabala in part.

Drudhabala

Drudhabala, son of Kapilabala and resident of Pancanadapura reconstructed the Charaka Sanhita, which was deficient in its one-third part *e.g.* 17 chapters in Chikisitaasthana and entire sections of Kalpa and Siddhi. He completed the Sanhita in these respects by taking relevant materials from several treatises (then available).

There is difference of opinion as to which Drudhabala reconstructed 17 chapters of Chikitsitasthana. Bengal and Bombay editions of the text represent two prominent views. Chakrapani says that the eight chapters upto yaksma, arsa, atisara, visarpa, dwivnaniya and madatyaya were of Charaka and the remaining seventeen chapters were completed by Drudhabala.

Kapilabala, Drudhabala's father is quoted in Vagbhata's Ashtanga Sangraha (6th Cent A.D.). Drudhabala is quoted by Jejjata (9th Cent A. D.). Chakrapani has quoted both Kapilabala and Drudhabala in one context. Chakrapani also says Vagbhata as following the views of Kapilabala. This shows that Kapilabala preceded Vagbhata and was renowned at the latter's time. Hence Kapilabala and his son Drudhabala may be placed in 4th Cent. AD, during the Gupta period.

Bhaav Mishra

Bhaav Mishra, the son of Latkan Mishra was a Brahmin by cast. Keeping similarity to ancient Sanhitas, he introduced new thoughts and Dravyas. It is the last and important Novel of Laghutrai. It is popular in critics. Besides existing knowledge, addition of latest knowledge has made ayurveda more comprehensive and rich.

The writings that relate to Ashtang Ayurveda provide an introductory information of the eight branches of Ayurveda, namely: Surgery (Shalya), ENT with Ophthalmology (Shalakya), Medicine (Kaya), Bhoota, Gynac, Obstretics, Paediatrics (Kaumara), Medical jurisprudence and toxicology (Agada), Gerontology (Rasayana) and Science of Aphrodiasic (Vajikarana).

Shloka

Shusruta Sutra 1/6 - 7 Pg. 30 (Ayurveda Hitopadesha) Ayurveda, Upaveda of Atharvaveda, written by Brahma is named as the Brahma Sanhita. It envelops in utself one million shlokas in form of rhymes written in one thousand chapters.

Ayurveda was rewritten in the in eight parts popularly known as the ashtang ayurveda due to the decline in the intellectual level and life span of human beings.

Shalya

Shloka: Ref Shusruta Su. 7/4 the word shalya refers to the things that cause discomfort to the body and the mind. Shalya is of two types, namely: shaarir (within the body) and agantuj (outisde the body).

The shalya present inside the body is considered as the Shaarir Shalya. Example: Unhealthy teeth, Hair, Nails, imbalanced doshas, dhatus and mala, abscess, tumor, fetus *etc*. Whereas agantuj are the shalya presen toutside the body. Example: Thorns, Stone pieces, iron pieces, dust particles, worry, *etc*.

Shalya Tantra (Surgery)

Shloka : Shusruta, 1/8

It deals with the means such as Yantra (Tools), Shastra (Instruments), Kshara (Alkalies) and Agni (Fire) to remove the shalya in the body by different methods.

Shalakya

Shloka: Ref Shusruta Su. 1/8

It deals with the diseases related to Nose, Ear, Throat and Eyes. Inother words it deals with the disease of Urdhva Jatru region *i.e.* diseases in the organs above the Clavicle (Jatru) and their treatment.

Kaya Chikitsa (Medicine/Therapeutics)

Shloka : Shusruta Su. 1/8

It is the branch of Ayurved that deals with internal medicine. The treatment involved is called "Kayachikitsa", where Kaya means "Agni "and Chikitsa means "treatment".

It is noteable that the entire Ayurvedic therapeutics is based on this concept of Agni. The concept of Kaya (Agni) is unique and is responsible for bio- transformation.

As it is known that energy can neither be created nor it can be destroyed. In human body Kaya provides the necessary energy for all bodily activities. As energy can be changed from one form to another the living body derive energy from the food eaten and breathing air. Biological Kaya transforms this energy to the energy, which is utilized by our cells.

In simple words, the vitamins, minerals, carbohydrates, fats *etc.* eaten are bio transformed by this Kaya to the bodily substances. As long as Kaya is proper all the activities in body are carried out smoothly. Any disturbance in Kaya causes imbalance in the homeostasis (equilibrium) and disturbs physiology which is nothing but the disease. In ayurveda therapeutics devotes to correction and maintenance of biological Kaya through the means of Mantra, Mani and Aushadhi.

Ayurveda is a completely herbal-based medical treatment using only the purest of elements. As such, there are no side effects whatsoever. For the truly health conscious, natural herbal therapy is the best way to pamper the body and mind.

In today's fast paced world, Ayurveda is your means to vibrant good health. It eliminates all toxins from the body to enhance immunity and vitality. At the physical level, it rejuvenates the senses and vital organs and heals every part of the body. On the mental plane, it instils new energy, releases stress and provides total relaxation.

According to Charaka Samhita, "life" itself is defined as the "combination of the body, sense organs, mind and soul, the factor responsible for preventing decay and death."

According to this perspective, Ayurveda is concerned with measures to protect "ayus", which includes healthy living along with therapeutic measures that relate to physical, mental, social and spiritual harmony.

According to tradition, Ayurveda was first described in text form by Agnivesha, named - Agnivesh tantra. The book was later redacted by Charaka, and became known as the Charaka Samhita.

Another early text of Ayurveda is the Sushruta Samhita, which was compiled by Sushruta, the primary pupil of Dhanvantri, sometime around 1000 BC. Dhanvantri is known as the Father of Surgery.

There are Eight Branches (Ashthanga) of Ayurveda:

1. Internal medicine - Kayachikitsa Tantra
2. Surgery - Shalya Tantra
3. Ears, eyes, nose and throat - Shalakya Tantra
4. Pediatrics - Kaumarabhritya Tantra
5. Toxicology - Agada Tantra
6. Purification of the genetic organs – Bajikarana Tantra
7. Health and Longevity - Rasayana Tantra
8. Spiritual Healing/Psychiatry - Bhuta Vidya

Basic Principles of Ayurveda

Ayurveda is based on the following theories :

1. Pancha Mahabhuta theory (Five Elements)
2. Tridosha theory (Three Body Humors)
3. Saptadhatu theory (Seven Body Tissues)

Panchamahabhoot Theory

The basic premise of Ayurveda is that the entire cosmos or universe is part of one singular absolute. Everything that exists in the vast external universe (macrocosm), also appears in the internal cosmos of the human body (microcosm). The human body consisting of 50-100 million cells, when healthy, is in harmony, self-perpetuating and self-correcting just as the universe is. The ancient Ayurveda text, Charaka, says, "Man is the epitome of the universe. Within man, there is as much diversity as in the world outside. Similarly, the outside world is as diverse as human beings themselves." In other words, all human beings are a living microcosm of the universe and the universe is a living macrocosm of the human beings. In other words, everything is composed of the five elements (earth, fire, water, air, and space). In the human body, these elements combine to make three doshas or constitutional types. Your dosha is determined by the predominant energies that were present during your birth. If air and space were predominant, then your nature is Vata. If fire and water were predominant, then your nature is Pitta. If earth and water were predominant, your nature is Kapha. In Ayurveda, awareness of your dosha can help you to target potential problems with your health and will allow you to choose foods and activities to balance your lifestyle and support wellness.

The Pancha Mahabhuta, or "five great elements", of Ayurveda are:

1. Prithvi or Bhumi (Earth),
2. Ap or Jala (Water),
3. Agni or Tejas (Fire),
4. Vayu or Pavan (Air or Wind),
5. Akasha (Aether).

Concept of Elements Theory in Other Medicines and Religions

Classical Elements : Panchamahabhoot

Western : Air, Water, Aether, Fire, Earth

Chinese (Wu Xing): Water, Metal, Earth, Wood, Fire

Japanese (Godai): Earth, Water, Fire, Air, Wind, Void, Sky, Heaven

Hinduism (Tattva) and Buddhism (Mahābhûta):

Vayu/Pavan -Air/Wind, Agni/Tejas - Fire, Akasha -Aether,Prithvi/Bhumi - Earth, Jala - Water

Bön: Air, Water, Space, Fire, Earth

New Zealand: Air, Water, Flora, Fire, Earth

The Greek classical elements are Fire, Earth, Air, and Water. They represent in Greek philosophy, science, and medicine the realms of the cosmos wherein all things exist and whereof all things consist. According to Galen these elements were used by Hippocrates in describing the human body with an association with the four humours: yellow bile (Fire), black bile (Earth), blood (Air), and phlegm (Water). Some cosmologies include a fifth element, the "Aether" or "quintessence." These five elements are sometimes associated with the five platonic solids.

The Pythagoreans added idea as the fifth element, and also used the initial letters of these five elements to name the outer angles of their pentagram.

Man inscribed in a pentagram, from Heinrich Cornelius Agrippa's Libri tres de occulta philosophia. The five signs at the pentagram's vertices are astrological.

Tridosha Theory

The central concept of Ayurvedic medicine is the theory that health exists when there is a balance between three fundamental bodily humours or doshas called Vata, Pitta and Kapha.

☆ Vata is the air principle necessary to mobilize the function of the nervous system

☆ Pitta is the fire principle which uses bile to direct digestion and hence metabolism into the venous system.

☆ Kapha is the water principle which relates to mucous, lubrication and the carrier of nutrients into the arterial system.

In Ayurvedic philosophy, the five elements combine in pairs to form three dynamic forces or interactions called doshas. Dosha means "that which changes".

It is a word derived from the root dus, which is equivalent to the English prefix 'dys', such as in dysfunction, dystrophy, *etc.* In this sense, dosha can be regarded as a fault, mistake, error, or a transgression against the cosmic rhythm. The doshas are constantly moving in dynamic balance, one with the others. Doshas are required for the life to happen. In Ayurveda, dosha is also known as the governing principles as every living thing in nature is characterized by the dosha.

Dosha Related Elements

Vata: Air and Ether

Pitta: Fire and Water

Kapha: Water and Earth

Effect of Constitution Type On Body or Microcosm

	Vata	*Pitta*	*Kapha*
Functions of the Doshas:	Movement Breathing Natural Urges Transformation of the tissues Motor functions Sensory functions Ungroundedness Secretions Excretions Fear Emptiness Anxiety Thoughts Life force Nerve impulses	Body heat Temperature Digestion Perception Understanding Hunger Thirst Intelligence Anger Hate Jealousy	Stability Energy Lubrication Forgiveness Greed Attachment Accumulation Holding Possessiveness
Manifests in living things as:	The movement of: nerve impulses air blood food waste thought	Pitta controls the enzymes that digest our food and the hormones that regulate our metabolism	Cells whichmake up our organs and fluids which nourish and protect them
Characteristics:	Cold light irregular mobile rare fied dry rough	Hot light fluid subtle sharp malodorous soft clear	Oily cold heavy stable dense smooth

Contd...

Contd...

	Vata	Pitta	Kapha
Aggressive Dosha can result in:	Nerve irritation high blood pressure gas confusion	Ulcers hormonal imbalance irritated skin (acne) consuming emotions (anger)	Mucous build up in the sinus and nasal passages, the lungs and colon
Too little dosha force can result in	Nerve loss congestion constipation thoughtlessness	indigestion inability to understand sluggish metabolism	Experiencesa dry respiratory tract burning stomach
Predominant during the life stage of	Old age as we get older, we virtually shrink and dry out	Teen and Adult. During this stage, our hormone changes transforms us into adults	Childhood years. During this period, we grow or increase in substance of the body

Significance of Balanced Doshas

Every person (and thing) contains all three doshas. However, the proportion varies according to the individual and usually one or two doshas predominate. Within each person the doshas are continually interacting with one another and with the doshas in all of nature. This explains why people can have much in common but also have an endless variety of individual differences in the way they behave and respond to their environment.

Sapta-Dhatu (Seven Body Tissues)

Rasa - Final Metabolic Juice and Plasma

(Digestive System)

Rakta – Blood

(Blood Circulatory System)

Mamsa – Muscles and Tendons

(Muscular System)

Med – Fat

Majja - Marrow

Asthi – Bone

(Skeleton)

Shukra – Semen Fluied

(Reproductive System)

Srotas or Channels

Ayurveda refers the meaning of "Srotas" is the body channels consist of one or more then one system carries certain liquids, impulsions, and actions.

Such are as examples:

Pranvaha Srotas (Respiratory Channel)

Rasavaha Srotas (Metabolism)

Raktavaha Srotas (Blood circulatory channel), *etc.*

Concept of Prakruti and Vikruti

According to Ayurveda, basic constitution is determined at the time of conception. This constitution is called Prakruti.

The term Prakruti is a Sanskrit word that means, "nature," "creativity," or "the first creation." One of the very important concept of Ayurveda is that one's basic constitution is fixed throughout his lifetime. The combination of Vata, Pitta, and Kapha that was present in the individual at the time of conception is maintained throughout his lifetime.

Different persons can have different combination of Vata, Pitta and Kapha as their basic constitution or Prakruti. This is how Ayurveda can explain the subtle differences between individuals and explains why everyone is unique and that two persons can react very differently when exposed to the same environment or stimuli. Your Prakruti is unique to you just as your fingerprint and DNA. Thus, in order to understand a person, it is necessary to determine his or her Prakruti.

Diagnostic Tests based on Prakruti

A couple of parameters are explained in ayurveda to determine the Prakruti of a human, *i.e.*, whether or not the person is Flexible/optimistic/Ambitious/practical/intense/ Calm/peaceful/solicitous. Such type of questionnaire is described in Ayurveda to evaluate the Prakruti of a person. After finding the right Prakruti than an ayurvedic physician goes for finding Vikruti (Pathology).

Vikruti

For finding the pathogenesis of any ailment the parameters are given in Ayurveda in questionnaire form. Questions may varied according to type of disease. As example:

Symptom	0 = Does not apply 3 = Strongly apply			
Poor appetite	0	1	2	3
Worried	0	1	2	3

Branches of Ayurveda

Bhoota (Graha)

it is the branch of Ayurveda that deals with diseases acquired or inherited from apparently unknown causes. In modern terminology it can be considered as idiopathic diseases in which the exact cause of disease is unknown.

According to Ayurveda, diseases are caused by affliction due to Deva, Asura, Gandharva, Yaksha, Rakshasa, Pitara, Pishacha, Naga and other bad demons or evils. The exact patho-physiology of these disorders is to be extensively researched.

Bala/Kaumara

it is the branch of ayurveda that deals with the diagnosis and treatment of diseases related to preconception, childbearing (Pregnancy), childbirth (delivery) and diseases of children (paediatrics).

Rasayana Tantra

It is the branch of ayurveda that deals with various aspects of preventive health care. Without rasayana it is possible to gain neither oratory nor the desirable aura. It includes longevity, improved memory, health, youthfulness, glow, complexion, generosity, and strength of body and senses. Rasayana improves the metabolic activities and results in best possible bio- transformation leading to health.

Vajikarana

it is the branch of ayurveda that deals with the sexual aspects. It includes medications for diseases related with reproduction namely spermatogenesis, aphrodisiacs *etc.*

Basic Principles of Ayurveda

Everything in this universe is composed of five elements. The human body is also made up of the five elements and the soul. These five elements are:

1. Prithvi or earth
2. Apa or water
3. Tejas or fire
4. Vayu or air
5. Akash or space

These five elements are the basic constituents and everything can be explained in terms of these five elements. For example, the bulky and solid part is earth, the digestive enzymes which are responsible for cooking or digesting is fire, the hollow, empty parts and the big beer belly is space.

These five elements in their biological form in the living body are of three different kinds. These are three primary life forces or three biological humors. The Ayurvedic term for these forces is Dosha. As they are three in number they are called as Tridosha (Tri means Three). In Sanskrit they are known as : Vata, Pitta and Kapha.

As these are specific terms or names of the three types of humors in the body it is not possible to translate them into English. They can be understood, experienced or felt only from their qualities, behavior and actions in the body.

Vata is compared to air, Pitta is compared to fire and Kapha is compared to mucus and water.

Other important basic principles of Ayurveda which are briefly mentioned here are:

Dhatus – These are the basic tissues which maintain and nourish the body. They are seven in number, namely – chyle, blood, muscles, fatty tissues, bone, marrow and semen. Proper amount of each dhatu and its balanced function is very important for good health.

Mala – These are the waste materials produced as a result of various metabolic activities in the body. They are mainly urine, feces, sweat, *etc.* Proper elimination of the malas is equally important for good health. Accumulation of malas cause many diseases in the body.

Srotas – These are different types of channels which are responsible for transportation of food, dhatus, malas and doshas. Proper function of srotas cause many diseases.

Agni – These are different types of enzymes responsible for digestion and transforming one material to other.

All these factors need to function in a proper balance for good health. They are inter-related and are directly or indirectly responsible for maintaining equilibrium of the Tridosha.

Ayurveda believes that specific disease conditions are symptoms of an underlying imbalance. It does not neglect relief of these symptoms, but its main focus is on the big picture: to restore balance and to help you create such a healthy lifestyle that the imbalance won't occur again.

Living in health and balance is the key to a long life free from disease. Perhaps the most important lesson Ayurveda has to teach is that our health is in our own hand it is our own choice and wisdom. Every day of our lives, every hour of every day, we can, and do, choose either health or illness.

When we choose wisely, nature rewards us with health and happiness. When we persistently choose unwisely, nature, in her wisdom, eventually sets us straight: She makes us sick and gives us a chance to rest and rethink our choices.

The theoretical side of Ayurveda provides insights into how to live one's life in harmony with nature and natural laws and rhythms. Its practical side - specifically its guidelines for an intelligently regulated diet and daily routine, its techniques for stress management, and its exercises for increased fitness and alertness-help us take control of our lives and develop radiant health.

The unique principle in Ayurveda is this: Each and every person has his/her own individual body constitution according to three elements Vata, Pitta and Kapha.

Vata person has known as Vatic (controls movement).

Pitta person has known as Paitic (controls metabolism).

Kapha person has known as Kaphic (controls structure).

According to Avurveda, everything is composed of the five elements (earth, fire, water, air, and space). In the human body, these elements combine to make three doshas or constitutional types. Your dosha is determined by the predominant energies that were present during your birth. If air and space were predominant, then your nature is Vata. If fire and water were predominant, then your nature is Pitta. If earth and water were predominant, your nature is Kapha. In Ayurveda, awareness of your dosha can help you to target potential problems with your health and will allow you to choose foods and activities to balance your lifestyle and support wellness.

The best way to determine your dosha is to be examined by an Ayurvedic doctor. You can get an idea of your constitution by taking this self-test. Put a check beside each statement which describes you. Your predominant dosha will be the one with the most checks. It's common for a person to have a combination constitution, especially to favour two of the doshas over the third.

Vata - Air and Space

Thin build and delicate bone structure (tall or short)

Prominent joints

Doesn't gain weight easily

Dry skin which chaps easily and may be rough

Cool, prominent veins

Dry, coarse hair

Small, active, dark eyes

Very physically active

Active mind

Thin, dry lips which chap easily

Nails may be brittle, ridged, or cracked

Little strength

Little endurance - tires easily

Variable appetite, but can get very hungry

Bowel movements irregular, may be dry and hard

Tendency toward stress-related anxiety

Good short term memory, poor long-term memory

Light sleeper

Cold hands and feet

Minimal perspiration

May dream of flying, movement, fears

Tendency towards constipation, anxiety, nervousness, poor sleep

Vata types tend to be quick, wiry, and creative. The Vata's primary organ is the colon. Vatas can benefit from a regular routine to ground high levels of moving energy. Travel, especially air travel, can imbalance a Vata. Vata types are aggravated by cold, frozen, or dried foods. In order to be balanced, a Vata should favour moist, warming foods and spices over raw, frozen, or dried foods.

Pitta - Fire and Water

Medium build

Medium strength

Moderate body weight, may be athletic and muscular

Skin is oil, warm, and sensitive, may be ruddy or inflamed

Hair is fine and oily, may be balding or prematurely grey

Medium, penetrating, photosensitive eyes

Lips are soft, medium-sized

Soft, flexible nails

Strong appetite, irritable if meals are missed

Enjoys physical activity, especially competitive

Bowel movements easy and regular, may be soft, oily, or loose

Possessed of mental focus and clarity

May become angry or frustrated when under stress

Excellent memory

Usually sleep well

Good circulation, perspire frequently

Dreams may be angry, violent, passionate, or fiery

Tendency towards inflammation, high blood pressure, hypersensitivity, aggression

The primary organs of the Pitta are the small intestine and stomach. Pitta types, therefore, have good digestion as well as intestinal fortitude or strong will and determination. Pitta is associated with the fire element and Pitta types tend toward heat, oiliness, and lightness. Imbalance in a Pitta type may manifest as skin rashes, burning, inflammation, fever, ulcers, anger, jealousy, or excessive urination. Pitta types benefit from staying cool and dry. Pittas should avoid oily or fried foods as well as caffeine, alcohol, red meat, hot spices, and salt. Balancing choices include fresh fruits and vegetables and whole grains. Pitta types should try to get plenty of fresh air and express their emotions.

Kapha - Earth and Water

Thick or stocky build. Larger bone-structure.

Tendency toward overweight

Thick cool skin, prone to acne

Thick, shiny hair

Large, round eyes with thick eyelashes

Lips are large, full, and smooth

Strong, thick nails

Strong, good endurance

May tend toward physical inactivity

Bowel movements thick, may be oily, heavy, or slow

Mind is calm and slow

Tendency to avoid conflict and difficult situations

Memory is slow but reliable

Elemental Balance Reference Chart

Category	Ether	Air	Fire	Water	Earth
Characteristics	Vacuum	Mobile/Dynamic	Transforming	Transportation	Structure
	Light	Cold	Hot	Cool	Cool
	Openness	Dry	Spreading	Heavy	Dry
	Immeasurable	Rough	Liquefying	Liquid	Heavy
	Omnipresent	Movement	Intense	Flowing	Nurturing/Nourish.
	Subtle	Erratic	Sharp	Cleansing	Tough/Gross
	Expanding	Subtle	Light	Nourishing	Stable
	Formless	Light	Subtle	Connected	Constant/Slow
	Soft	Clear	Dry	Oily/Slimy	Reliable
	Smooth	Soft		Slow	Rough
Dominant body system	Etheric Body, Higher Intellect	Nervous System, Mind	Liver, Math and Scientific Intellect	Plasma, Lymph	Bones, Cranium, Muscles
Ideal Function	Descent of intelligence into the heart	Movement in a particular direction	Brilliance and Luminosity	Nourishment and Cleansing	Gravitation and Structure
Human Creative Expression	Spirit energy, infinite potential, voice	Direction, Breath, Touch	Visual, colour, action	Love, connection, birth and sensuality	Building, planting seeds, growing trees. Long term expressions
Type of energy and direction	Nuclear energy, inwards	Electrical energy, the 5 prans of direction	Radiant energy, upwards and outwards	Chemical energy, downward movement	Physical and mechanical energy. Downward movement
Psychology	Non-attachment, Aloof	Multiplicity, Fear, Worry	Joy, Focus, Anger, Impatience	Connection, Love, Attachment, Sadness	Relaxed, Content, Stubborn, Fixed
Exercise, Movement	Awareness, stillness, meditation	Aligning, breathing, walking, jogging	Cardio, heating, sprinting, karate	Elasticity, fluidity, asana, stretch	Toning, weights, resistance
Nutrition to increase element	Philosophy, Fasting	Oxygen, Freshness	Herbs, Spices	Carbohydrates, Fats	Proteins, Minerals

Sound, heavy sleeper

Moderate perspiration

Dreams may be peaceful and romantic and may involve the water or ocean

Tendency towards respiratory congestion, water retention, lymph congestion, cystic acne, tiredness.

The primary organ of the Kapha is the chest. Kapha may be associated with the production of mucus. Kapha types have good strength and endurance. Kaphas naturally prefer to follow routine, but breaking from routine is beneficial. Kapha types tend to attach to people or things. Food and security are important. Kapha types have a tendency towards weight gain, congestion, sinusitis, sluggishness, diabetes, and water retention. Kaphas gain balance from engaging in physical activity and avoiding fried or fatty foods, icy drinks, sweets, or excessive amounts of starchy foods. Fresh vegetables, change, and challenge are all of benefit to Kapha types.

Same principle applies to each and every food or drink item that we take. So, food also can be Vatic, Paitic and Kaphic. By the same principle, if a "Vatic person" eats or drinks a "Vatic item" more and more: that could be the cause of a "Vatic disease" because each and every disease is also classified by the same principle. You will be surprised to know that in Ayurveda it is already established that each and every herbo-mineral medicine has also the same characteristics based on the same principle and that is why by using a "Paitic medicine" you can cure a "Vatic disease" and so on.

Ayurvedic Medicine works to restore the balance of the three factors on the same principle. There are 16000 plus beneficial herbs, only available in India that contribute to the medical system called Ayurveda. Ayurveda is not an invasive system of medicine. It has no side effects, only side benefits. Moreover, it is inexpensive.

Ayurvedic Home Remedies

Constipation: Copper pot, fill it with drinking water, keep it overnight (room temperature), next morning warm up the water and drink it first in the morning.

Gas – Abdominal: Drink Ginger juice 1 ml to 2 ml with warm water. And/or drink Club soda with some salt and Black pepper powder.

Common Cold: Ginger powder, Ghee (Clarified Butter) and Jaggery (Brown Sugar) mix them together in same quantity and have it first thing in the morning.

Indigestion: Ginger powder, Ghee (Clarified Butter) and Jaggery (Brown Sugar) mix them together in same quantity and make small balls size of small marbles and have it after each meal.

Poor – Metabolism: Ginger powder, Ghee (Clarified Butter) and Jaggery (Brown Sugar) mix them together in same quantity and make small balls size of small marbles have it after each meal.

Cough: Apply Vicks Vaporub (or similar products) on the chest and gently massage clockwise. Also take a 1 glass of milk, add 1 tsp. of Turmeric powder, warm it up, and drink it. You can repeat it several times of day if you wish or according to the condition of person.

Vomiting: ½ gram of ground cardamom, mix with 1 tsp. Honey and have it slowly. And/ Or Drink Club Soda with some salt and black pepper powder added.

Worms: First thing in the morning, Eat 1 small piece of Jaggery (Brown sugar) then wait for 5 minutes and eat 1 tsp. of Caraway Seeds or Caraway powder.

Acidity: Drink 1 glass of milk with 1 tsp. of Ghee (clarified butter).

Colic Pain (Stomach Pain): Take a ½ gram of Asafetida (Hing Powder/Indian Spice), then mix it with little water make a paste of it and fill up your Belly button (umbilicus), lie down for 15-30 minutes and you will release some gas for few times, after that you will be fine.

Fat Remover/Weight Loss: Boil 1 glass of water, cool it down then add 1 tsp. of Honey and drink it. Do this every morning. Or take 5 – 10 gms. of Triphala Powder boil it with 1 ½ glass of water for 5 – 10 minutes, then cool it down little as the temperature you can handle to drink, drink it everyday.

Paralysis: Black pepper powder mix it with Sesame seed oil, warm it up and massage on effected areas regularly.

Debility (weakness): Take a 1 glass of Milk, add 1 tsp. of Ghee (clarified butter), add little bit of Ginger powder, warm it up and drink it regularly. And/Or Eat Chick peas, split chick peas, roasted chick peas. And/Or Eat 5 Dates with Ghee (clarified butter), everyday in the morning.

Eye Problems: 1 tsp of Triphala powder, soak it whole night in the one glass of water, next morning filter it with cheese cloth and wash your eyes with it several time a day. Do it regularly, it makes big difference.

Ear Problems: If you have whistling sound eat Pecan nuts (walnuts), And/Or Sesame seed oil or Ghee (clarified butter) warm it up on low temperature and pour 2-3 drops into ears regularly.

Tooth/Gums Problems: Sesame seeds oil or Clove oil, massage gently with your fingers on teeth and gums every morning. Always use soft brush, don't brush your teeth more then one time, but if you will like to then use your finger then brush. There is some great Ayurvedic toothpowder/toothpaste available at Ayurved Centre, 416-778-9341

Sleeplessness/Insomnia: Massage Ghee (Clarified Butter) or Castor oil on the feet before going to the bed daily for 5 to 10 minutes.

Mental Disorders: Massage Sesame seed oil/Ghee (Clarified Butter) gently on the forehead, scalp regularly any time of the day or do it at night, next morning do shampoo. And/Or Soak 2-5 almonds over night in the water, next morning, remove skin of almonds and have it.

Piles: Take cotton and soak it with castor oil and apply and placed on effected areas and keep it for while.

Bed Wetting: Do massage with Ghee (clarified Butter) on the abdomen (clockwise) for 5 to 10 minutes, before going to the bed on regular basis. And/Or Chew Black Sesame seeds (5 gms), before going to the bed And/Or Coriander powder with Sugar (10 gms. Each), mix it together and have it 2-3 times a day in summertime only.

Night Discharge: Small towel soak it with the cool water and placed on the abdomen and then go to the bed. And/Or First thing in the morning, swallow 3 pieces of Black Pepper, then eat 2 tsp of Ghee (Clarified Butter) regularly.

Sexual Weakness: 1 or 2 tsp. of Ashwagandha Powder (Indian Ginseng), 1 tsp. of Ghee (Clarified Butter), 1 cup of milk and sugar for your taste, boil it for 5-10 minutes then drink it with temperature you can handle as hot or warm. Do it regularly in the morning or once a day. Also eat 1 tsp. Chyavanprash twice a day, regularly. And/Or Eat Urid Dal (lentils) in your food. Also have warm milk and butter in your meals. If you have less Sperm Counts then fry 5 to 10 g. Onions into the Ghee (clarified Butter), twice a day.

Skin problems/Prickles/Eczema/Acne/Pimples: Gently massage affected areas with Ghee regularly then after 2-3 hours take a shower or sleep over with it at night. Take a shower only with Sandalwood soap or Neem Soap only. And/Or make a paste with Milk, Sandalwood powder and Rose water and apply on effected areas, when its dry take a shower with warm water not a hot water and again only use above mentioned soaps.

Hair Loss: Regularly do massage on the crown area of the head with Ayurvedic Medicated oil and use Ayurvedic Shampoos (but when you do massage always use your palm not a fingers because your nails will cut your hair more). Take internally Bhringaraja, Manjishta and Mukta Shukti Pishti.

Eye problems: Massaging Castor oil on your feet regularly not only improves your vision but may be help you reduce numbers on the glasses.

Low Blood Pressure: First thing in the morning, 3 pieces of Whole Black Pepper swallow it regularly.

Keeping in mind all the benefits of ayurveda it must be kept in mind that anything and everything that is herbal can not naturally become Ayurvedic medicine, but only the therapy which considers the above mentioned concepts of Ayurved qualifies to be called as Ayurvedic medicine.

8

Bach Flower Medicine

Bach Flower Remedies

Flowers have been used for healing since ancient times and their use much pre-dates homoeopathy. Many cultures, including Egyptian, Malay, African and aboriginal Australian, used flowers to treat emotional states and imbalances. In Europe folklore on the healing power of flowers goes back to medieval times at least, but the earliest recorded use of flower essences occurred in the sixteenth century, when Paracelsus, the great physician and healer, collected the dew from flowers to treat his patients' emotional states.

Flower essences do not work from biochemical interaction with the body, but they contain the life force of the plant as a specific energetic pattern which works on the human energy fields which in turn influence mental, emotional and physical well-being.

Dr. Edward Bach (1886 – 1936) is well known for bringing the use of flower essences into our modern times. He was an English surgeon, bacteriologist and later pathologist. In spite of his success in orthodox medicine, he felt that doctors concentrated too much on disease and not enough on the patient. He was drawn to homoeopathy but felt that there was another way. In 1930 he left his Harley Street practice and decided to research natural remedies from flowers which addressed particular mental and emotional states. He was sure that by treating patients in this way, their own natural healing energy could be unblocked and used to address their physical symptoms.

As an holistic physician, he was very much aware that a person's attitude and personality has an effect on their health and well-being. This is why there is a strong link with colour Therapy since colour also has a profound effect on our emotions as well as our physical and spiritual well-being. The other obvious reason for including this piece on Bach flowers is, of course, the link with nature.

With the emotional aspect of illness in mind, Dr.Bach went on to develop 38 flower remedies plus the combination remedy known as Rescue Remedy. The remedies are all made using the flowers of plants, bushes and trees growing in the English countryside. The flowers are then combined with pure water and placed in full sunlight. The only exception is Rock Water which is made from pure natural spring water which is said to have healing properties.

In 1930 Dr.Bach wrote a small book called "Heal Thyself". Published in 1931, this book is still in print and is a very useful book.

In the 1930's, Dr. Edward Bach used his knowledge of homeopathy to devise a plant-based remedy to treat a particular set of negative feelings. The Bach Flower Remedies are thought to work by stimulating the body's capacity to self-heal and by balancing negative feelings. Remedies are selected which most closely correspond to a person's basic personality type or the particular emotional stress that is being experienced. The Bach Remedies are generally considered to be non-toxic, non-addictive, and safe to use with other medications. Bach flower essences are widely available from suppliers who carry homeopathic or naturopathic remedies.

Flower	*Indication*
Agrimony	Repressed worries, for one who has a cheery outward appearance that conceals internal fears and concerns
Aspen	Apprehension, fear of the unknown, anxiety, foreboding
Beech	Intolerance, being critical of others, narrow-mindedness, feeling annoyed by others
Centaury	Inability to say 'no', trying to please others, easily exploited
Cerato	Lack of trust in own decisions, always turning to others for decisions
Cherry Plum	For some compulsions and obsessions, fear of impulsively doing something known to be wrong, impulsiveness
Chestnut Bud	Failure to learn from mistakes, destructive patterns of behavior
Chicory	Possessive love, needing to be involved in the lives of others, meddling
Clematis	Daydreaming, withdrawing into a fantasy world
Crab Apple	Feelings of self-hatred, poor self image, shame of physical appearance
Elm	Feeling overwhelmed by responsibility, feeling unequal to a task
Gentian	Discouraged by a setback, making 'mountains of molehills'
Gorse	Hopelessness, despair, feeling nothing can be done
Heather	Self-centered, needs to talk about one's self, unhappy when alone
Holly	Hatred, jealousy, envy
Honeysuckle	Dwelling on the past, expecting to never be happy again
Hornbeam	Procrastination, fatigue due to boredom
Impatiens	Impatience, irritability with slowness in others

Flower	Indication
Larch	Lack of confidence, expectation of failure
Mimulus	Fear of known things or things encountered in everyday life, such as fear of the dark, fear of growing old, *etc.*
Mustard	Gloominess, feeling overshadowed by a cloud
Oak	For one who continues past the point of exhaustion, workaholic, one who continues to fight a battle that cannot be won
Olive	Exhaustion following mental or physical effort, lacking vitality
Pine	Guilt, perfectionist, dissatisfaction with the efforts of others
Red Chestnut	Too much concern for the welfare of loved ones, always anticipating the worst
Rock Rose	Fright, terror, extreme fear in the face of an emergency or accident
Rock Water	Self-denial, repression, being too hard on oneself
Scleranthus	Indecision, mood swings, unable to achieve balance
Star of Bethlehem	Shock, refusing to be consoled, for trauma following receipt of bad news or loss
Sweet Chestnut	Extreme mental anguish, when it seems no hope remains, when you have reached the limits of your endurance
Vervain	Overly enthusiastic, feeling the need to convert others over to your way of thinking
Vine	Inflexibility, dominance, domineering, seeming to have too much self-assurance
Walnut	Protection from change and unwanted influences, for periods of transition and adjustment to new beginnings, protection from peer pressure and negativity from others
Water Violet	Pride, arrogance, for loners who appear aloof, for those who seem unapproachable or distant
White Chestnut	Unwanted thoughts and mental arguments, for when the mind replays the same 'broken record' to distraction
Wild Oat	Uncertainty over one's direction in life, for feeling lost
Wild Rose	Apathy, resignation, accepting what life has in store without an effort to influence it
Willow	Self-pity, resentment, for unjust suffering and the feelings that tend to accompany it, for feeling unfortunate
Rescue Remedy	Mix of cherry plum, clematis, impatiens, rock rose and star of Bethlehem, used to help deal with any emergency or stressful event.

Since Bach's time there has been increasing interest in flower essences from all over the world. They can be remarkably physical in their effects as well as treating emotional states. Their action can be compared to the effects we experience on hearing a particularly moving piece of music or seeing something which stirs us such as a work of art or a beautiful landscape. The light or sound waves reaching us may evoke profound feelings in our soul which indirectly affect our breathing, pulse rate and other physical activity. The essences can be combined to produce remedies for specific problems. Common problems such as anxiety or lack of confidence can be treated effectively. Furthermore flower essences can be combined for a particular person and their own individual problems. They are completely safe and have no side-effects.

9

Biochemic Medicine

Biochemic Tissue Salts are inorganic minerals which are important for daily functioning, maintaining, rebuilding, and detoxifying the cell. The illness of the cell originates through the loss of inorganic minerals. Biochemic Tissue Salt therapy affects the body, and even the mind, on multiple levels. Biochemic Tissue Salts, also known as Cell salts or Schuessler Salts, are minerals in an energy form. Mineral energies form the functional part of every cell, and ideally are derived from organically grown plant foods eaten in an uncooked state. Unfortunately, due to incorrect treatment of the soils, growing conditions and added chemicals, the mineral and vitamin content of our plant foods has diminished to levels unable to sustain cellular resonance.

Although the importance of vitamin supplementation in healthcare has been recognized, mineral replacement in the same form as found in plant foods has been underestimated, even though it was recognized and developed over a hundred years ago by Dr. Wilhelm Heinrich Schuessler in Germany. With the rediscovery of the Biochemic System of Medicine, speedy recoveries to health and maintenance of wellbeing have become possible. The first signs of Tissue Salt deficiencies show in the face, long before physical symptoms occur. You are encouraged to familiarize yourself with the facial signs and body symptoms related to the Tissue Salt deficiencies. In this way you are able to take the first step toward working with your own body, assuming responsibility for your wellbeing. You will also find that you slow down the aging process and start looking younger too!

Biochemic Tissue Salts are extremely safe and gentle remedies, which, apart from the very rare case of lactose intolerance, have no risk of side effects. They are normal naturally occurring substances found in the body so they will not interfere with or counteract the effect of other medication or treatment. After the first sufficient intake of Tissue Salts there is an immediate absorption through the membranes of the mouth into the bloodstream, becoming available to the cells of the body within a short time. This gives rise to an increasing feeling of wellbeing with the relief of pain and symptoms of discomfort as the cellular imbalances are corrected over a period of time.

As the stresses on the physical body decrease and feelings of healthy wellbeing are restored, the energetic pattern is also shifted: cellular vibrations are corrected and emitted from the dense physical body into the surrounding energy field, aligning its frequency with the tissues and cells.

The twelve tissue salts with curative properties as introduced by Schuessller are described below:

CALCAREA FLUORICUM

For glandular tumors, varicose veins, prolapsus uterus, bone affections *etc.*

This salt is indicated in all diseases affecting the substance forming the surface of the bone, the enamel of the teeth and part of elastic fibres, no matter in what structure they may be found. A deficiency of the elastic fibres in the muscular tissue causes a relaxation, which is a primary condition in many diseases. The facial sign for Calcium Floride deficiency is oblique or vertical lines at the inner corners of the eyes. Calc Fluor maintains the essential quality of elastic tissue and is chiefly indicated for conditions resulting from relaxed muscular or supportive tissue. These include diluted blood vessels, varicose veins, poor circulation, haemorrhoids, constipation, strained tendons and ligaments, prolapsed organs and chapped or hardened skin. Calc Fluor helps reduce the formation of stretch marks.Indicated in all ailments that can be traced to a relaxed condition of the elastic fibres, including dilatation of blood vessels such as blood-tumors, piles, enlarged and varicose veins, indurated glands, enlargement of the heart, relaxed uvula, prolapsed of the womb, *etc.* Also useful in the diseases of the respiratory organs with difficult expectoration of small yellow lumps. The symptoms are all worse in damp weather and are relieved by fomentation and by rubbing.

CALCAREA PHOSPHORICUM

For anaemia, chlorosis, deficient development, dentition disorders, bone affections, dyspepsia *etc.*

Pale, waxy looking areas and facial skin are indicative of a Calcium Phosphate deficiency. Calc Phos is the most abundant tissue salt in the body. It assists with the digestion and absorbtion of food and is vitally important for the building of sturdy bone and body structures and a robust constitution. Its excellent restorative powers help speed up convalescence and to replenish the body's reserves. Calc Phos is indicated for blood and bone disorders including anaemia, oestoporosis and growing pains, and is also benefitial for fear and anxiety. It is excellent for growing children and the elderly and is a good supplemental tonic with any treatment.

The sphere of the Phosphate of Lime includes all bone diseases, whether due to some inherited dyscrasia or to defective nutrition in osseous and other structures dependent upon a proper distribution of lime molecules in the body. It is the bone cell-salt. Without this element, no bone can be formed and hence it is a valuable remedy in childhood when the bones are in the condition of being formed. Cal. Phos. is also found in the gastric juice and plays an important part in the process of digestion and assimilation. It is of importance in anaemia, in chlorosis, in convulsions and spasms, in weak scrofulous subjects; during dentition when the teeth are slow in making their appearance and decay too rapidly; also in other teething disorders. In convalescence after acute and chronic wasting diseases, it acts as a tonic by building up new red blood corpuscles and by restoring the lost vitality. Deficient development of children and young people emaciation without apparent cause onanism, suppuration of bones, spinal weakness and curvature. Aids the union of fractured bones. Rheumatism of the joints, chronic enlargements of the tonsils and goiter. Cholera infantum, faeces hot and offensive. Dyspepsia; food seems to lie in a lump; vomiting after cold drinks. Cold, change of weather, motion, getting wet, generally aggravates symptoms. Relieved by rest, warmth and by lying down.

CALCAREA SULPHURICUM

Calcium Sulphate deficiency shows in the face as acne in adults, and severe cases of acne in adolescents. Calc Sulph is a blood purifying tissue salt that helps the liver to remove waste products from the blood stream. It is indicated for conditions associated with the discharge of pus, such as pimples, boils, abscesses and also for ulcerations. It is particularly effective at drying up matter that continues to ooze after the bulk of the pus has been discharged such as in a slow healing or weeping wounds.

For suppurations, boils, carbuncles, gonorrhoea and chronic catarrhal conditions.

This is a most excellent remedy in suppurations to which it is closely related. It is an eliminator of effete matter. A deficiency of this salt causes long protracted suppurations. It is indicated in swelling of the soft parts in connective tissue with threatened suppuration or where suppuration has already taken place. Abscesses either aborting or hastening suppuration. Suppurative stage of eruptive diseases of the skin. Boils, carbuncles, felons, pimples and pustules on the face; suppurative stage of gonorrhea; 'in fact in all ailments where pus-formations are liable to develop or have already developed. Chronic catarrhal conditions with purulent secretions. Advanced stages of lung diseases. The true indication for this remedy is a thick, heavy yellow pus, sometimes streaked with blood. Symptoms are aggravated by getting wet or by washing and working in water,

FERRUM PHOSPHORICUM

Facial Signs of Ferrum Phosphate deficiency include redness of the forehead, outer ears, or entire face looking flushed. When very deficient, dark circles under the eyes appear. Ferrum Phos an oxygen carrier and anti-inflammatory, and is the principle biochemic first aid remedy. It should be used at the onset of all inflammatory conditions, *i.e.* disorders ending with 'itis'. Key indications for Ferrum Phos are shortness of breath (including asthma), weak immunity, colds and influenza, viral, bacterial, fungal and parasitic infections, pain, fever, anaemia and heavy menstuation.

For acute congestions, inflammations, fevers, *etc.*

Ferrum Phos. is given in the initial stage of all congestions, inflammations and fevers. This salt possesses an affinity or attraction for oxygen and carries it to all parts of the body and hence is an important agent in all diseases in which the blood and the corpuscles are involved. Consequently, the first remedy in all cases depending upon the relaxed condition of the muscular tissue and in abnormal conditions of the corpuscles of the blood themselves. It gives strength to the circular wall of the blood vessels. It supplies the colour to the blood corpuscles. A deficiency of this cell salt is the cause of fevers and inflammatory conditions and so indicated in all febrile disturbances and in inflammatory and congestive diseases at the beginning, especially before exudation has commenced. The principal accompaniments of these conditions calling for Ferrum Phos are manifested by flushed face, fever with full pulse, hot, dry skin, thirst, pain and redness of the part. An excellent remedy in anaemia, chlorosis, pneumonia, inflammatory rheumatism, tonsillitis, diphtheria, cystitis, nosebleed, incontinence of urine, apoplexy, gastritis *etc.* The symptoms calling for this remedy are always aggravated by motion and ameliorated by cold.

KALI MURIATICUM

Kalium Muriaticum deficiency is indicated facially by a bluish-whitish area under the eyes, and when very deficient the entire face appearing pale and white. This deficiency may present as allergies, hayfever, sinusitis, snoring or nausea (including morning sickness). Kali Mur is the principle biochemic remedy for catarrhal conditions accompanied by the discharge of thick, white mucous or phlegm from the skin or mucous membranes.

For second stage of inflammatory diseases, diphtheria, croup, catarrh, pneumonia, glandular swellings, deafness *etc.*

Kali Mur. is given for the sequel of all inflammations, for exudations and infiltrations, especially of a fibrous character and in inflammations of serous membranes, when the exudation is plastic in nature. A most excellent remedy in the later stages of all catarrhal states. This cell-salt works with the fibrin, which is found in every tissue of the organism except boric. In inflammatory exudations we find fibrine in the serous cavities and on the mucous membranes. The use of Kali Mur. usually follows Ferrum Phos. in inflammatory disorders. The characteristic symptoms of Kali Mur which is always indicated in the second stage of the inflammatory diseases, are glandular swellings, discharges or expectorations of a thick, white fibrinous consistency, white or grey exudations and a white or grey coating of the tongue. The efficacy of this remedy is demonstrated in chronic catarrhal conditions, croup, diphtheria, dysentery, pneumonia *etc.* Diarrhoea with pale, yellow faces. With Ferrum Phos is should be given in coughs. Useful in deafness from the catarrh of the Eustachian tubes, in skin eruptions with small vesicles containing yellowish secretions, ulcerations, ulcerations with swelling and white exudations, in rheumatism with swelling of the parts, in leucorrhoea and gonorrhea with characteristic discharge. Symptoms in general are worse from motion, while gastric and abdominal symptoms are worse after taking pastry, rich and fatty foods.

KALI PHOSPHORICUM

Kalium Phosphate deficiency is indicated by a bluish-greyish undertone around the mouth and chin, or a generally greyish complexion. Kali Phos is a brain and neve cell nutient and acts a s stong antispectic. Its deficiency may present in symptoms of lack of concentration, hyperactivity in children, memory loss, insomnia, any offensive body odour, mental or physical exhaustion and nervous conditions.

For neurasthenic condition in general, sleeplessness.

Kali Phos. is a constituent of the brain, nerves, muscles and blood vessels. It unites with albumen and creates nerve fluid or the gray matter of the brain. Nervous symptoms indicate a deficiency of this cell-salt, which must be supplied to restore the normal condition. Whatever disease can be traced to the nerve degeneracy, we enter the field of Kali Phos. Nervous condition known as neurasthenia is a field in which this salt has become prominently curative.

The results of the want of nerve power, such as prostration, loss of mental vigor, depression, brain-fag, softening of the brain, insanity, paralysis, epilepsy, hysteria, locomotor ataxia and when there is rapid decomposition of the blood. It is curative in septic haemorrhages, scorbutus, gangrene, stomatitis, offensive carrion-like diarrhoea, dysentery, a dynamic or typhoid conditions, incontinence of urine, urticaria, predisposition of epistaxis in children, dizziness and vertigo from sleeplessness, nervous exhaustion, nervous dyspepsia, asthma, neuralgia, nervous headaches, tongue coated as if with dark liquid mustard. Symptoms are aggravated from noise, by mental or physical exertion, by rising from a sitting position. Pains worse in cold and ameliorated by gentle motion

KALI SULPHURICUM

Kalium Sulphate deficiency shows in the face as pigmentation marks including chloasma of pregnancy, or an overall bronzy undertone to the face. Kali Sulph is important for the maintenance of healthy skin, hair, mucous membranes, and is an oxygenerator and kidney conditioner.

For third stage of inflammatory or catarrhal condition, leucorrhoea, diarrhoea, skin diseases, rheumatism, ulcerations, dyspepsia.

Kali Sulph is a carrier of oxygen. It furnishes vitality to the epithelial tissues. Light yellow, sticky and watery secretions are the characteristics of this remedy. The chief indication for this remedy is yellow deposit on the tongue. It has been given in bronchitis with yellow, slimy or thin, watery expectoration, in whooping cough, pneumonia pthisis, skin diseases with a sticky yellowish secretion and peeling off of the epidermis. Epithelioma, rheumatism, retrocession of eruptions in measles. Dyspepsia with slimy yellow coating of the tongue, catarrh of the bowels, diarrhoea, leucorrhoea, ophthalmia etc; with light yellow, watery secretions; dandruff; yellow scales in skin diseases. Amenorrhoea, menstruation too late and too scanty with weight and fullness in the abdomen *etc*. All ailments are worse in a warm room and toward evening and better in cool, open air.

MAGNESIA PHOSPHORICUM

Magnesium Phosphate deficiency can be seen as a slighly pinkish to dark red area close to the nose. Mag Phos is the anti-spamodic, analgesic tissue salt. Symptoms of deficiency may include flatulence, nervous tension, hunger pains and muscle cramps. It is indicated for all spasmodic pains, hiccups, spasmodic palpitations, cramping, mentrual pain and headaches accompanied by shooting or stabbing pains.

Note: Mag Phos will often act more rapidly when tablets are taken with a sip of hot water.

For convulsions, epilepsy, cramps and spasmodic affections, neuralgia and sciatica.

It is a constituent of muscles, nerves, brain, bone, spine, teeth and blood corpuscles. It helps to create the white fibres of the muscles and nerves. It uses water and albumen to form the transparent fluid, which nourishes the white threads and fibres of nerves and muscles. The deficiency of this salt allows the fibres to contract, causing spasms, cramps, convulsions and other nervous phenomena. It is a remedy to be thought of in all maladies having their origin in nerve cells and in the muscular tissue. It is particularly indicated in lean, thin, emaciated persons of a highly nervous temperament. Magnesia patient is always languid, tired and easily exhausted. The value of this remedy is shown in all forms of spasms, cramps, tetanus.

St. Vitu's dance, epilepsy, spasmodic retention of urine, paralysis agitans, hiccup, whooping cough, chorea. It is the remedy for the neuralgic pains in the head, face, teeth, car, stomach, abdomen *etc.* spasmodic palpitation of the heart, dysmenorrhoea. Attacks are often attended with great prostration, sometimes with profuse sweat. Valuable remedy for flatulent colic forcing the patient to bend double. Dysentery with cramp-like pains, angina pectoris, sciatica with excruciating spasmodic pains, angina pectoris, sciatica with excruciating spasmodic pain writer's and piano-player's cramps. All pains are lightening-like, shooting or boring, and change their location frequently and are worse on the right side, from cold air, washing in cold water, and from touch. Relieved by warmth pressure, friction and by bending double.

NATRUM MURIATICUM

Natrium Muriaticum deficiency reveals itself as puffiness under the eyes and as a strong nasolabial fold - a line running between the edge of the nose and the corner of the mouth. Nat Mur is the water balaning tissue salt. Symptoms of Nat Mur deficiency will be characterised by excessive moisture or dryness suzh as dry lips and a running nose. Itchyness of the skin or eyes, lack of vitality, tearfulness, feeling of depression, swelling of any area of the body, proneess to mosquito bites and constipation or diarrhoea are all indicative of Nat Mur deficiency.

For catarrhal affections with thin, clear, transparent, watery secretions, influenza, hay-fever, diarrhoea, constipation, leucorrhoea, intermittent fevers and skin diseases.

This salt is a constituent of every liquid and solid part of the body. Its function is to regulate the degree of moisture within the cells. It properly distributes the water through the organism.

Deficiency in this salt causes the disturbance of water in the human organism. An unequal distribution of water in the system sometimes causes excessive dryness of mucous membranes in some part, while there may be a copious, watery fluid from another part. Naturm Mur. will set things right and will establish equilibrium. It acts upon the lymphatic system, the blood, liver, spleen and upon the mucous lining of' the alimentary canal. It is indicated in headache, toothache, faceache, stomachache and also where there is either salivation or hyper secretion of tears or vomiting of water and mucus. Catarrhal affections of mucous membranes with secretion of transparent, watery, frothy mucus. Small watery blisters, breaking and leaving a thin crust. Skin diseases with watery symptoms. Diarrhoea with transparent, frothy, slimy stools. Constipation with dull headache and profuse tears. Conjunctivitis with clear watery discharge. Tongue clean, slimy, small bubbles of frothy saliva on sides. Leucorrhoea with watery, smarting or clear, starch-like discharge, gleet, coryza with loss of smell and taste, follicular catarrh of the pharynx, thrush with salivation, diphtheria with drowsiness, inflammation of uvula with elongation, catarrh of the bladder, impotence, prostatic fluid, paralysis, chorea, chlorosis, insomnia, pneumonia with loose rattling phlegm, intermittent fever, especially after abuse of quinine. Anasarca, gout, chronic swellings of lymphatic and sebaceous glands. Hemicrania.

NATRUM PHOSPHORICUM

Natrium Phosphate deficiency shows up either as a greasy skin, especially the T-bar, enlarged pores, blackheads and pimples, or extremely tight, almost invisible pores. If there is a Nat Mur deficiency as well, the cheeks will appear to droop. Nat Phos is the acid neutralising tissue salt. It is the principle biochemic remedy for ailments associated with an accumulation of acid in the body. It is indicated for heartburn, acid indigestion, gout, cholesterol problems, sweet or chocolate cravings, smelly feet or body odour, constipation and any yello coloured body discharge.

For gastric and intestinal disorders with sour and acid symptoms, worms, acidity and diarrhoea.

Through the presence of this salt, lactic acid is decomposed in carbonic acid and water. It eliminates the carbonic acid through the lungs. It has also the power to eliminate any excess of sugar from the blood.

This is the remedy in all cases where there is an excess of lactic acid. It acts also upon the bowels, glands, lungs and abdominal organs. Deficiency of this salt in the gastric juice will bring about acid and sour fermentations, and so it cures sour belching and sour rising of fluids. Sour vomiting, sour eructation with flatulence, greenish, sour smelling, diarrhoea, colic, spasms due to acid conditions of the stomach. Fever from acidity of stomach in children. Febrile condition with acid and sour smelling perspiration. Ague with characteristic coating of the tongue. Thick, moist, golden yellow coating of the tongue and palate and also a creamy yellowish exudation are the characteristic symptoms calling for this remedy. Eye complaints discharging a creamy yellow matter. Intestinal worms. Incontinence of urine in children with acidity. Atony of bladder, seminal emissions without dreams, leucorrhoea. with creamy or honey colored or acid and watery discharge, uterine displacements accompanied by rheumatic pains. Goiter, scrofulosis with acid symptoms. Hives, crusta lactea.

NATRUM SULPHURICUM

Natrium Sulphate deficiency shows first on the outer edge of the ear. Should the deficiency become servere, the tip of the nose may become red to purple in appearance, or the entire face can take on a greenish-yellow undertone. Nat Sulph promotes the elimination of excess water from the body and is the chief biochemic remedy for water retention. Nat Sulph contributes to the health functioning of the liver and is reccomended for disorders associated with the liver and bile such as biliousness, alcohol overindulgence, jaundice and hepatitis.

For hepatic diseases, bilious attacks, intermittent fevers, headaches, diarrhoea, diabetes, asthma, la grippe, dropsy.

This salt aids and regulates the excretion of the superficial water. It eliminates and excess of water from the blood. It keeps the bile in normal consistency. The three natrum salts work in the body as follows. Natrum Phos creates the water by splitting up of lactic acid; Nat Mur distributes the water through the system. Nat. Sulph. regulates the quantity of water in the organism. Nat. Sulph. is preeminently a liver remedy. It exhibits a marked similarity to the uric acid diathesis and is certainly a valuable remedy in combating or numerous phases of that malady. Dirty, brownish green or grayish-green coating on the root of the tongue. Dark greenish stools from excess of bile, bitter taste, vomiting of bile, diabetes from jaundice, bilious headache, dropsy, diseases of the liver. Excellent in influenza, intermittent fever and ague in all stages with vomiting and other bilious symptoms. Enlargement of the liver, erysipelas, smooth, red shiny eruptions coming in blotches with swelling of skin. Gravel, sand in urine, gout, fig warts, asthma and dyspnoea aggravated by change to damp weather, edematous inflammations of the skin, skin diseases with moist, yellowish scales, yellow watery secretions in all vesicles. Vomiting in pregnancy with bitter taste, fistulous abscesses discharging watery secretions, tendency to warts around the eyes, scalp, face, chest and arms *etc*. Symptoms are worse in the morning, in damp, wet weather and from lying on left side. Relief in warm dry weather and in the open air.

SILICEA

The first signs of Silicea deficiency show in the face as wrinkling, especially around the eyes. Other signs are hair loss including balding, a shiny, glossy nose tip, dull, fluffy hair, skin problems, and when extreme the eyes may appear deeply sunken into their sockets. Silicea is a connective tissue cleanser and conditioner. It is an important anti-stress tissue salt. A deficiency results in symptoms that could include irritability, noise or light sensitivity, feelings of agression, clamp-like headaches and involuntary twitching of the eyes or facial muscels. It also removes degenerative matter from the body and is the biochemic remedy for abscesses, sties, boils and pimples.

However,due to its ability to remove foreign matter from the body, Silicea should not be used for extended periods by people with implants or foreign object in their body.

For chronic suppurations and ulcerations, carbuncles, tonsillitis, glandular swellings, indurations, scrofula, foot sweats, uterine disorders.

It acts more upon the organic substances of the body, involving prominently bones, joints, glands, skin and mucous surfaces, producing malnutrition and corresponding to the scrofulous diathesis. Its action is deep and long lasting. It is especially suited to imperfectly nourished constitutions. It is a remedy for deep-seated chronic and acute suppurations and ulcerations, affecting the tendons, periosteum and bone. promotes suppuration. Indicated when there is hardness and indurations with suppuration. Infiltrations terminating in suppuration. It ripens abscesses by promoting suppurations. Valuable also in diseases resulting from suppressed foot sweats. Metrorrhagia in menstruation, associated with icy coldness over the whole body. Constipation and fetid foot-sweats. Ozena, carbuncles, tonsillitis, scrofula, syphilitic indurations, glandular swellings which become hard and threaten to suppurate. Styes on eyelids. Epilepsy especially nocturnal arid with change of moon. Gouty deposits in large joints of fingers. Chronic dyspepsia, copious night sweats especially in phthisis. Third stage of tuberculosis. Symptoms are always worse at night, during full moon, in the open air and from suppressed foot sweats and are better by the application of heat and warmth.

10
Biofeedback

Biofeedback is a treatment technique in which people are trained to improve their health by using signals from their own bodies. Physical therapists use biofeedback to help stroke victims regain movement in paralyzed muscles. Psychologists use it to help tense and anxious patients learn to relax. Specialists in many different fields use biofeedback to help their patients cope with pain.

Chances are you have used biofeedback yourself. You've used it if you have ever taken your temperature or stepped on a scale. The thermometer tells you whether you're running a fever, the scale whether you've gained weight. Both devices "feed back" information about your body's condition. Armed with this information, you can take steps you've learned to improve the condition. When you're running a fever, you go to bed and drink plenty of fluids. When you've gained weight, you resolve to eat less and sometimes you do.

Clinicians rely on complicated biofeedback machines in somewhat the same way that you rely on your scale or thermometer. Their machines can detect a person's internal bodily functions with far greater sensitivity and precision than a person can alone. This information may be valuable. Both patients and therapists use it to gauge and direct the progress of treatment.

For patients, the biofeedback machine acts as a kind of sixth sense which allows them to "see" or "hear" activity inside their bodies. One commonly used type of machine, for example, picks up electrical signals in the muscles. It translates these signals into a form that patients can detect: It triggers a flashing light bulb, perhaps, or activates a beeper every time muscles grow more tense. If patients want to relax tense muscles, they try to slow down the flashing or beeping.

Like a pitcher learning to throw a ball across a home plate, the biofeedback trainee, in an attempt to improve a skill, monitors the performance. When a pitch is off the mark, the ballplayer adjusts the delivery so that he performs better the next time he tries. When the light flashes or the beeper beeps too often, the biofeedback trainee makes internal adjustments which alter the signals. The biofeedback therapist acts as a coach, standing at the sidelines setting goals and limits on what to expect and giving hints on how to prove performance.

The word "biofeedback" was coined in the late 1960s to describe laboratory procedures then being used to train experimental research subjects to alter brain activity, blood pressure, heart rate, and other bodily functions that normally are not controlled voluntarily.At the time, many scientists looked forward to the day when biofeedback would give us a major degree of control over our bodies. They thought, for instance, that we might be able to "will" ourselves to be more creative by changing the patterns of our brainwaves. Some believed that biofeedback would one day make it possible to do away with drug treatments that often cause uncomfortable side effects in patients with high blood pressure and other serious conditions.Today, most scientists agree that such high hopes were not realistic. Research has demonstrated that biofeedback can help in the treatment of many diseases and painful conditions. It has shown that we have more control over so-called involuntary bodily function than we once though possible. But it has also shown that nature limits the extent of such control. Scientists are now trying to determine just how much voluntary control we can exert.

Clinical biofeedback techniques that grew out of the early laboratory procedures are now widely used to treat an ever-lengthening list of conditions,these include:

- ☆ Migraine headaches, tension headaches, and many other types of pain
- ☆ Disorders of the digestive system
- ☆ High blood pressure and its opposite, low blood pressure
- ☆ Cardiac arrhythmias (abnormalities, sometimes dangerous, in the rhythm of the heartbeat)
- ☆ Raynaud's disease (a circulatory disorder that causes uncomfortably cold hands)
- ☆ Epilepsy
- ☆ Paralysis and other movement disorders

Specialists who provide biofeedback training range from psychiatrists and psychologists to dentists, internists, nurses, and physical therapists. Most rely on many other techniques in addition to biofeedback. Patients usually are taught some form of relaxation exercise. Some learn to identify the circumstances that trigger their symptoms. They may also be taught how to avoid or cope with these stressful events. Most are encouraged to change their habits, and some are trained in special techniques for gaining such self-control. Biofeedback is not magic. It cannot cure disease or by itself make a person healthy. It is a tool, one of many available to health care professionals. It reminds physicians that behavior, thoughts, and feelings profoundly influence physical health. And it helps both patients and doctors understand that they must work together as a team.

Biofeedback places unusual demands on patients. They must examine their day-to-day lives to learn if they may be contributing to their own distress. They must recognize that they can, by their own efforts, remedy some physical ailments. They must commit themselves to practicing biofeedback or relaxation exercises every day. They must change bad habits, even ease up on some good ones. Most important, they must accept much of the responsibility for maintaining their own health.

Scientists cannot yet explain how biofeedback works. Most patients who benefit from biofeedback are trained to relax and modify their behavior. Most scientists believe that relaxation is a key component in biofeedback treatment of many disorders, particularly those brought on or made worse by stress.

Their reasoning is based on what is known about the effects of stress on the body. In brief, the argument goes like this: Stressful events produce strong emotions, which arouse certain physical responses. Many of these responses are controlled by the sympathetic nervous system, the network of nerve tissues that helps prepare the body to meet emergencies by "flight or fight." The typical pattern of response to emergencies probably emerged during the time when all humans faced mostly physical threats. Although the "threats" we now live with are seldom physical, the body reacts as if they were: The pupils dilate to let in more light. Sweat pours out, reducing the chance of skin cuts. Blood vessels near the skin contract to reduce bleeding, while those in the brain and muscles dilate to increase the oxygen supply. The gastrointestinal tract, including the stomach and intestines, slows down to reduce the energy expensed in digestion. The heart beats faster, and blood pressure rises.Normally, people calm down when a stressful event is over especially if they have done something to cope with it. For instance, imagine your own reactions if you're walking down a dark street and hear someone running toward you. You get scared. Your body prepared you to ward off an attacker or run fast enough to get away. When you do escape, you gradually relax.

If you get angry at your boss, it's a different matter. Your body may prepare to fight. But since you want to keep your job, you try to ignore the angry feelings. Similarly, if on the way home you get stalled in traffic, there's nothing you can do to get away. These situations can literally may you sick. Your body has prepared for action, but you cannot act. Individuals differ in the way they respond to stress. In some, one function, such as blood pressure, becomes more active while others remain normal. Many experts believe that these individual physical responses to stress can become habitual. When the body is repeatedly aroused, one or more functions may become permanently overactive. Actual damage to bodily tissues may eventually result.

Biofeedback is often aimed at changing habitual reactions to stress that can cause pain or disease. Many clinicians believe that some of their patients and clients have forgotten how to relax. Feedback of physical responses such as skin temperature and muscle tension provides information to help patients recognize a relaxed state. The feedback signal may also act as a kind of reward for reducing tension. It's like a piano teacher whose frown turns to a smile when a young musician finally plays a tune properly.

The value of a feedback signal as information and reward may be even greater in the treatment of patients with paralyzed or spastic muscles. With these patients, biofeedback seems to be primarily a form of skill training like learning to pitch a ball. Instead of watching the ball, the patient watches the machine, which monitors activity in the affected muscle. Stroke victims with paralyzed arms and legs, for example, see that some part of their affected limbs remains active. The signal from the biofeedback machine proves it.

This signal can guide the exercises that help patients regain use of their limbs. Perhaps just as important, the feedback convinces patients that the limbs are still alive. This reassurance often encourages them to continue their efforts.

If you think you might benefit from biofeedback training, you should discuss it with your physician or other health care professional, who may wish to conduct tests to make certain that your condition does not require conventional medical treatment first. Responsible biofeedback therapists will not treat you for headaches, hypertension, or most disorders until you have had a thorough physical examination. Some require neurological tests as well.

How do you find a biofeedback therapist? First, ask your doctor or dentist, or contact the nearest community health centre, medical society, or State biofeedback society for a referral. The psychology or psychiatry departments at nearby universities may also be able to help you. Most experts recommend that you consult only a health care professional a physician, psychologist, psychiatrist, nurse, social worker, dentist, physical therapist, for example who has been trained to use biofeedback.

11

Chakra Healing

The word chakra is derived from the Sanskrit word meaning "wheel". If we were able to see the chakras (as many psychics, in fact, do) we would observe a wheel of energy continuously revolving or rotating. Clairvoyants perceive chakras as colorful wheels or flowers with a hub in the center. The Chakras begin at the base of the spine and finish at the top of the head. Though fixed in the central spinal column they are located on both the front and back of the body, and work through it.

Each chakra vibrates or rotates at a different speed. The root or first chakra rotates at the slowest speed, the crown or seventh chakra at the highest speed. Each chakra is stimulated by its own and complimentary color, and a range of gemstones for specific uses. The chakra colors are of the rainbow; red, orange, yellow, green, blue, indigo, and violet. The size and brightness of the wheels vary with individual development, physical condition, energy levels, disease, or stress.

If the chakras are not balanced, or if the energies are blocked, the basic life force will be slowed down. The individual may feel tired, out of sorts, or depressed. Not only will physical bodily functions be affected so diseases may manifest, but the thought processes and the mind may also be affected. A negative attitude, fear, doubt, *etc.* may preoccupy the individual.

A constant balance between the chakras promotes health and a sense of well being. If the chakras are opened to much, a person could literally short circuit themselves with too much universal energy going through the body. If the chakras are closed, this does not allow for the universal energy to flow through them properly which may also lead to disease.

Most of us react to unpleasant experiences by blocking our feeling and stopping a great deal of our natural energy flow. This affects the maturation and development of the chakras. Whenever a person blocks whatever experience he is having, he in turn blocks his chakras, which eventually become disfigured. When the chakras are functioning normally, each will be open, spinning clockwise to metabolize the particular energies needed from the universal energy field.

As already mentioned, any imbalances that exist within any chakra may have profound effects upon either our physical or emotional bodies. We are able to use our quartz crystals and gemstones to re-balance all our chakric centers and once the chakra has been properly balanced then our body will gradually return to normal.

The reason why crystals and gemstones are wonderful and powerful healing tools are because of what science calls its piezoelectric effect. (One can see this effect in the modern quartz watches). Crystals and gemstones respond to the electricity that is coursing through our body, and if the energy is sluggish, the constant electrical vibrations of the stones will help to harmonize, balance, and stimulate these energies.

The Seven Major Chakras

First Chakra - Root

Studying the individual chakras begins with the root chakra, called Muladhara in Sanskrit. The root chakra is located at the base of the spine at the tailbone in back, and the pubic bone in front. This center holds the basic needs for survival, security and safety. The root chakra is powerfully related to our contact with the Earth Mother, providing us with the ability to be grounded into the earth plane. This is also the center of manifestation. When you are trying to make things happen in the material world, business or possessions, the energy to succeed will come from the first chakra. If this chakra is blocked an individual may feel fearful, anxious, insecure and frustrated. Problems like obesity, anorexia nervosa, and knee troubles can occur. Root body parts include the hips, legs, lower back and sexual organs.

NOTE: A man's sexual organs are located primarily in his first chakra, so male sexual energy is usually experienced primarily as physical. A women's sexual organs are primarily in her second chakra, so female sexual energy is usually experienced primarily as emotional. Both chakras are associated with sexual energy.

Second Chakra - Belly (Sacral)

The next chakra or second chakra is often referred to as the belly or (sacral). It is located two inches below the navel and is rooted into the spine. This center holds the basic needs for sexuality, creativity, intuition, and self-worth. This chakra is also about friendliness, creativity, and emotions. It governs peoples sense of self-worth, their confidence in their own creativity, and their ability to relate to others in an open and friendly way. It's influenced by how emotions were expressed or repressed in the family during childhood. Proper balance in this chakra means the ability to flow with emotions freely and to feel and reach out to others sexually or not. If this chakra is blocked a person may feel emotionally explosive, manipulative, obsessed with thoughts of sex or may lack energy. Physical problems may include, kidney weakness, stiff lower back, constipation, and muscle spasms Belly body parts include sexual organs (women), kidneys,bladder, and large intestine.

Third Chakra - Solar Plexus

The third chakra is referred to as the Solar Plexus. It is located two inches below the breastbone in the center behind the stomach. The third chakra is the center of personal power, the place of ego, of passions, impulses, anger and strength. It is also the center for astral travel and astral influences, receptivity of spirit guides and for psychic development.

When the Third Chakra is out of balance you may lack confidence, be confused, worry about what others think, feel that others are controlling your life, and may be depressed. Physical problems may include digestive difficulties, liver problems, diabetes, nervous exhaustion, and food allergies. When balanced you may feel cheerful, outgoing, have self-respect, expressive, enjoy taking on new challenges, and have a strong sense of personal power. The body parts for this chakra include the stomach, liver, gall bladder, pancreas, and small intestine.

Fourth Chakra - Heart

The fourth chakra is referred to as the heart chakra.It is located behind the breast bone in front and on the spine between the shoulder blades in back. This is the center for love, compassion and spirituality. This center directs one's ability to love themselves and others, to give and to receive love. This is also the chakra connecting body and mind with spirit. Almost everyone today has a hard, hurt, or broken heart, and it is no accident that heart disease is the number one killer in America today. Deep heart hurts can result in aura obstructions called heart scars. When these scars are released, they raise a lot of old pain, but free the heart for healing and new growth. When this chakra is out of balance you may feel sorry for yourself, paranoid, indecisive, afraid of letting go, afraid of getting hurt, or unworthy of love. Physical illnesses include heart attack, high blood pressure, insomnia, and difficult in breathing. When this chakra is balanced you may feel compassionate, friendly, empathetic, desire to nurture others and see the good in everyone. Body parts for the fourth chakra include heart, lungs, circulatory system, shoulders, and upper back.

Fifth Chakra - Throat

The fifth chakra is referred to as the Throat. It is located in the V of the collarbone at the lower neck and is the center of communication, sound, and expression of creativity via thought, speech, and writing. The possibility for change, transformation and healing are located here. The throat is where anger is stored and finally let go of. When this chakra is out of balance you may want to hold back, feel timid, be quiet, feel weak, or can't express your thoughts. Physical illnesses or ailments include, hyperthyroid, skin irritations, ear infections, sore throat, inflammations, and back pain. When this chakra is balanced you may feel balanced, centered, musically or artistically inspired, and may be a good speaker. Body parts for the fifth chakra are throat, neck, teeth, ears, and thyroid gland.

Sixth Chakra - Third Eye

The sixth chakra is referred to as the Third Eye. It is located above the physical eyes on the center of the forehead. This is the center for psychicability, higher intuition, the energies of spirit and light. It also assists in the purification of negative tendencies and in the elimination of selfish attitudes. Through the power of the sixth chakra, you can receive guidance, channel, and tune into your Higher Self. When this chakra is not balanced you may feel non-assertive, afraid of success, or go the opposite way and be egotistical. Physical symptoms may include headaches, blurred vision, blindness, and eyestrain. When this chakra is balanced and open you are your own master with no fear of death, are not attached to material things, may experience telepathy, astral travel, and past lives. Sixth chakra body parts include the eyes, face, brain, lymphatic and endocrine system.

Seventh Chakra - Crown

The seventh chakra is referred to as the Crown. It is located just behind the top of the skull. It is the center of spirituality, enlightenment, dynamic thought and energy. It allows for the inward flow of wisdom, and brings the gift of cosmic consciousness. This is also the center of connectedness with the Goddess (God), the place where life animates the physical body. The silver cord that connects the aura bodies extends from the crown. The soul comes into the body through the crown at birth and leaves from the crown at death. When this chakra is unbalanced there may be a constant sense of frustration, no spark of joy, and destructive feelings. Illnesses may include migraine headaches and depression. Balanced energy in this chakra may include the ability to open up to the Divine and total access to the unconscious and subconscious.

Activation of the Energy Centers

The three lower chakras co-relate to basic primary needs- those of survival, procreation and will. The four higher chakras relate to our psychological makeup. They define love, communication and knowledge and can also provide a spiritual connection to this universe and beyond. Seven chakras in our body play a vital role in our total wellbeing. Our chakras get disturbed due to stress and negative emotions, creating disease and disharmony, which prevent us from functioning at our most vibrant and joyful level.

The details of Chakras with related values:

1. Power Activates Muladhara (Root) Chakra

The root chakra not only sends energy to bones and muscles of the body but also acts as a sex center. It is the source of creative energy and power. We are both male and female in our energies, but society does not let us express both.

Most societies repress the feminine energy: the left side of the body and the right side of the brain, which is the abode of beauty, stillness, poetry and art. Until the age of seven, your energy is total. Notice how children are beautiful in their beings. But beyond seven, children are forced to suppress their other side and thus wound their consciousness. In their search for the other half, girls adore their father and boys their mother.

Between the age of 7 and 14, the wound can be healed if parents remain most of the time with their children, but if the parents are not there, children look in vain in the outside world for that feminine and masculine energy.

Children in the 14-21 age groups are ready to engage with the opposite sex- to marry- but they are rarely allowed. So they collect and cherish imagination and dreams of the opposite sex through the media. Manufacturers exploit this by selling dreams and sex appeal through their products. And these play a vital role in deforming and disabling the astral and the mental body of youths, which further causes several diseases in the physical body at a youthful age of their life. This all leads to further degeneration of their character and conduct.

The root chakra is a continuous fountain of positive energy power, while our imaginations and expectations are like a huge stone blocking this marvelous energy fountain. When we burnaway all our unwanted imaginations and memories and open ourselves to reality, the creative energy of this chakra flows.

2. Purity activates Swadhisthana (Spleen) Chakra

This chakra is located two inches below the navel and is the place where fear attacks us. The average person has six to 12 fear strokes every 24 hours, while in awaken and sleeping time. This weakens our immune system and leads to depression, illness, aging, infertility, and impotency. It also creates diseases of the kidney.

Fear primarily falls into four categories:

☆ Fear of loss of status, wealth or comfort.

☆ Fear of disease or loss of a bodily part.

☆ Fear of loss of family or friends.

☆ Fear of death

Purity leads to fearlessness. It is not the absence of fear but it is the courage to face the fear and experience it completely. If we can truly enter into the space of fear and experience it with power of purity, it can never affect us again and will activate Swadhisthan Chakra.

3. Happiness Activates Manipura (Navel) Chakra

Worry and suppressed emotions sit in the navel center and block this chakra. "Worrying is nothing but a constant repetition of certain words in the mind". This drains away our energy and takes us out of the total awareness and bliss of this moment. Worry is a terrible wastage of time, thought and energy. 99 per cent of our worries never come true and the one percent that do, end up being good for us. Many stomach problems; skin diseases and pains are related to the navel center. Obesity is another byproduct of worry and depression. Happiness and contentment activate this chakra.

4. Selfless Love Activates Anahata (Heart) Chakra

Unconditional expression of love and affection expands this chakra. Constant need for other's attention and approval blocks it. "Life is a long signature campaign." We are constantly working for social success, rather than individual success. Make a list of all those people who have the power to upset us and realise that we are psychological slaves to these people. The conscious realisation itself will bring about a freedom.

We never know the people who are close to us. We simply form an image of them and continue to relate to that image through out, while the person is in an ever-changing multifaceted centre of consciousness.

5. Peace Activates Visuddhi (Throat) Chakra

"Three layers of energy merge in the throat chakra; ordinary physical energy, reserve energy to be used in emergencies and spiritual energy." We usually use only the first level and forget that there are two more levels behind it; in order to access these levels we need to get rid of our ego, which results in either a superiority or inferiority complex. Comparison and jealousy lock this chakra. When we run in the rat race, the visuddhi closes. Understanding the value of peace and appreciating our uniqueness open it.

We usually create an idea about ourself and compare it to others, which is senseless. If we live our life, we will grow uniquely and will have profound satisfaction, whether we have money, fame or not *etc*. When we feel tired in our body, we can say "no" and go to the next layer of energy, which will provide us with all the energy, and vitality that we need.

6. Truth and Knowledge Activates Ajna (Brow) Chakra

"Ajna"(Agya) means will power. We usually spend 80 per cent of our energy what we desire and 20 per cent of our energy what we want to create. These percentages get reversed when this chakra is opened; resulting in fewer desires but more power to create them into reality. Conditioning, labelling and judgements contaminate this chakra. All judgements are prejudiced; they draw conclusions from a few observations (observations are contaminated by our conditioning). "I am this, you are that, life is this, *etc*." are statements of the ego.This mental chatter reduces all our experiences merely to words.

To live spontaneously and intensely we need to deal with every person and situation without a script. If we trust our being we will have an enormous leap of growth through such spontaneous interaction with the world. Facing life without a script, we will realise that we know much more than we imagine. Whether we believe it or not, like it or not, the fact is that we are God's children and we are same as our father is. This is the truth of life. True knowledge and wisdom enlighten us to open brow chakra.

7. Bliss Activates Sahastrara (Crown) Chakra

"Sahastrara" means the thousand petals lotus. When this chakra opens, we are flooded with ecstasy and eternal bliss. Discontentment and negativity towards life block this chakra.

Gratitude for life and all that God has given us opens this chakra. So many blessings are showered upon us and we take them for granted. Our life and body are amazing gifts from God- absolutely incredible. Living out of deep joy and bliss is the product of simply showing gratitude and respect for the divine. Loving the whole world is easy; loving our neighbour is difficult because it requires actions through acts of gratitude, compassion and affection. Feel gratitude for every small thing in life.

"Gratitude is the greatest attitude which determines our altitude of success in life."

12

Chiropracty and Osteopathy

Chiropractic and osteopathy are both manipulation-based therapies used for treating problems associated with bones, joints and the back. The two therapies have much in common, but chiropractors tend to focus on the joints of the spine and the nervous system, while osteopaths put equal emphasis on the joints and surrounding muscles, tendons and ligaments.

How they Work

Chiropractors and osteopaths share conventional medicine's view that the human body is like a machine and that any disease is due to a breakdown of part of the machine. However, they also believe that many health problems can be traced to poor posture and to misalignment of muscles and joints (and, with chiropractic, particularly the spine). They suggest that, if the structure of the body is improved and the spine put back into alignment, its function will improve, problems will be alleviated and good health restored. Misalignment is usually thought to be due to an external cause, such as a fall or other accident (even one that happened years before), or to long-term poor posture.

What they are for

Osteopaths and chiropractors both mainly treat back and neck pain. But they are also treat other health problems, such as headaches, migraines, vertigo and tinnitus (ringing in the ears).

Some practitioners also treat a wider range of diseases, including:

☆ Heart and circulatory problems

☆ Arthritis

☆ Sports injuries

☆ Digestive problems

☆ Asthma, and

☆ Period problems

Are they Effective?

While osteopathy and chiropracty are well accepted by many conventional medical practitioners, the scientific evidence for them is relatively sparse, and very few medical studies have compared the two therapies with each other. However, the evidence is stronger for chiropractic than osteopathy, with a few studies showing that chiropractic is effective in treating lower back pain.

There have also been several studies of spinal manipulation and mobilisation techniques (by osteopaths, chiropractors, physiotherapists and doctors) for lower back pain. They suggest that the techniques, whoever does them, do provide short-term relief from pain, and improvement in mobility. There have not been as many studies of manipulation and mobilisation techniques for neck pain.

At a Consultation

A first consultation with a chiropractor or osteopath generally takes about an hour. The chiropractor or osteopath will ask detailed questions about your general health, lifestyle, emotional state, and medical and family history. He or she will also perform a physical examination, with you sitting, standing, walking and possibly carrying out other movements. He or she may test your reflexes with a reflex hammer, and measure your blood pressure. Some chiropractors (but not usually osteopaths) also use X-rays and other conventional medical tests to help them make a diagnosis.

Later sessions usually last about 30 minutes. The number of treatments you will need and how often you need them will depend on your problem. However, the practitioner should give you an idea of this at the first treatment session.

Treatment is usually carried out with you lying down in various positions. Chiropractors mainly use a manipulative technique on the spinal column and pelvic area consisting of short, rapid forceful movements called high-velocity thrusts. These are designed to realign and mobilise the spine, and may result in an audible sound - a clicking similar to knuckles being stretched.

Osteopaths also carry out high velocity thrusts on the spine, but these play a much smaller part in treatment than in chiropractic.

Osteopaths also use a wide range of other techniques, from stretching of soft tissues and massage to rhythmic joint movements and manipulation, on other parts of the body as well as the back. And, they may use only gentle "release" techniques with some people, particularly children and older people. These are called muscle energy techniques and are used to release tension in specific muscles. Some osteopaths also do cranial manipulation.

Related Therapies

Physiotherapists and doctors sometimes use spinal manipulation and mobilisation techniques similar to those used by chiropractors and osteopaths.

Some osteopaths do cranial manipulation, also called cranial osteopathy. This consists of gentle manipulative techniques on the cranium (skull). A belief underpinning cranial osteopathy is that childbirth, an accident or long-term muscle tension can cause compression of the cranium. This, in turn, can affect how fluid called cerebrospinal fluid flows in the spine and around the brain, and so can result in disease. Practitioners claim that gently manipulating the bones of the cranium can correct the flow of cerebrospinal fluid, by restoring the skull to its natural shape.

Craniosacral therapy is similar to cranial osteopathy. It differs in believing that the flow of cerebrospinal fluid affects every cell in the body.

Side Effects and Risks

The most serious potential risks of chiropractic and osteopathy are spinal cord injury or stroke after manipulation of the neck.

These are rare, though there have been calls for research to establish how significant the risk is.

There is a general consensus that osteopathy is less risky in terms of spinal injury because osteopaths usually use less forceful manipulation techniques on the spine.

Less serious, but more common, side effects include discomfort or mild pain at the point of manipulation, mild headaches or tiredness, which should disappear within 24 hours of treatment.

There are certain situations where forceful manipulation can be dangerous and should never be done. These include if you are pregnant or if you have osteoarthritis of the neck, or osteoporosis of the spine.

However, chiropractors and osteopaths are trained to check patients for these and other risk factors.

13

Dowsing

Dowsing therapy uses the art of dowsing to locate and treat disease and faulty organs. The theory is that all substances emit radiation waves that can be sensed or picked up and amplified with the aid of an indicator, such as a dowsing pendulum or rods. Much as a radio receives and amplifies sound wave frequencies. The dowsing therapy is the diagnostic part and once it is made, the client can then be treated with appropriate remedies such as herbal or homeopathic. It's believed that dowsing works because it helps you tap into energy fields –human, animal and geological. Dowsing is also known as 'reading biofields' and a form of dowsing called 'muscle testing' is sometimes used by alternative health practitioners to determine the needs of their patients. Long distance dowsing can be performed with a lock of the clients hair or sample of urine or blood, in place of holding the pendulum over the client.

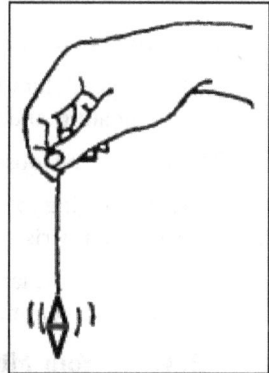

Dowsing is where quantum physics meets with mysticism, still impossible to measure with scientific instruments, only accessible to our intuition. Imagine a device where pendulum movement in the hand of a dowser is an equivalent of a needle on a meter used to measure subtle energy not detected by other senses.

Self Hypnosis

It sounds like a pretty easy task, simply sit with your eyes closed, relax, connect with your breath and guide your thoughts towards a specific intention, but, in reality, it can be much more challenging than this instruction. This is because the conscious mind often has another agenda. One that involves distracting you with a billion thoughts so you're unable to truly connect with that all-important unconscious mind.

It happens inevitably when you're starting with, it's important to find a way to bypass the critical factor, *viz.*, the conscious mind, so that you can begin. One of the most effective and easiest ways to do this is using a pendulum.

The History of the Pendulum

People first started using the pendulum for water dowsing in Europe back at the turn of the 19th century. But it wasn't until a Frenchman Michel Eugene Chevreul started experimenting with it. He found that the pendulum had no magical properties on its own, although, rather interestingly, it did give an ideodynamic response when interacting with the unconscious mind, causing it to move when it receives a signal.

In other words, when the unconscious mind expects something to happen, it gives little micro-muscular twitches that cause the pendulum to move one way or another.

While these signals are very subtle, they can easily be read using a pendulum, helping you to communicate with your unconscious mind.

It's always advisable to carry your pendulum around for a few days so it attunes with your energy.

Using the Pendulum

Step 1: Familiarize yourself with the Pendulum

Hold the pendulum loosely between your thumb and index finger. Make sure your elbow is free floating, and not resting on a table or locked in place.

Stay relaxed and loose, and start familiarizing yourself with the feel of your pendulum.

Start by swinging your pendulum, making large swings and then smaller swings. Notice how it feels in your wrist, arm and fingertips when it swings.

Once you become familiar with the weight and the feel, you can then begin practicing using it as a tool to communicate with the unconscious mind.

Step 2: Clear Your Mind

Start by connecting with your breath by doing some simple breathing exercises which you normally would do during meditation.

Step 3: Imagine The Pendulum Moving

"Tell" the pendulum to start moving backwards and forward.

Keep repeating this command, and then start telling it to swing in different directions, for example, clockwise, counter-clockwise, or back and forth and side-to-side.

Once it starts to move on your request, you can also start to ask it direct questions that have an obvious answer; depending on which way the pendulum swings, you'll know what a "yes" or "no" answer is.

Types of questions you can ask include:

"Is my name… ?"

"Is my age… ?"

"Is my profession… ?"

"Do I live in…?"

Typically the pendulum will swing forward and backward for a "yes" response, and side-to-side for a "no" – however, your experience may be different.

It's a good idea to always calibrate your pendulum so you know what a "yes" versus a "no" response looks like.

The Five Unconscious Answers

The next step is to have different swings for the Five Unconscious Answers. These answers will help you get further clarity when you ask a question.

The Five Unconscious Answers are:

Yes

No

Please rephrase the question

I don't want to answer that

I don't know

It might take some time to get clear responses to each of these, but keep commanding your pendulum to show you responses until you're confident with your signalling system.

Once you're happy with the above, start asking your pendulum some "truth" questions.

For example:

Are you going to give me the absolute truth to the questions I ask now and in the future?

Asking these kinds of questions is important as you'll always want to make sure your unconscious mind is committed to giving you the truth before you start any serious work. However, keep in mind that every time you work with your pendulum, it's important to reset it or calibrate it with those "yes" or "no" questions and with the five unconscious answers.

Getting to the Root of the Problem

Using a pendulum during self-hypnosis is especially useful when you need to get to the root of a particular problem.

For example, let's say you have to give a speech and you feel nervous about standing in front of a large group of people and your regular self-hypnosis practice isn't working.

To help you dig a little deeper so you can solve this problem, ask your pendulum a series of questions relating to the issue.

For example:

Did this issue begin when I was a child?

Was I younger than 18?

Was I younger than 12?

Did this issue stem from one incident?

Did this incident occur while I was in school?

Did this incident occur some other time?

The trick here is to keep drilling down until you get to the root cause of the issue. This may take as little as 10 minutes, or as long as a few days of regular practice.

When you discover the root cause, you can then create self-hypnosis suggestions to counteract it, allow time and space for your mind to work through the issue.

Additional Tips

Keep a record of your questions

Prepare questions ahead of time

Learn from past questions

Work with your pendulum for a few minutes each day

Never loan your pendulum out or borrow someone else's (it's important it stays attuned to your energy only)

The pendulum is a very powerful tool that can be used to connect with your unconscious mind and help you solve problems. So why not give it a go to see how it can deepen your self-hypnosis practice.

Here is a short list of areas that dowsing can be used:

1. Finding water - this is the most traditional way of using dowsing, also known as witching or divining. Search for gold and minerals - this also include oil and gas dowsing and is quite widely used. Some oil companies wont admit they hire dowsers to locate rich oil deposits.

2. Finding lost objects or missing people.

3. Detecting geopathic zones. Again - this is a growing field. Several European countries won't issue building permit without testing for geopathic zones.

4. Finding water or gas mains or finding leaks

5. Sexing unhatched eggs - this one requires some skills and a way to neutralize the dowsing tool and the spot to avoid the shadow effect that can skew your results.

6. Dowsing stock market - a bit controversial - from what I gathered you can get better results dowsing trends. It's hard to find out the truth - I don't think successful stock market dowsers would share their techniques with the public for obvious reasons-

7. Dowsing fruit and vegetables for freshness and contaminants - irradiated, sprayed or otherwise treated food will give very low life force reading.

8. Testing supplements for compatibility with your body.

9. Testing for mineral deficiencies or surplus.

10. Adjusting charkas - for spin rate, direction, opening ratio, *etc.*

How to Use Dowsing or Divining Rods

Before technology came about that would allow us to see into the ground, people depended on dowsing to find water wells, metals, gemstones, and even missing people and unmarked graves. Although dowsing has never been scientifically proven to work in a controlled setting, the practice remains popular in many parts of the world. It's been suggested that humans may be able to sense electric and magnetic energy that's invisible to the eye (as many animals can) and subconsciously manipulate the dowsing rods or pendulum to reflect that information (the ideomotor effect). Whether you're a stout defender of dowsing or you think it's hogwash, doing your own experiment can be both educational (from a historical perspective) and fun.

Dowsing Using a Dowsing or Divining Rod

Find a forked ("Y"-shaped) branch from a tree or bush. Hold the two ends on the forked side, one in each hand. You may want to experiment with holding it with your palms facing up or down; one may be more effective than the other. Hazel or willow branches were commonly used because they were light and porous, and were believed to better absorb vapours rising from buried metals or water, thus weighing down the un-forked end and pointing towards the source.

You may bend two identical pieces of wire into an "L" shape and hold one in each hand by the short part of the "L" so that the long part is parallel with the ground and so they can swing freely from side to side. You can use coat hangers to make these rods. Some dowsers claim certain metals, such as brass, to be more effective.

You may also make a pendulum by suspending a weight (such as a stone or crystal) by a string or chain. Pendulums are used with maps or to answer yes/no questions, rather than to guide the dowser on unfamiliar terrain; instructions for using a pendulum are given in a separate section below.

Relax. Whether you're priming yourself to receive paranormal insight, or you're relaxing your muscles so they can better transmit the ideo-motor effect, or you're just experimenting with this for fun, relaxing will make it a more effective or enjoyable experience. Take a few deep breaths or meditate for a minute or two.

Calibrate your dowsing rods, lay out cards numbered 1-5 face up and in a line with about 1-2 feet (1/2 meter) of separation between each card. Start at one end, holding your dowsing rods, and make a request, like "Show me where the card labeled 4 is". Close your eyes and visualize the card you want the rod(s) to find for you. Then open your eyes and walk slowly next to the line of cards with your dowsing rod(s) over them, pausing over each one, and see what happens when you go over the card you requested. You may find that the wooden rod points downward, or the metal rods cross each other.

Test your dowsing abilities. Repeat the previous step, but this time, shuffle the cards and put them on the ground facing down, so you don't know which is which. Make your request and see if you can correctly identify the card you requested by dowsing. If you can't, either you're a bad dowser (you're not focused or relaxed enough, you're psychically challenged, you're holding the rod(s) incorrectly, or you're too skeptical to allow dowsing to work for you) or dowsing is nothing more than superstition punctuated by coincidence.

Pendulum Dowsing

Calibrate the pendulum. Hold it perfectly still over a bare surface, then ask a specific question to which you know the answer is "yes". Does it go in a circle (if so, clockwise or counter-clockwise?), swing right to left, or swing up and down? This is your "yes" answer. Repeat to find a "no" answer. If your goal is to find a lost person or object, hold the pendulum over a picture of that person or object and see what the pendulum does.

Hold the pendulum over an object or person and make a request. The simplest way to use a pendulum is to ask a specific, yes or no question and see what the pendulum does. A dowsing pendulum can also be used in other ways:

For map dowsing, hold the pendulum still over the map and make a request, *e.g.*, "Show me where this object or person is". Move the pendulum slowly over all areas of the map until you see activity that coincides with your calibration. An alternative to this is to hold your pendulum in your 'dominant' hand and use a pencil, pen, or other pointed object in your other hand to navigate over the map or chart, seeking activity.

Write several answers on a piece of paper, leaving the centre blank. Hold the pendulum over the centre and ask a question. Watch the pendulum carefully to see in which direction it swings. Which answer does it point to.

Radiesthesia is the practice of using dowsing to make a medical diagnosis. A common technique is to hold the pendulum over a pregnant woman's stomach to identify the gender of the child. It's not wise, however, to depend solely on a pendulum for medical advice.

Questions and Answers

Be the first to ask a question about this topic:

What do you need to know? We'll do our best to find the answer.

Submit

Allow 3 minutes of knowledge! Answer should follow by yes or no.

Sample Questions:

Can you tell us about Graphic designing and computer art?

Can you tell us about First impressions?

Can you tell us about Humidifiers?

Can you tell us about Travel packing?

Note

The "L"-shaped rods will only perform well when parallel to the ground. Do not let the rods droop towards the ground. Whatever it is that you're trying to find through dowsing, visualize it as clearly as you can. Once you've found a water source with rods, you may be able to use a stiff pendulum (a floppy horizontal wire with a weight at the end) to determine how deep the well is by counting how many times it bobs.

It is tempting to stare at the rods while you are working. Please be mindful of where you are walking though so you do not trip on something or fall in a well.

Make sure your dowsing area is clear of other people. Do not use the rods in a crowded place or in close proximity to other people because someone could get poked and injured. Besides, the rods might pick up energy fields from other people and they won't work.

How it Works

There are different theories about how dowsing works although it is not known exactly how, but the information received is outside the five senses so it appears that we can use it to tap into our hidden powers by bringing together the rational and intuitive faculties, the left and the right sides of the brain to access knowledge outside normal conscious awareness.

Early explanations suggested that there was a sympathetic attraction between specific minerals and rods cut from certain trees and that underground water, gold or oil emit a strong natural energy which the diving tool picks up like an aerial and transmits to the dowser.

Some say that the pendulum has a mind of its own and as an inanimate object has statues inside waiting to be freed. Divination traditionally meant getting in touch with a god or goddess for inspired direction for a future course of action.

14

Electrohomeopathy

Electrohomeopathy: This term is derived from three words-

ELECTRO comes from the electro bio-energy content extracted from plants and the way that energy is extracted and made into remedies. The force of electro bio-energy connects itself to any disturbance within the body and expels it from the system. It thereby restores the organic tissues and nervous system to their proper balanced state of health.

HOMOEO refers to the equilibrium or balance between the blood and lymphatic systems and works towards "Homeostasis" which is the correct physical and chemical consistency of the body's cells. This ensures that each and every cell interrelates as it should so that the body functions at its best.

PATHY basically means a system of treatment. In short, Electro Homeopathy is the system of natural medicines, made exclusively from plant extracts, by which a healthy balance is achieved and maintained throughout the whole body.

Electro Homeopathy has four fundamental laws:

1. LAW OF POLARITY
2. CONSTITUTION and TEMPERAMENT
3. LAW OF DOSOLOGY
4. SELECTION OF REMEDY

Who Discovered Electrohomeopathy?

The practice of Electrohomoeopathy grew with the Italian Noble, Count Cesare Mattei who founded true Electrohomoeopathy in the latter part of the 19th century in Bologna, Italy. Being strongly influenced by the philosopher Paulo Costa, Count Mattei was motivated to devote his time and wealth to the service of humanity. He built on the works of Indian metallurgists and Ayurvedic/Siddha practitioners, also on Paracelsus and Balroot but he focused on preparing remedies extracted exclusively from plants. He called these remedies Electrohomoeopathic medicines. He called these remedies Electrohomoeopathic medicines on own his experiments of 30 years which show in his plants extract as vegetable electricity.

Later on this system was further developed by Mattei adherent Theodore Krauss who have given the manufacturing method known as Krauss method (also mentioned in German Pharmacopoeia HAB) for making electro homeopathic remedies. He also added 22 more remedies in Mattei rest of remedies which was 38. This system became Electro complex homeopathy in his life time worldwide. After death of Count Mattei the progress and development was stopped during the First World War, during this period some so called perfectionists like Dr.A.Sauter in Geneva,Switzerland, and Father Augustus Muller in Mangalore, South India, took charge and established the so called true pharmacy of Electrohomoepathy, just as Minerva came out armed from the head of Jupiter(quote from Maria Venturoli Mattei).

Electrohomeopathic Remedies: How Do They Work?

They work on the principle that diseased organisms are far more sensitive than healthy organisms. Much like the area of skin around a cut or a graze is more sore and tender than its surrounding areas. A remedy is never given to just remove the symptoms from the body, but to work wholly on the disease and remove its cause. It therefore works on the body as a whole.

Each remedy is derived from the active enzymes of several plants. When you take this complex remedy it completely covers and controls the body. It will then move through the body via the blood and lymphatic systems and target only those areas that require it. Once the diseased part is sufficiently saturated with the remedy, it stops absorbing it. When all the parts of the body have been healed and a healthy balance restored, the medicine itself is no longer absorbed. Therefore, there are absolutely no risks of over-dosing, side effects or drug dependency.

Electrohomoeopathic remedies purify the lymph and blood systems and can find and destroy the gravest of disorders. They can also help prevent disease by keeping the body pure and giving it a healthy balance.

Principle of Electrohomeopathy

In sick person, the body fluids, lymph and blood become polluted and all the parts become abnormal state, which is why the disease develops in the body. Electrohomoeopathy is system of dynamic form of healing force extracted from the plants. It preserves and controls the organism Lymph and Blood, consisting of atoms, molecules, cell tissues, organs etc, by purifying them and returning them back to the normal state. The sick person recovers speedily and completely.

Why was the Name given Electrohomeopathy?

The name Electrohomoeopathy comes from the Electro content extracted from the plants, and the way this energy is extracted and made into the remedies. Electrohomoeopathy is a name given by the founder, on the basis that those remedies are class of energy conforming to the law of similar and they posses a power and promptness and speed of action of the remedy which allows them to be compared to electricity. Electro is not related to the electricity current and not using any type of electric machine in the treatments.

Theodore Krauss was the disciple of Count Mattei. Mattei made only 38 remedies in his life time but Krauss added more 22 remedies in his life time. After death of Count Mattei the progress and development was stopped in First World War, so therefore Krauss took charge and established the pharmacy of Electro Homeopathy in Germany which is why this pathy also known as spagyric method of Krauss.

The Proving of Remedies

Electrohomoeopathic remedies work on the principle that diseased organisms are far more sensitive than healthy organisms and change the condition of the bodily fluids. A remedy should not be given to remove the symptoms of the body but to work on the cause of the disease.

What is the difference between Electrohomeopathy and other Therapies?

Electrohomoeopathy is a system of medicine that draws all of its remedies from non-poisonous plants. It does not use sources from the mineral or animal kingdom. The remedies work with nature and help the healing force in the shape of Electro element.

Who can Use Electrohomeopathic Remedies?

Electrohomoeopathy can be used by anyone of any age group who suffers from acute to chronic illnesses, for example, depression, arthritis, migraine, ulcers or ME.

What are the Side Effects?

These remedies are all extracted from non-poisonous plants and do not create drug addiction or side effects. All remedies use the physical properties and not the chemical ones and are therefore ideal to combat the human body disorders without any side effects.

How do Electrohomeopathic Remedies Work?

The manner of preparing these compound remedies is the method of "cohobation". This method uses plant extracts in the higher energies (vital force). This is a very potent invisible force, abstract in the natural form and transforms every constitution of the living being as an 'Od Force' (vital force). Each remedy is derived from the active enzymes of several plants. When the complex remedy is administered to the patient it completely covers and controls the body. The remedy will then move through the body acting on the parts that require it. As soon as the diseased part is sufficiently saturated with the remedy, it stops to absorb it. When all the parts of the body have been healed, the medicine itself is no longer absorbed.

What Does the Treatment Involve?

This will depend on the individual and their personal health fitness. The holistic approach to illness needs to be given time and many ailments will not disappear overnight. Therefore a course of treatments is recommended to get the best result. Electrohomeopathy treatment is always based on the Law of Temperament. The patient may be lymphatic, sanguine, mixed,bilious or of a nervous temperament. The person belonging to a specific temperament will show symptoms based on that constitution. In order to learn the constitution of the patient, a full background into their history any inherited family health matters is required. A remedy is then given accordingly.

What to Expect when, Visiting a Practitioner?

The patient will be asked questions relating to their medical history and perhaps that of their immediate family. Lifestyle questions, such as stress levels, diet; exercise regimes, and sleep patterns will be asked to ensure that a holistic approach is taken. Their own expectations of the treatment and what they hope the therapy will achieve will also be discussed. It is necessary for the practitioner to know that the patient comes under which temperament then the treatment can be assessed. The disease or condition they are displaying will also indicate what medicines should be taken. The patient will need to take the medicines, which come in the form of pill and liquid. You will be asked to attend another consultation when your prescription may change.

About Spagyric Therapy

Nagarjuna an Indian metallurgist and alchemist, born at Fort Daihak near Somnath in Gujarat in 931. He wrote the very first treatise Rasaratnakara that deals with alchemical preparations.

Spagyric medicine is ancient natural medicine the has undergone 500 years of trial, research and practice. Paracelsus the famous Swiss physician, Philosopher and alchemist is regarded as the founder of spagyric, but it was almost forgotten until the 19th century.

Dating from thee times of Paracelsus, the terms spagyric and spagyrism are derived from the Greek verbs "span (to seperate) and ageirein (to unify). They designate the art of creating medications of enhanced efficacy by performing the two fundamental alchemical operations. the valuable portions are first separated from those which are impure processed and subsequently reunited to yield an improved medicinal form (HELMSTADTER, 1990). The source materials of spagyric medicines are exclusively from botanical origin. Depending upon the manufacturing process, differentiation is made among spagyrism according to Krauss.

Around the turn of the last century the Italian herbalist Count Cesare Mattei (1809-1896) rediscovered the lost spagyric art and founded the system known as Electrohomoeopathy in Italy. Mattei borrowed from Paracelsus the process of preparing the vegetable substances by means of a more or less complicated mode of fermentation, called cohobation, and also the final combination of he number of ingredients with similar or supplementary effects to form a complex medicinal unity. Mattei has kept the formulation secretly until the age of 78 and he gave all his work to his adopted son in laws Mario Ventrouli Mattei in 1887. After Mattei death in 1896 one of his disciple Theodor Krauss (1864-1924) has established his system in more advanced way. When the real supply of these remedies stopped in 1917 due to World war 1, Krauss reinvented the method in more scientific process known as Krauss method. Today the pharmaceutical company ISO, GmbH Germany is continuing his tradition via JSO Komplex heilwisen. First of all, these preparations enjoy success and worldwide appreciation because of their carefully chosen ingredient. Based on his key research with the founder of system count Mattei, Krauss added a few more combination remedies to the system, characterized by their broad therapeutic spectrum, they are highly acclaimed spinal cord of the system by experts from all over the world specially Germany since 1920.

Spagyric Therapy According to Krauss

The central pillar exists in spagyrism according to Krauss: phytotherapy, within the present context, phytotherapy is to be understood as the traditional art of healing with plants. The spagyric medications are administered in either their 4th or 10th decimal potency. The clinical propertise of the active medicinal substances within each preparation help determine the selection of the appropriate medication. In the Krauss' school of spagyric therapy, pathological processes are not viewed simply as isolated instances of dysfunction in individual organs or within the mind. Therefore, the objective is to detect and treat the course of disease in the interactions with all the body processes.

In order to adequately cover the multi-level interconnections of an illness, this type of therapy employs complex medications consisting of a combination of numerous single-ingredient spagyric remedies. Selection of the appropriate spagyric agents to be included within a complex is based on the physiological and anatomical relationships existing among individual body functions, organs, and organ groups. The single-ingredient agents are combined into such a manner to influence a group of organs closely associated with one another in either a physiological or anatomic-histological manner. Contained within one Electro homeopathic medication are single-ingredient agents with similar fields of actions as well as those that influence the pathological condition from a variety of aspects.

The constituents are combined to target the illness synergistically while mutually supplementing their effects. Thus they selectively direct their effects within the diseased organism at multiple levels, each in correspondence with its individual rang of action, thereby collectively creating stimulation of the body's intrinsic defensive and regenerative powers (Bruch 1939, Krauss 1989).

In the book "The Hidden Roots: A History of Homeopathy in Northern, Central and Eastern Europe" by Prof. Dr. Robert Jütte, Director of the Institute of History of Medicine Foundation Robert Bosch in Germany is written that Dr Alexander Rosendorff (1871 - 1963) became "acquainted with Electrohomoepathy" when he met Cesare Mattei. The sign which he had installed in 1920 outside his practice in Tallinn, Waldstraße, however, said: 'Homeopath'.

In the book "The essay on history of Homeopathy in Latvia" Marina Afanasieva, Medical Dept., Vice-President of LMHI of Latvia, published by the "Liga Medicorum Homeopathica Internationalis," France has written a significant part on Mr. Feliks Lukin (1875-1934) that he learned Elettromeopatia from Dr.Rosendorf in Tallinn (Estonia) and practiced in France but with the name Homeopathy.

In Farmacopia Omeopatica Germany, published by the federal government, the formulas are described as delll'Elettromeopatia formulas Omeopatiche.

Bärbel Tschech (Dip. biologist) in her article : A Discussion of Complex Homeopathic Remedies,translation by Siegfried Letzel, published in Hpathy Ezine - July, 2007 Zimpel and Mattei have been described as homeopaths.

Description of the Remedies

- ☆ Scrofoloso remedies, act on the scrofulous disorders and the metabolism.
- ☆ Linfatico remedies act on both systems of lymph, and blood especially the white corpuscles.
- ☆ Angioiticos remedies, which act on the blood vessels the arteries and the veins, and the circulatory system.

☆ Canceroso remedies, act on cellular construction and the chronic degeneration of Lymph.

☆ Febrifugo remedies, act on the fevers and all types of intermittent diseases, as well as disorders of the spleen and liver.

☆ Vermifugo remedies act on the intestines, but also on the other parts of our organism, and then also destroy all worms.

☆ Pettorale remedies, act on the respiratory system and bronchial tubes.

☆ Venereo remedies have a general constitutional effect.

Electricities or Fluids

☆ Red Electricity is an arterial and nerve stimulant. This increases the vitality of the body.

☆ Blue Electricity has tonic effects on the circulatory system, when used in a diluted form it calms the organs which is over stimulated.

☆ Green Electricity promotes venous system and helps with chronic diseases and arthritic conditions.

☆ Yellow Electricity relaxes muscular or nervous tension, diminishing organic function and excitement.

☆ White Electricity exerts a gentle, calming influence over the nervous system.

☆ APP water for skin keeps the skin fresh and smooth as cosmetic effect.

Advantages of Electrohomeopathic Remedies

These remedies are all extracted from plants, having energetic properties. They produce energy in the body; its action is gentle and sometimes instantaneous, but more generally gradual of such a nature that the results can only be perceived after some minutes.

They correct the deficiency or excess to natural proportions and recover the disease speedily, gently, completely and permanently.

They do not simply cure the particular disease for which they are prescribed, but at the same time good to the constitution of the patients who uses them.

The true remedy gives the extra something, which the body needs. They are non-poisonous and do not create drug disease or leave any bad effects on the body.

It is not extremely facilitated and simplified in this way, in the most complicated illness as well, but also it is more economical of what is so important in the present period of necessity.

It is especially adapted for universal use as household remedies. No mother dread her child's life has been endangered because she has taken some of the medicines in error, nor need she ever be haunted by the fear, so often experienced by those who give bromides and other hurtful drugs that may turn out to be worse than the disease.

Electrohomoepathy medicines are purely herb based medicines and 114 plants are used in the preparation of these complex medicines. The medicines are prepared by following the three major steps of spagyrism: purification, separation and cohobation. The prepared medicines are prescribed homeopathically to the patient. In this way, Electrohomoepathy is a complex art of healing with the effects of both healing systems - spagyrism and homeopathy. Mattei gave eight specific groups of medicine in addition to five "electricities", also known as the electric fluids, which are mostly used in making a compress for the body points. Mattei classified the medicines in groups as follows:

Sl.No.	Name of Medicine	Action
1.	Antiscrofoloso Group	For purification of the lymph
2.	Antiangitico Group	For purification of the blood
3.	Antilymphatico Group	For lymph metabolism disorders
4.	Pettorale Group	For respiratory problems
5.	Vermifugo Group	For all kinds of infective germs as well as worms
6.	Antivenereo Group	For constitutional disorders as ell as for venereal disorders
7.	Febrifugo Group	For all kinds of fevers as well as for use as a nerve remedy
8.	Anticanceroso Group	For all kinds of cancers (benign as well as malignant)
9.	Five Electric Fluids	
	a) Red Electricity	Stimulant
	b) White Electricity	Sedative
	c) Blue Electricity	Antihaemorrhagic
	d) Green Electricity	Pain killer
	e) Yellow Electricity	Intestinal remedy

$\boxed{15}$
Feng Shui

Feng Shui (pronounced fung-shway) was developed 3000 years ago in China and is regarded as an ancient art and science. 'Feng' means 'wind' and 'shui 'means 'water'. It assures the health and good fortune of people and reveals how to balance the energies of any given space. Not just the Chinese but millions of people all over the world repose their faith in and live by this natural force to enhance and attract good health, wealth and relationships. 'Feng' means 'wind' and 'shui 'means 'water'. According to the Chinese culture, good harvest and health are associated with 'gentle wind and clear water'. Thus,' Good Feng Shui' is synonymous with good livelihood and fortune and whereas 'bad feng shui' came to mean hardships and misfortune. Your health is connected to the Feng Shui of your environment. Good Feng Shui may not heal all that ails you, but it can help you become as healthy (both physically and mentally) as you can be. Bad Feng Shui can cause, exacerbate, or increase susceptibility to illness, stress, and depressed energy (chi).

Feng shui means 'wind, water' and traditionally symbolizes the space between heaven and earth — the environment where we live. The underlying philosophy recognizes that we and our environment are sustained by an invisible, yet tangible, energy called chi. The skill of a feng shui consultant lies in recognizing where chi is flowing freely, where it may be trapped and stagnant, or where it may be excessive. These skills and work are applied together with a harmonious re-balancing of yin and yang, the dark and light of all situations.

Good health is the most important thing anyone amid want. If good health is there, good luck follows. According to Feng Shui, east is the direction associated with health. If the eastern sector is rightfully activated the family members will enjoy good health and enjoy a healthy long life.

The element of this direction is wood. Wood is denoted by plants and trees or the green color. According to the destructive cycle in Feng Shui, metal destroys wood and fire burns wood. So avoid keeping any metallic or fire elements in the east,

However water nourishes wood, so water features are welcome in this sector. Placing healthy green plants in this sector is the best. However avoid using dried or artificial plants. This suggests stagnation of energies as well as negative energies. Bamboo is the best suited plant for this sector as it is known to absorb negative energies and symbolizes 'longevity, strength and growth. One can also use a crystal or a wooden turtle to energize this sector.

A Green dragon is considered very auspicious when kept in this sector, If there is someone in the family who is suffering from some ailment, then one needs to cleanse the negativity from the eastern sector of their room.

Do this by placing two small wooden containers filled with unrefined salt and change the salt daily. This will help absorb the negativity from the room. Also lighting up this sector with green lights is helpful.

Cranes are a symbol of longevity in Feng Shui. Their picture can be placed in the east of the room or east of the room of an ailing person, Cannons and arrows can prove to be very harmful for health according to Feng Shui.

A feng shui practitioner suggests ways to bring harmony and improve the environment in any home or business through the use of the 5 elements.

Some Useful Feng Shui Tools and Tips to Improve your Life

FLAT BAGUA MIRROR

It guards and wards off negativity and harm as it balances and adjusts the Chi or the energy. The flat Bagua is an adequate one except for, if there are conditions that necessitate the use of the convex and concave types.

CONCAVE BAGUA MIRROR

The negative energy, known as the Shar Chi or killing energy is deflected and is prevented from entering the given space-whether home or office.

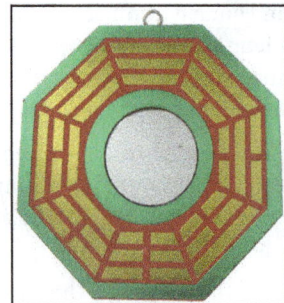

FU DOGS

Fu Dogs are seen as protectors. They should be placed on the each side of the front door, on the outside and if that's not possible they can be positioned on the inside of the front door.

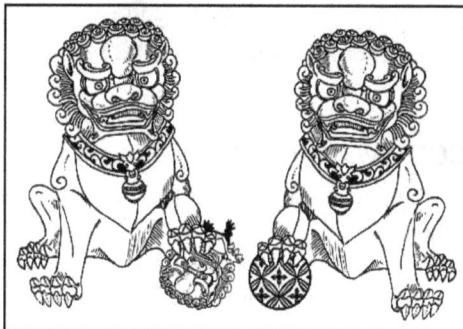

LAUGHING BUDDHA

The laughing Buddha is regarded as one of the many Gods of wealth. It garners prosperity, success and financial gains to the keeper. Its position must directly face the main door. The Laughing Buddha greets the energies that come in from the main door.

FUK LUK SAU

According to the Chinese, placing the Three Star Gods - Fuk Luk Sau, the Gods of Health, Wealth and Prosperity, in the dining and/or living room, symbolizes good fortune for the household.

THREE LEGGED FROG

It is seen as to augment to and protector of wealth and prosperity. It should be placed inside the main door of House/Office, facing indoors thereby representing bringer of health and wealth.

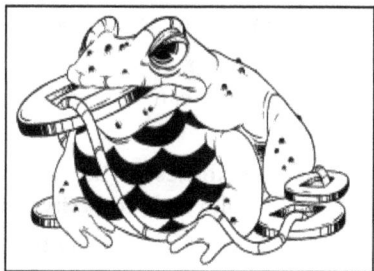

AROWANA FISH

Fishes symbolize boundless wealth and generate positive energy in Feng Shui. According to the Chinese, the Arowana fish is a Feng Shui fish and many people concur that their good fortune is because of keeping live arowanas in their offices. The arowana figurine, swimming on a seabed of coins conveys

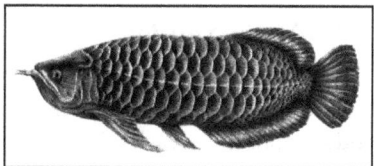

auspiciousness and garners wealth! This fish is also known as Dragon Fish. It should be kept in North, East or South east of the house or Office.

COIN WITH BELLS

This symbolizes good luck and auspiciousness and brings well being and positivity. These bells with coins should be hung on the inside of the main door of your house. The coins bring financial luck while the vibrations produced by the bells energize the positive Chi (energy).

SWASTIK - OM - TRISHUL SYMBOL

The Chinese believed in displaying pious and religious symbols outside their homes to ward off evil spirits and attract good luck and fortune. In India the three most auspicious signs are of- "SWASTIK - OM and TRISHUL ". Therefore, exhibiting these three signs together outside the main door, brings luck and also protects the people living inside the house from all kinds of evil. This sign known as "THE TRISHAKTI" can be used in the office or it can be kept in the purses, bags, briefcases *etc.*

GOLDEN PLATED INGOT

Golden Ingots signify wealth, material assets and they represent savings. It can be placed as a table top or on the cabinet to bring good chi luck and wealth to you.

The ingots can be placed in the following directions:

☆ **North-west** – it helps to enhance the family, wealth and prosperity.

☆ **South-east** – it helps to generate money and wealth

☆ **East** – it brings good health, longevity and descendant luck.

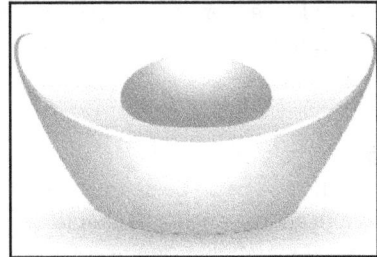

YIN-YANG SYMBOL

The yin and yang are the basic energies of the universe. These two energies represent the female and the male energies and are opposite to each other. The well being of human beings depends upon a perfect balance and harmony between these

two positive and negative energies. Placing this symbol in a room or a work place helps in balancing the yin - yang energy of that place thereby making that place better for living or working.

METAL TURTLE

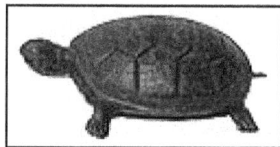

Turtle is one of the four celestial animals in Feng Shui. A turtle represents longevity and good luck. It should be placed in the north or north-east orientation of the house preferably in a plate containing some water. However, if your bedroom is in the north and north-east direction, then the turtle should be placed without water.

☆ Let the first impression of entering your home be warm and welcoming.

☆ Keep a fish pond with 10 fishes, with 9 goldfishes and 1 black fish. The black fish absorbs all the negativity. And death of a fish or fishes denotes that some bad luck has been warded off and taken up by the fishes which could otherwise be meant for the resident. Fishes should not be kept in the bedroom, bathroom or kitchen.

☆ The front and backyard can have items that catch one's attention-items like pinwheels, bright colored figurines, garden gazing balls *etc.* the mail box must be brightly colored and eye-catching, it'll lure more mails especially if your business is home based.

☆ Flowering plants (red colored being the best) should be placed along the walkway in groups of 3 up to the front door.

☆ Keep the alley or walkway to your door lighted at night, it'll keep the path to your wealth and prosperity illuminated.

☆ Dead and dried flowers, which symbolize dead wealth, should be replaced by new or silk or artificial flowers.

☆ The yard should be clutter-free and the plants must be well trimmed and not over grown. Any cut-down trees or shrubs should be replaced.

☆ Try to keep as many plants as possible in as many rooms. Better still if their leaves are coin-shaped.

☆ It is advisable to change the positioning of the furniture regularly as it'll help to change the flow of CHI (energy). And make sure there is no clutter, when it comes to furniture-less is more, as more furniture hampers the flow of energy.

☆ Filled up kitchen cupboards provide a sense of plenty. It is advisable to keep it well stocked. Also keep closets and cupboards clean.

☆ The brooms, mops and other such items should be out of sight. The livelihood of the family is prevented to be swept away.

☆ To soothe the working or home atmosphere play tranquil and calming music like forest sounds, rain falling *etc.*

☆ Make sure your work desk is adorned with fragrant flowers as they'll keep you in high spirits. However, they shouldn't be kept in the corner of love and marriage as it clashes with the element of romance-the earth.

☆ If the room is small in size, mirrors can make it appear as larger. Spacious room denotes endless opportunities.

☆ Keep all your watches in working condition (*i.e.* wrist, wall *etc.*) according to feng shui, non-functional watches are a barrier to growth and fortune.

Tips to Improve Love and Romance

Feng shui practitioners believe that the best way to attract love is to adjust the environment accordingly. Dingy surroundings, piles of clutter, and self-absorbed artwork can actually drive out Cupid from your door. Lone/solitary figures should be replaced with Pictures of Happy Couples as artwork has an incredible impact on the subconscious. Surrounding one with photos, paintings, sculptures and knick-knacks of solitary figures, will let you carry yourself accordingly. Replacing such images with representations of happy couples will make you more receptive to love.

Create Cozy Seating Arrangements

Back off-that's the message sent to the prospective suitors and spouses by single chairs. If you're looking for love, create cozy seating arrangements from put chairs at comfortable angles to each other signalling that you're ready for a relationship.

Keep the Bedroom Sans the Television

Nothing kills romance like the buzz of late night television. If you have trouble falling asleep, try unwinding by reading love poetry or romantic novels. Your subconscious will shift accordingly, making you a virtual love magnet.

Treat Yourself to Sensual Bedding

That old tatty pillow from college and those serviceable sheets aren't doing your sex life any favors. Go ahead and splurge on pillows, blankets and sheets that invite you to linger in bed.

Prepare for Company

When you demonstrate that you are willing to share your space with someone special, your romantic prospects will soar. You can do so by-clearing out one of your dresser drawers, leaving some empty hangers in the closet, and keeping an extra toothbrush in the bathroom.

Downsize from a King to a Queen

Sleeping on a huge mattress won't nurture intimacy. Invest in a comfortable queen-sized bed if you'd like to find the love of your life. If you simply can't make the switch, make sure to sleep on patterned sheets to add some zest to your sex life.

Pull Your Bed Away from the Wall

If you push your bed against the wall it crowds out any chance for love to enter your life. Therefore, arrange your bed so that it has enough space to walk on either side. Before you even realize it, you'll be cuddling close with someone special, instead of hugging your pillow for comfort.

Make Your 'Relationship Gua' a Shrine to Love

Using the front door of your home as a reference point, the far right corner of your home represents relationships and is known as the "Relationship Gua" in Feng Shui. Therefore, you need to keep this area intimate and inviting. A love seat illuminated with adjustable lighting is ideal for this special spot.

Keep the Family at Bay

The bedroom represents romantic life. Therefore, the last images that should be placed here are of your dear old Mom and Dad. Also Children's artwork and toys will undermine your sex life. Celebrate these relationships in other areas of your household, but keep your bedroom as a private retreat.

Think Pink!

Warm colors like pink and red can liven up your love life significantly. Soft shades like rose, salmon and coral can attract a gentle partner who is considerate to your needs, while bold colors like scarlet, crimson and burgundy will draw a passionate adventurer to your side.

Crystals in Feng Shui

Crystals are widely used in feng shui as they bring specific energy, or vibrations to the space. For example, rose quartz is used to attract love and romance, as well as to heal a broken or scarred heart because rose quartz has specific frequencies that promote healing of the heart.

- ☆ Two rose quartz hearts are often placed in the Southwest area of the space to symbolize happy hearts in a love relationship; sometimes a full bowl of rose quartz is placed for healing, for grounding or centering, as well as protective purposes tumbled hematite is used.

- ☆ If your child tends to get overexcited easily and has trouble focusing/concentrating, hematite will bring a grounding energy which might be very helpful. It can be either placed in your child's room, or in the West area of your home.

- ☆ Since hematite stone has protective qualities, it is used at the main door for feng shui purposes.

- ☆ Crystals should be cleansed when first bought or received as a gift. Cleansing them resets their vibration to be receptive to the new home and owner. They should be cleansed regularly, as well as treated in a respectful and loving manner.

16

Gem and Crystal Therapy

Gems are wonderful creations of nature. Crystals have many special properties and find applications in scientific areas including electronics. Gemstones are found in natural form in many shapes and colors. They differ in physical and chemical properties from one another. The stones are cut and polished to give the required beauty and shine. From ancient times scholars have associated gems with planets, based on the color and the effect that is produced by wearing them. Not only the Indians but Chinese and Egyptians believed in the mystical and therapeutic qualities of gems.

Crystals are solids that form by a regular repeated pattern of molecules connecting together. In some solids, the arrangements of the building blocks (atoms and molecules) can be random or very different throughout the material. In crystals, however, a collections of atoms called the Unit Cell is repeated in exactly the same arrangement over and over throughout the entire material. Because of this repetitive nature, crystals can take on strange and interesting looking forms, naturally.

Crystal healing is a form of healing that uses crystals or gemstones. The crystals are mainly placed on specific areas of the body called "chakras" meaning spiritual energy. According to this teaching there are seven basic energy centers in the body, each having a color associated with it. Some crystal healers place the same color crystals as the color of the chakras on the person to enhance the flow of energy. Crystals are said to direct the flow of energy to the person in a particular part of the body and bring balance to a person's energy. Ultimately, they are used to cleanse the person from bad or negative energy believed to cause an illness. Clearing out the bad spiritual energy alleviates the physical ailment. Crystals are used for physical, mental, emotional and spiritual healing. Additionally, crystals can be worn, placed next to a person's bed as they sleep, and in some cases placed around a person's bath. It is an alternative healing technique for strengthening the body and resolving issues and patterns using various forms of natural crystals. The earliest records of crystal healing have been traced in Ancient Egypt. India's Ayurvedic records and traditional Chinese medicine also claim healing with the use of crystals, dating back to 5000 years.

Effects of Crystal Therapy

Crystal therapy aims to restore the balance and well being of a person. In this therapy assorted stones are placed on or around the body that resonate with chakras of the body. This helps to release impending blockages, refining the flow of energy. The recipient is able to experience a deep state of tranquility, letting go of attitudes and habits that are not beneficial, assisting in tuning ourselves to the rhythm of life. Crystals work on the subtle energy levels and in our auras to rebalance energies and improve our well-being. This therapy promotes peace and tranquility and reduces mental and nervous stress.

Ailments and Crystal Healing

Inherent properties of a crystal make them beneficial and invaluable in the healing of ailments Keep your stone with you as much as possible for a week. Place it next to your bedside when you sleep at night, wear it or put it on your desk during the day. Pay attention to your moods, your interactions with others and your dreams.

The body needs all 7 color rays: red, orange, yellow, green, blue, indigo, and violet. Each color ray is associated with one or more of the chakras. By using the color ray that supports a particular chakra, you can speed healing of the associated areas and organs.

The 12 Crystal Healers

With just 12 stones in your crystal toolkit you can heal dis-ease at a physical emotional, mental and spiritual level or transform your environment. The stones in this directory bring together seven crystals for cleansing, energizing and activating the traditional seven chakras together with five 'master healer' stones, for healing the body (Bloodstone) mind (Clear Quartz) emotions (Rose Quartz) Spirit (Labradorite) and environment (Smoky Quartz). The master healers also work on higher chakras that are now a days coming into play. However all 12 specially selected stones are extremely versatile and will have a healing effect at all levels.

You can place your stone in a metal spiral and wear this around your neck-over the thymus gland about a hand's breadth below your collarbones is an ideal place, or you can tape the stone over the organ or body part concerned-tumbled stones have no rough edges and are comfortable against your skin. If you are placing the stones on any part of your body for a shorter time, it is usually sufficient just to lie still. Most stones work through your clothes although there are occasions when the stone is more effective against your skin.

Crystals for Cleansing and Recharging your Chakras

People who are intuitive can see or feel, chakras spinning, looking like whirling Catherine wheels of light. Dull or black patches or a spin that'wobbles' or is too fast or slow, indicates dis-ease at the physical, emotional mental or spiritual level according to the chakra concerned. Fortunately you do not need to 'see' such dis-ease because a crystal will pick up any disharmony, put it right and re-energize the chakra.

Strengthening your Aura

Your aura, also known as the etheric body or biomagnetic sheath, is a subtle energy field that surrounds your physical body. Made up of several diffuse layers, the outer layer relates to spirit, the next layer to mind, the next to emotion and the one closer to the physical body relates to the physical level of being.

Crystal for your Body

Detoxification

If your body is toxic, it cannot maintain health. Stimulating your liver with the detoxification layout release toxins and encourages your lymphatic system to remove them, bringing about a physical cleansing. A Yellow jasper at your solar plexus stimulates an emotional detox as well.

Grounding

One of the most widespread causes of feeling vaguely unwell and out of sorts is being ungrounded. Ungroundedness means you are not fully in your physical body. You are not in the present moment and your energy is scattered in all directions.

Vitality

Red and orange are excellent colours for instilling vitality and for overcoming energy depletion.

Crystals for your Mind

Crystals have a powerful effect on your mind, calming an overactive mind and stimulating a sluggish one. They promote concentration, focus and creativity.

Crystal to Improve Concentration, Focus, Memory and Learning

Problem-solving and Creativity

Your creativity and capacity to solve problems can be expanded by uniting the two different hemispheres of your brain. The left side of the brain is the analytic, rational side, concerned with sequential logical thought process and the right side is the illogical, intuitive part that makes random leaps to find creative answers, Crystals connect the two hemispheres and harness their powers.

Healing Mental Problems

Crystals can be an enormous support if you suffer from mental problems or have minor psychiatric conditions. They can stabilize your mood or assist you to overcome addictions, particularly as they help you deal with the underlying cause of you problem. Wear your stone constantly to support your mind and your intention to heal.

Crystal for your Emotions

Wearing or surrounding yourself with crystal helps you keep your emotions in balance. The crystals gently release the emotional blockages or suppressed feelings that cause your moods to fluctuate.

Crystal for Emotional Balance

Healing Depression

Depression is an extremely debilitating emotional state which manifests some what differently according to its cause. Wear a crystal according to the type of depression you are experiencing.

Healing Negative Emotions

Crystals draw out difficult emotions. They can be placed over the appropriate chakras. As the crystal is placed, visualize it pulling the negative emotion from your physical body and the subtle bodies and then filling the space with the crystal's unique healing vibration.

Crystals for Anger (Base Chakra)

Crystals for Healing Past Hurts (Base Sacral, Solar Plexus, Heart and Throat Chakras)

Crystals for Guilt (Base, Solar Plexus and Brow Chakras)

Crystal for Forgiveness (Higher Heart Chakra)

Revitalize your Sex Life

Sexual activity is linked to the lower chakras and when these are cleansed and energized, libido is able to flow freely. Crystals activate these centers and also those of the emotional body so that love can be given and exchanged freely. Your sex life can be gently enhanced if you and your partner share the sexual recharge ritual placing the stones on each other as part of your foreplay or enfolding them between your bodies if this is comfortable for you.

Crystals for your Spirit

As they are full of spiritual energy and light, crystals work brilliantly to heal or support your spirit and to open your intuition. The high-vibration crystals such as the Quartz and Labradorite work through the spiritual body to ground spiritual energy into the physical body and to bring about profound insights.

Crystals for Intuition and Insight

Healing Dreams

Crystals can be programmed to bring you healing dreams. Simply hold your crystal in your hands for a few moments, visualizing it surrounded by light, firmly state your intention that your crystal will being you a healing dream that you will remember and understand on waking. Then place the crystal under your pillow before you go to sleep. Keep a pad and pen by the bed to write down your dream.

Crystal Meditation

Meditation switches off the mind and puts you in touch with spiritual reality Gazing into the depths of a clear crystal enables you to quickly enter a meditative state. Crystals have a natural affinity with meditation as they calm your mind and open it to receive spiritual energy.

Connecting with the Divine

Connection with the divine energies and the spiritual level of being often occurs during meditation. When it is known as bliss consciousness or enlightenment, but you can be connected to divine energy all the time with the assistance of your crystal. Connecting to this energy and anchoring it into your crystal means that you will always be connected to the divine light at the heart of the universe.

Crystals for Relaxation

Regular relaxation conveys an enormous benefit to your health and plays a major part in preventing disease. Crystals quieten your body, mind and emotions, bringing about a deep sense of calm and centredness. If you feel uptight, crystals will help you to unwind; if you feel physically jittery and cannot stay still, crystals will relax your muscles; and if your thoughts are racing, they will shut down your mind. Taking 15 or 20 minutes to relax with your crystals will bring you inner peace that radiates out into your daily life.

Stress Relief

Stress whether physical, emotional or mental, takes an enormous toll on your body causing headaches, aches and pains, hot seats, fatigue, insomnia, stomach ulcers and various auto-immune disease. When you are under stress your adrenal glands, which sit on top of your kidneys, go into 'fight' or 'flight' mode, over producing adrenaline. Lying quietly for 20 minutes with the appropriate crystal helps you to release stress, but you can also carry the crystal in your pocket and use it as a hand soother to play with whenever you need it.

Insomnia

Insomnia can arise from various causes and there are several crystals that will assist you to sleep more deeply. The crystal can either be placed under your pillow or worn around your neck.

Crystals for Clearing and Protection

If you feel drawn to wear a particular crystal, it may well be because you are subconsciously registering a need for its protective vibes. Crystal vibrations are excellent for counteracting negative energy of all kinds, gently deflecting negative energy, they stabilize excess energy and absorb toxicity, creating harmony within your home, workplace or external surrounding.

Protecting your Home

Crystals are beautiful, decorative objects that can enhance your home but they have a practical purpose too : they can keep the vibes harmonious and the energies vibrant. Crystals will protect against environmental pollution or electromagnetic smog (caused by the likes of microwaves, radio waves and radar radiation) Remember to cleanse the crystals in your home regularly.

Enhancing your Workspace

Crystals are discreet tools for enhancing your workspace and for encouraging co-operation between co-workers. You can use a crystal as a paperweight, slip one into a plant pot or place one on your computer. A small crystal can have just as powerful effect as a large one. A small Orange Carnelian, for instance, energizes a whole room.

Crystals and Electronics

The subtle emanations from electrical equipment can seriously damage the biomagnetic sheath surrounding your body, Kirlian photography can take pictures of the energy around your body, and can demonstrate, for example, that the emanations from a mobile phone create gaps in the biomagnetic sheath around the head and neck, and also around the reproductive organs where the aura virtually disintegrates. This occurs when the phone is simply switched on.

Crystal Healing: Use of Crystal for Curing some Diseases

☆ Abuse - Pink Tourmaline, Rose Quartz, Thulite, Larimar

☆ Acne - Cuprite, Larimar, Rose Quartz

☆ Addiction - Avalonite, Malachite-Azurite, Peridot

☆ Adrenal Glands - Kansas Pop Rocks, Sulfur, Kyanite, Black Tourmaline

☆ Ageing - Emerald, Rose Quartz, Rutilated Quartz, Boulder Opal

☆ Alcoholism - Avalonite, Golden Calcite, Peridot, Phenacite

☆ Allergic Reaction - Bloodstone, Hematite

☆ Altitude Sickness - Hematite, Jet

☆ Alzeihmers - Blue Chalcedony, Fluorite

☆ Anemia - Hematite, Garnet, Bloodstone

☆ Anger - Black Tourmaline, Chrysocolla, Peridot, Oligoclase

☆ Anorexia - Rose Quartz, Thulite, Golden Calcite, Malachite

☆ Anxiety - Aventurine, Lepidolite, Amethyst, Amazonite

☆ Arthritis - Boji Stones, Chrysocolla, Lapis Lazuli, Imperial Topaz

☆ Assertiveness - Chrysoprase, Golden Calcite, Tigers Eye

☆ Asthma - Malachite, Vanadinite

☆ Attention Defect Disorder - Fluorite, Selenite, Azurite

☆ Backache - Amber, Green Tourmaline, DT Quartz

☆ Bacterial Infection - Anhydrite, Malachite, Sulfur

☆ Bi Polar Disease - Kunzite, Rutilated Quartz

☆ Birthing - Peridot, Shiva Lingham,

☆ Bladder and Kidneys - Cuprite, Prehnite

☆ Bleeding (Decrease) - Bloodstone, Malachite, Sodalite, Sapphire

☆ Blisters - Anhydrite, Sulfur, Rose Quartz

☆ Blood (Purifyin, Disorders, Circulation) - Garnet, Malachite, Ruby

☆ Blood (Pressure, Balance) - Chiastolite, Malachite, Ruby

☆ Bone Marrow - Malachite, Lapis Lazuli

☆ Bowels - Lepidolite

☆ Breast Milk - Carnelian, Chalcedony, Chiastolite, Turquoise

☆ Breathlessness - Amber, Vanadinite

☆ Broken Bones - Axinite, Calcite, Green Tourmaline, Topaz, Selenite

☆ Bronchitis - Aquamarine, Chrysocolla, Amazonite, Aventurine

☆ Burn Out - Garnet, Ruby, Vanadinite, Zincite

☆ Burns - Boji Stones, Chrysocolla, Blue-Lace Agate

☆ Bursitis - Amber, Blue-Lace Agate

☆ Cancer - Lepidolite, Sugilite, Red Jasper, Watermelon Tourmaline

☆ Cataracts - Green Apophyllite

☆ Chemotherapy - Malachite, Sulfur, Anhydrite

☆ Chickenpox - Childrenite

☆ Child Abuse Recovery - Manganocalcite, Kunzite, Green Fluorite

☆ Childbirth - Shiva Lingham, Hematite, Ruby, Bloodstone, Opal

☆ Chronic Fatigue Syndrome - Ruby

☆ Circulation - Bustamite, Fire Quartz, Tektite

☆ Computer Stress - Smoky Quartz

☆ Colic - Boji Stones, Amber

☆ Colds/Sinus - Azurite, Amethyst, Fluorite, Kyanite

☆ Communication - Amazonite, Blue Calcite, Turquoise, Oregon Opal

☆ Conception - Quartz Scepter, Rhodochrosite, Shiva Lingham Stone

☆ Confidence - Golden Calcite, Orange Calcite, Malachite

☆ Confusion - Charoite, Fluorite, Howlite

☆ Constipation - Ruby, Smoky Quartz, Black Tourmaline

☆ Cough - Aquamarine, Turquoise, Chrysocolla

☆ Courage - Charoite, Chrysoprase, Golden Calcite, Tigers Eye, Sunstone

☆ Cramps - Lepidolite, Hematite

☆ Creative Blocks - Orange Calcite, Chiastolite, Golden Calcite

☆ Degenerative Nerve Disease - Malachite, Azurite, Rhodochrosite, Lapis, Sapphire, Sodalite

☆ Depression - Amethyst, Angelite, Elestial Quartz, Holley Blue, Lepidolite, Smoky Quartz, Sugilite

☆ Detoxification- Malachite, Anhydrite, Sulfur

☆ Diabetes - Citrine, Anhydrite, Jade

☆ Diarhhea - Black Tourmaline, Smoky Quartz

☆ Digestion - Amber, Citrine, Sulfur

☆ Dreamwork - Bustamite, Herkimer diamond, Holley Blue

☆ Drug Issues - Smoky Quartz, Jet, Tigers Eye, Rutilated Quartz

☆ Dyslexia - Sugilite

☆ Earache- Amber, Amazonite

☆ Eating disorders- Rose Quartz, Avalonite, Thulite

☆ Eczema- Blue Sapphire

☆ Emotional Healing - Isis Crystal, Malachite

☆ Endocring Glands - Pietersite, Topaz,

☆ Environmental Pollution - Anhydrite, Sulfur, Zincite

☆ Epilepsy - Red Jasper, Carnelian, Malachite, Coral, Emerald

☆ Eye Problems - Obsidian, Topaz, Tigers Eye, Jade, Opal, Obsidian

☆ Fatigue - Barite, Red Calcite, Vanadinite, Zincite, Peru Opal

☆ Fear - Angelite, Kunzite, Sunstone, Golden Calcite, Holley Blue, Jet, Lepidolite, Tigers Eye

☆ Fertility (improving) - Garnet, Shiva Lingham Stone, Quartz Scepter

☆ Fevers - Brazilianite, Opal, Pietersite, Ruby, Tektite

☆ Fibromyalgia - Amethyst, Aventurine, Blue Lace Agate, Citrine, Quartz Crystal, Rose Quartz

☆ Focus - Yttrian Fluorite, Fluorite slices,

☆ Forgiveness - Angelite

☆ Gall Bladder - Carnelian, Citrine, Malachite, Emerald, Green Tourmaline

☆ Goiter - Amber

☆ Grief - Apache Tears, Sugilite, Smoky Quartz, Jet, Elestial Quartz, Angelite, Kyanite

☆ Grounding - Hematite, Jasper, Mochi Marbles, Obsidian, Smoky Quartz

☆ Guilt - Larimar

☆ Gums - Agate (all varieties)

☆ Hay Fever - Blue Lace Agate

☆ Headaches - Aquamarine, Chrysocolla, Turquoise, Sodalite

☆ Hearing Loss - Lapis Lazuli

☆ Heart Attack - Garnet

☆ Heartburn - Dioptase, Peridot, Sulfur

☆ Heart Disease - Garnet, Dioptase

☆ Heart Healing (emotional) - Chrysoprase, Danburite, Kunzite,

☆ Heat Stroke - Sunstone

☆ Herpes - Jadeite, Lapis Lazuli

☆ Hip Pain - Jadeite

☆ Hormone Production - Pietersite, Amethyst

☆ Hot Flushes - Moonstone, Chrysocolla, Gem Silica

☆ Hypertension - Chrysocolla

☆ Immune System - Angel Hair Quartz, Herkimer diamond, Rutilated Quartz,

☆ Impotence - Varascite, Shiva Lingham

☆ Incest Recovery - Avalonite, Thulite, Rose Quartz

☆ Infections - Sulfur, Anhydrite, Malachite

☆ Infertility - Shiva Lingham Stone, Quartz scepters

☆ Inflamations - Pyrite, Chrysocolla

☆ Insomnia - Amethyst, Labradorite, Smoky Quartz, Celestite

☆ Intellect - Imperial Topaz, Hiddenite

☆ Intestines - Thulite, Brown Tourmaline

☆ Intuition - Amethyst, Azurite, Selenite

☆ Irritability - Amethyst, Blue Lace Agate, Turquoise

☆ Itching - Chrysocolla

☆ Jealousy - Peridot

☆ Jet Lag - Hematite, Black Tourmaline

☆ Joint Soreness - Green Calcite, Hiddenite

☆ Joy - Blue Sapphire, Sunstone, Vanadinite

☆ Lactation - Carnelian, Chalcedony, Chiastolite, Turquoise

☆ Laryngitis - Stilbite

☆ Learning Difficulties - Fluorite, Azurite

☆ Leg Cramps- Jadeite, Lepidolite

☆ Leukemia - Chrysocolla

☆ Liver - Danburite, Imperial Topaz, Opal

☆ Loneliness - Shiva Lingham, Thulite

☆ Love - Thulite, Rose Quartz, Manganocalcite, Stellar Ice Calcite,

☆ Lungs - Bustamite, Garnet, Pyrite

☆ Measles - Turquoise

☆ Meditation - Amethyst, Fluorite, Holley Blue, Lapis Lazuli, Moonstone,

☆ Memory Problems - Yttrian Fluorite

☆ Meniere's Disease - Dioptase, Diopside

☆ Menopause - Moonstone, Carnelian

☆ Menstrual Cramps - Carnelian, Cuprite, Moonstone,

☆ Mental Confusion/Disorders - Apophyllite, Fluorite, Danburite

☆ Migraines - Aquamarine, Chrysocolla, Turquoise, Sodalite and Blue Tourmaline

☆ Miscarriage (prevention)- Carnelian, Ruby, Aquamarine

☆ Miscarriage (recovery)- Chrysocolla

☆ Morning Sickness - Moonstone, Red Jasper, Malachite, Sodalite,

☆ Mumps - Aquamarine, Topaz

☆ Muscle Aches - Diopside, Jadeite, Lepidolite

☆ Multiple Sclerosis - Lapis, Sodalite, Blue Sapphire, Moonstone, Amethyst

☆ Nail Problems - Apatite

☆ Nausea - Green Calcite, Peridot, Golden Calcite

☆ Nervousness - Barite, Sodalite, Watermelon Tourmaline

☆ Neuralgia - Lapis, Carnelian, Amber

☆ Nose Bleed - Carnelian

☆ Nurturing - Shiva Lingham, Thulite, Manganocalcite

☆ Pancreas - Herderite

☆ Parkinson's Disease - Opal

☆ Past Life Regression - Kunzite, Apophyllite

☆ Peace - Angelite, Blue Calcite, Celestite, Kyanite

☆ Pneumonia - Fluorite

☆ Poison Ivy - Serpentine

☆ Pre-menstrual Syndrome - Moonstone, Jade, Chrysocolla, Turquoise

☆ Problem Solving - Moonstone, Scolecite,

☆ Protection - Amber, Black Tourmaline, Chiastolite, Hematite, Hawks Eye, Mochi Marbles, Obsidian

☆ Prostate Gland - Zincite

☆ Psoriasis - Sulfur, Green Tourmaline, Rhodochrosite

☆ Rape Recovery - Lapis, Chrysocolla, Black Tourmaline

☆ Rash- Larimar, Blue Calcite, Blue Lace Agate, Angelite

☆ Reproductive Organs - Orange Tourmaline, Rubellite Tourmaline,

☆ Relationships - Brazilianite, Peridot, Manganocalcite, Kyanite,

☆ Relaxation - Blue Lace Agate, Turquoise

☆ Respiratory System - Coral, Topaz, Amber, Tigers Eye, Zircon, Agate

☆ Rheumatism - Boji Stones, Imperial Topaz

☆ Sciatica - Green Tourmaline, Kyanite, Smoky Quartz

☆ Self Esteem/Worth - Golden Calcite, Isis crystal, Oregon Opal

☆ Self-sabotage - Golden Calcite, Rose Quartz, Thulite, Avalonite

☆ Shock - Bloodstone, Larimar, Sunstone

☆ Shyness - Golden Calcite, Orange Calcite, Hessonite Garnet Sight - Obsidian, Topaz, Tigers Eye, Jade, Opal,

☆ Obsidian

☆ Sinus Problems - Azurite, Blue Lace Agate

☆ Skin-Nails-Hair - Bustamite, Lepidolite

☆ Smoking-Quitting - Smoky Quartz, Staurolite

☆ Sore Throat - Blue Lace Agate, Blue Calcite, Angelite

☆ Sprains - Sphene, Green Calcite, Aventurine, DT Quartz

☆ Spine - Tigers Eye, Labradorite, Selenite, Calcite, Magnetite

☆ Spiritual Developement - Amethyst, Apophyllite, Azurite, Celestite, Epidote, Fluorite, Holley Blue, Iolite, Hour Glass

☆ Selenite, Selenite,

☆ Stage Fright - Golden Calcite, Aquamarine, Amethyst

☆ STD'S - Tektite, Chrysoprase

☆ Stomach - Holley Blue, Sunstone, Jet, Peridot, Varascite

☆ Stress - Amethyst, Angelite, Aventurine

☆ Surgical Recovery - Amber, Smoky Quartz, Jasper, Chrysocolla

☆ Swelling - Peridot, Emerald, Green Tourmaline, Aquamarine, Sapphire

☆ Teeth - Calcite, Selenite, Sphene

☆ Throat - Blue Lace Agate, Citrine, Turquoise

☆ Thyroid Balance - Citrine, Halite

☆ Trauma - Aventurine, Citrine, Blue Lace Agate, Malachite

☆ Trust - Axinite, Hiddenite, Lepidolite, Oligoclase

☆ Tuberculosis - Morganite

☆ Tumours - Malachite, Bloodstone, Smoky Quartz

☆ Ulcer - Sunstone, Peridot, Lepidolite, Smithsonite

☆ Urinary Tract Infection - Blue Sapphire, Kyanite

☆ Varicose Veins - Amber, Blue Lace Agate, Bloodstone

☆ Vertigo - Cuprite, Lapis Lazuli, Elestial Quartz,

☆ Vomiting - Lapis Lazuli

☆ Warming - Imperial Topaz, Vanadinite, Ruby, Garnet, Zincite

☆ Water Retention - Chrysocolla, Cuprite

☆ Whooping Cough - Amber, Blue Lace Agate, Topaz

☆ Wisdom - Encourage - Apophyllite, Avalonite, Holley Blue Agate, Oregon Opal, Sodalite

Crystals and Colour

Crystals can be very powerful and are very sensitive. Crystals, like all things, have an energy field - aura - and their different energies can be used to balance/enhance our own body's energies. As with many natural therapies, crystals have been used as a healing aid for centuries. Crystals may be helpful in relation to the seven main chakras and their colours.

Crystal Energy

Crystals have been used as a healing aid for centuries.Crystal energy, as colour energy, can help to awaken our awareness, not just of the physical imbalance/dis-ease but awareness of the psychological, spiritual and emotional aspects of ourselves. Some crystals suit some people better or in different ways than others, but this simply emphasizes our uniqueness and our different needs. As with many forms of healing, we take what we need. There are many ways of using crystals and they can be obtained in different forms: the natural unpolished pieces, tumbled stones, cut points, essences, and pieces of jewellery set with crystals. Essences are similar, in a way, to solarized water, in that the crystal is placed in water for a period of time and the specific energy of that crystal is transmitted to the water which is subsequently used for drinking or bathing.

Crystal Healing and Aura

The physical body has an "astral body" like an energy field around it. It is the "Aura" or the bioplasm. This bioplasm is present as long as there is life. After death, it separates from the physical body. The astral body or bioplasm is of great importance, as the origin of most diseases lies here. When external organisms cause an infection in our body, they first come in contact with our astral bodies, and then, the symptoms of infection are felt.

Seeing your Aura

When the astral body is full of positive energy, it easily wards off outer influence. At least 85 per cent of people can see their own aura. This can be done by standing in front of a mirror. There can ideally be a white background behind you. Now focus upon a point 6 inches above and 1 foot behind your head. You must fix your gaze upon this point. At the same time you can also see the outlines of your head and shoulders in the corner of your eyes. After 30 seconds to 1 minute you will see a prominent coloured light parallel to your body. As your gaze moves away from your body, this coloured light will become lighter and lighter. This is you aura or energy field. With practice, you will see this more easily and clearly. It has been proved with the help of the "oscilloscope", that anything or anyone coming within 20 feet of the human aura can cause disturbance in the brain waves of that person.

So you feel disturbed by some people coming near you, either negative or positive. People with similar energy fields, feel very comfortable with each other and may think alike. So we say that their "wavelengths" match.

We can cleanse the aura by spiritual means. This may be by chanting mantras, thinking positive. Even bathing with saline water, made from raw natural salt, cleanses and brightens the aura. So bathing in non polluted sea water is rejuvenating to many. Praying to the Sun, also gives us positive life energy.

Crystal healing is a useful method of cleaning a persons aura. This has to be done carefully and correctly; based upon the study of Chakras and their required colours. Otherwise, this may result in disturbing the aura rather than balancing it.

Detailed Informaton of Gem Stones

☆ **Adventurine** (Aventurine) - A good luck stone, especially in financial matters. Stimulates creativity, intelligence and perception. A great healing stone, it gives a sense of mental and emotional well-being.

☆ **Agate** - general protection and healing, increases courage, self-confidence, energy and promotes longevity. Also good for gardening.

☆ **Alexandrite** - renewal and regeneration, openness to higher self. Brings love, joy, and luck.

☆ **Amazonite** - helps to clarify and improve thinking.

☆ **Amber** - healing, protection from negativity, attractiveness and energy.

☆ **Amethyst** - increases vivid dreams, relieves depression, promotes calm, serenity and spirituality. Helps with addiction and stress.

☆ **Apatite** - helps in hypertension. Rids one of guilt and grief.

☆ **Aquamarine** - improves sight, sharpens the mind, calms and clarifies emotions.

☆ **Aragonite** - grounding and centering, facing truth and reality.

☆ **Azurite** - increases psychic ability and aids in meditation. Relieves arthritis and joint pain.

☆ **Bloodstone** - healing, especially related to the blood. Increases courage and charitability.

☆ **Beryl** - Laziness, hiccups.

☆ **Chrysoberyl** - opens and activates the crown chakra, stone of immortality, assisting one to overcome and to progress towards excellence.

☆ **Chalcedony** - touchiness, melancholy, stimulates maternal feelings and creativity, release.

☆ **Chrysolite** - Inspiration, prophecy.

☆ **Clear Quartz** - Transmitter and amplifier of healing energy and clarity, balance, channeler of universal energy and unconditional love, all purpose healer, programmable, brings good health and balance, use for healing, meditation, knowledge, clear thinking and protection, strengthens all other stones when together.

☆ **Copper** - carry or keep in house to attract money.

☆ **Carnelian** - general healing stone. Restores the natural energy flows of your body.

☆ **Chrysocolla** - soothes nervous tension.

☆ **Chrysoprase** - soothes heartache and helps depression, promotes emotional balance, wisdom and peace.

☆ **Citrine** - helps digestion and aids in mental and intellectual functions. Increases self-esteem and energy. Balances and dissipates negative energy.

☆ **Coral** - helps in childbirth and adoption. Wards off bad thoughts from others and promotes general well-being.

☆ **Diamond** - manifests abundance. Used in conjunction, it strengthens the properties of other stones.

☆ **Emerald** - a mind stone. Enhances memory, promotes clear and quick thinking and right action stemming from right thought.

☆ **Flourite** (multicolor) - enhances spiritual energy work, focuses the will and balances the psyche.

☆ **Garnet** - balances your natural energy. A highly protective stone. Increases confidence and security.

☆ **Goldstone** - uplifting, reduces tension and stomach problems.

☆ **Hematite** - great stone for grounding. Calms and soothes, reducing stress and blood pressure.

☆ **Howlite** - How lite reduces stress and anxiety. Quietly soothes and calms to bring gentleness and patience.

☆ **Jade** - Jade wisdom, mercy, humility, generosity, peace, and harmony. Protective and lucky when worn as jewelry. Place a bit under your pillow at night for restorative dreams.

☆ **Jasper** - calms a troubled mind. Helps with digestive problems and other ills.

☆ **Kunzite** - emotional equilibrium, self esteem.

☆ **Lead** - access to ones spiritual guides when questions related to direction and course of action are pending, can also facilitate a new start.

☆ **Labradorite** - strengthens the will and inner strength and convictions. Aids in combatting addictions.

☆ **Lapis Lazuli** - emotional healing and stability. High intensity stone which cleanses the spirit to bring out inner truth and peace.

☆ **Malachite** - strong stone which helps release negative emotions and protect from psychic attack.

☆ **Moonstone** - helpful in psychic work. Opens the spirit to the feminine aspect. Draws love and helps in interpersonal relationships.

☆ **Morganite** - love, empathy, patience and compassion. Heals the emotional self.

☆ **Pearl** - Eliminates emotional imbalances, helps one master the heart chakra, most powerful when worn at night, speed love vibrations and attract love and good fortune, attract money, health, courage and physical strength.

☆ **Red Jasper** - helps with conflict and aggression. Promotes grace and perserverence.

☆ **Sandstone** - assists one to maintain strength against the distractions of the mind, providing for balancing of ones reality, promotes clarity in thought and sight.

☆ **Onyx** - good protective stone. Also releases old relationships and keeps away general negativity.

☆ **Opal** - Awakens and helps understanding of psychic intuition and mysticism. Encourages living authentically in feeling and action, with faithfulness and loyalty. Helpful for eyesight and during childbirth.

☆ **Peridot** - balances your emotional mind and calms emotional storms. Aids digestion and insomnia and increases psychic powers.

☆ **Quartz** - the greatest of all healing stones. Acts as an amplifier for psychic energy and aids meditation and visualization.

☆ **Rose Quartz** - great for attracting love. Promotes self-loving and heals emotional wounds as well as promoting peace, forgiveness, and nurturing.

☆ **Smoky Quartz** - a grounding stone which helps reduce depression and negative emotions.

☆ **Rhodochrosite** - love drawing stone. Helps release past psychological issues. Helps eyesight.

☆ **Rhodonite** - self-affirmation and self-love. Fosters ability to remain calm in arguments and resolve disagreements in a loving way.

☆ **Ruby** - stimulates nurturing emotions and economic stability. Excellent for shielding from psychic attacks and gathering and amplifying energy. Helpful for blood-related health issues, such as anaemia, menstrual issues, and poor circulation.

☆ **Sapphire** - promotes light, pure emotions - purity of mind, serenity, joy, and peace. Opens the mind to beauty and love. Clears the mind of unwanted thoughts.

☆ **Sardonyx** - encourages self-control and courage in public speaking. Relieves anxiety and grief. Protection, luck, and peace.

☆ **Snowflake Obsidian** - protects against nightmares and emotional draining.

☆ **Sodalite** - purifies your aura and protects you from negative energy from people around you.

☆ **Tanzanite** - increases psychic abilities and enhances spiritual connectedness.

☆ **Tiger's Eye** - clear thinking, helps in manifesting will, courage, and confidence.

☆ **Topaz** - soothes physical pain, promotes peace and calms emotions, as well as promoting forgiveness. Promotes individuality, self-confidence, and creativity. Counteracts negative emotions.

☆ **Tourmaline (multicolour)** - calm, protection, focus, and balance. Promotes self-assurance and goodwill.

☆ **Black** - strong protection from negative energies. Reduces fears, obsessions and panic with a calm rootedness.

☆ **Pink** - promotes compassion, love and healing of emotional pain.

☆ **Green** - cleanses and detoxifies the spirit, brings prosperity.

☆ **Blue** - helpful with verbal, speech, and throat problems. Mental peace and patience.

☆ **Watermelon** - heals old emotional pains and brings emotional understanding.

☆ **Turquoise** - healing stone. Attracts friendships, luck, and happiness. Releases shame and guilt.

$\boxed{17}$
Herbology or Unani

Herbology or Unani system of Medicine owes its origin to Greece. It was the Greek philosopher - Physician Hippocrates (460-377 BC) who freed medicine from the realm of susperstition and magic, and gave it the status of Science. The theoretical framework of Unani Medicine is based on the teachings of Hippocrates. Unani Medicine as a healing system was founded by Hakim Ibn Sina. Yet, because it is a comprehensive system encompassing virtual all of the known healing systems of the world, the threads which comprise Unani Healing can be traced all the way back to Hippocrates.

After a number of other Greek Scholars enriched the system considerably. Of them Galen (131-210 AD) stands out as the one who stabilised its foundation on which Arab physicians like Rhazes (850-925 AD) and Avicenna (980-1037 AD) constructed an imposing edifice. Unani Medicine got enriched by imbibing what was best in the contemporary systems of traditional medicine in Egypt, Syria, Iraq, Persia, India, China and other Middle east and Far east countries.

In India, Unani system of Medicine was introduced by Arabs and soon it took firm roots in the soil. When Mongols ravaged Persian and Central Asian cities like Shiraz, Tabrez and Galan, scholars and Physicians of Unani Medicine fled to India. The Delhi Sultan, the Khiljis, the Tughlaqs and the Mughal Emperors provided state patronage to the scholars and even enrolled some as state employees and court physicians. During the 13th and 17th Century Unani Medicine had its hey-day in India. Among those who made valuable contributions to this system into period where Abu Bakr Bin Ali Usman Ksahani, Sadruddin Damashqui, Bahwa bin Khwas khan, Ali Geelani, Akbal Arzani and Mohammad Hashim Alvi Khan.

The scholars and Physicians of Unani Medicine who settled in India were not content with the known drugs.They subjected indian drugs to clinical trials. As a result of their experimentation added numerous native drugs to their own system further enriching its treasures.

During the British rule, Unani Medicine suffered a setback and its development was hampered due to withdrawal of governmental patronage. Since the system enjoyed faith among the masses it continued to be practised. It was mainly the Sharifi Family in Delhi, the Azizi family in Lucknow and the Nizam of Hyderabad due to whose efforts Unani Medicine survived. during the British period. An outstanding physician and scholar of Unani Medicine, Hakim Ajmal Khan (1868-1927) championed the caused of the system in India.

Fundamentals

Unani Medicine, as said earlier, is based on the principles put forward by Hippocrates. He was the first person to establish that disease was a natural process, that its symptoms were the reactions of the body to the disease and that the chief function of the physician was to aid the natural forces of the body. He was the first physician to introduce the method of taking medical histories. His chief contribution to the medical reals is the humoral theory.

The humoral theory presupposes the presence of four humors Dam (blood), Balgham (phlegm), Safra (Yellow bile) and Sauda (black bile) in the body. The temperaments of persons are expressed by the words sanguine, phlegmatic choleric and melancholic according to the preponderance in them of the respective humors blood, phlegm, yellow bile and black bile. The humors themselves are assigned temperaments : blood is hot and moist phlegm cold and moist, yellow bile hot and dry, and black bile cold and dry.

Every person is supposed to have a unique humoral constitution which represents his healthy state. And to maintain the correct humoral balance there is a power of self-preservation or adjustment called Quwwat-e-Mudabbira (medicatrix naturae) in the body. If this powder weakens, imbalance in the himoral composition is bound to occur. And this causes disease. In Unani Medicine, great reliance is placed in this power. The medicines used in this system, in fact, help the body to regain this power to an optimum level and thereby restore humoral balance, thus retaining health. Also correct diet and digestion are considered to maintain humoral balance.

The word 'Temperament' used as noun literally means "characteristic combination of physical, mental and more qualities, which together constitute the character of an individual and affect his manner of acting, feeling and thinking". Its adjective, 'Temperamental', besides relating to temperament means the variable or unaccountable moods. In modern medical terminology 'Temperament' means "the peculiar physical character and mental cast of an individual". Thus we see that the "moral qualities" are excluded from the total characteristic combination of a man in modern medical science. The concept of the 'Temperament' in all the eastern sciences, however, is 'holistic' and embraces all the qualities of a man, whether it be physical, mental, moral or spiritual.

Before discussing the temperamental theory in the Unani sciences of treatment, it is expedient to trace its origin and determine the actual purport of the term 'Temperament'.

The ancient Greeks held a variety of opinions about the elemental composition of nature. Anaximonas of Miletus (6th C.B.C.) believed that water was the only element. Hippocrates (C469 - C357 B.C.) is supposed to have believed in two elements - air and water, as he owes a treatise on "Air, Water and Places" as the first book on physiotherapy to his credit. But Empedocles (490-430 B.C.) who was born in Argentum, Sicily, and was the disciple of Pythaqoras and Permenides, had a philosophy that attempted to combine the technique of the Eleatic school with the doctrine of Heraclitus (573 B.C.). He believed that the universe is a compound of four basic elements - earth, air, fire and water, which combined in various proportions form different things. These elements are thought to be related to one another through four qualities, *viz.*, coldness, wetness, hotness and dryness, which also gave the elements their 'characteristics'. It was thought that one element could be transformed to another by changing their qualities. Thus the qualities or characteristics of the four elements can be termed as the basic concept of 'Temperament'.

But historically the concept of 'Temperament' originated with the Hippocratic school of Medicine and was based upon a theory of varying proportions of the four body humours or excretions in the body - (Latin "tempero" to mingle "compound in definite proportion"). These four fluids - blood, phlegm, black bile or melancholy and yellow bile... produce according to their degree of predominance the sanguine, phlegmatic, melancholic and choleric temperament. The various mixtures of these humours in different men determined their 'complexion' or 'Temperament'.

After the fall of the Roman Empire all knowledge and learning were transferred to the Arabs. But the journey of 'Temperament' again took its course in Europe after the 10th century, which continued even after the renaissance upto the 17th Century. During this period, a rapidly growing number of classical medical texts were translated from Arabic to Latin. Islamic works such as the "Canon" of Avicenna and clinical treatises of Rhazes were also translated. The theory of four 'humours' and 'qualities' remained the basis of explanation of health and diseases. John of Gaddesen (1280-1361 A.D.) the pattern of the "Doctor of Physick" in Geffrey Chaucer's (1340 -1400 A.D.) Canterbury Tales, was one of the English students there, and he relied upon 'astrology' and upon the 'humours', which is evident from Chaucer's description.

Well could he guess the ascending of the star,

Wherein his patient's fortunes settled were,

He knew the course of every malady

Were it of cold, or heat or moist or dry.

Also known as prophetic medicine it traditionally makes use of a variety of techniques including diet, herbal treatments, manipulative therapies, and surgery. Unani-tibbi is a complete system, encompassing all aspects and all fields of medical care, from nutrition and hygiene to psychiatric treatment.

Origins

The name unani-tibbi is something of a misnomer, as literally translated from the Arabic, it means Greek medicine. This is because the early Arab physicians took their basic knowledge from the Greeks. At the time, Greek medical knowledge was the best to be had, particularly from Galen, the renowned second-century Greek physician who treated the gladiators and Emperor Marcus Aurelius.

However, from that point onwards, Islamic medical scholars were responsible for many developments and advancements that, at the time, placed Arabic medicine firmly in the vanguard of medical science. There followed a steady stream of Muslim medical scholars, who not only upheld the high standards that came to be known of unani-tibbi, but carried on adding to and improving the basic pool of knowledge.

Some notable scholars of the science of unani-tibbi include:

Al Tabbari (838–870)

Al Razi (Rhazes) (841–926)

Al Zahrawi (930–1013)

Avicenna (980–1037)

Ibn Al Haitham (960–1040)

Ibn Sina (Avicenna), (980–1037)

Ibn Al Nafees (1213–1288)

Ibn Khaldun (1332–1395)

Medical innovations introduced by unani-tibbi physicians included:

Avicenna was the first to describe meningitis, so accurately and in such detail, that it has scarcely required additions after 1,000 years.

Avicenna was the first to describe intubation (surgical procedure to facilitate breathing)— Western physicians began to use this method at the end of the eighteenth century.

The use of plaster of Paris for fractures by the Arabs was standard practice—it was "rediscovered" in the West in 1852.

Surgery was used by the Arabs to correct cataracts.

Ibn Al Nafees discovered pulmonary blood circulation.

A strict system of licensing for medical practitioners was introduced in Baghdad in 931, which included taking the Hippocratic oath, and specific periods of training for doctors.

There was a system of inspection of drugs and pharmaceuticals—the equivalent of the Federal Drug Administration (FDA)—in Baghdad 1,000 years ago.

The European system of medicine was based on the Arabic system, and even as recently as the early nineteenth century, students at the Sorbonne had to read the canon of Avicenna as a condition to graduating.

Unani-tibbi hospitals were, from the beginning, free to all without discrimination on the basis of religion, sex, ethnicity, or social status.

Their hospitals allocated different wards for each classification of disease.

Hospitals had unlimited water supplies and bathing facilities.

Before the advent of the printing press, there were extensive handwritten libraries in Baghdad, (80,000 volumes), Cordova, (600,000 volumes), Cairo, (two million volumes), and Tripoli, (three million volumes).

Metaphysical and Philosophical Foundations

Unani medicine's philosophical underpinnings are tied to the Islamic spiritual tradition. In fact, the true meaning of philosophy as philo-sophia, or "love of wisdom-divine," has only accidental affinities to what is termed philosophy in the modern lexicon. Traditional philosophy as philo-sophia was, and still is, tied to a spiritual vision of humanity and the cosmos.

Epistemology and Ontology

Epistemology and ontology deal directly with the nature of knowledge-what something is, how it is obtained, and how it is used-and the study of existence or being. Since Unani medicine stems from the Islamic tradition, historians must search this spiri- tual source for the answers to questions regarding its theories on knowledge and existence.

Within the Islamic as well other traditional religious worlds, knowledge has a definite hierarchical structure.i This hierarchy maintains the essential links that exist between heaven and earth, between God and humankind. First, the hierarchy provides an acknowledgement of a presiding Divine Knowledge. Arising from that Divine Knowledge is the knowledge found in divine revelation, a perfect dispensation from heaven that enlightens the human intellect.j And as this revelatory knowledge sparks the heart-intellect, it necessarily begins to involve the mind in its quest for knowledge of this earthly domain (ilm al-dunya), the physical world.

Ontologically, the being-ness of humanity is intimately connected with the Being of God. God is seen as the Divine Being without which no thing or being can rightfully claim existence for itself. Thus, God's Being is termed Wājib al-Wujûd (Necessary Being). In this light, humanity-and everything else in the world of forms-is seen as borrowing its existence from the indispens- able nature of the Divine Being.

Cosmology and the Human Microcosm

The basic cosmological premise in Islam, as well as in other traditions such as Hinduism, Christianity, and Judaism, is that the Absolute Truth of Divinity (al-Haqq) manifests itself in the world of forms and within the human heart in a process called the "arc of descent."k A reciprocal "arc of ascent" of the human spirit1O occurs in response to the Divine's "arc of descent." These adwār (arcs or cycles) of descent and ascent relate directly to the interdepen- dence of all things on all levels of existence.

Indeed, these traditions perceive everything in creation as acknowledging, in one form or another, the all-pervading Divine Presence. The traditional view sees the human being as the most concentrated theophany (ie, locus of Divine Presence in the world of forms), especially as it concerns the qualities of an active intellect and free will. According to the Hermetic dictum "As above, so too below, "the human being acts as a mirror to the Divine. Furthermore, the Sufis (the mystics of Islam) have a popular saying that addresses this reality on the cosmic scale: al-insān qawn saghīr wa 'l-qawn insān kabīr, "Man is a small universe, and the universe is a large man."

The guiding principle of the dynamic interplay between the Divine Order and the rest of creation is called Tawhīd, "the principle of Divine Unity." All traditional sciences agree that creation depends on the Divine, with interdependence among all forms in the cosmos and intradependence of all elements within each specific form. These concepts offer an obvious analogy to the profound knowledge found in the divine Oriental symbol of the Taoist t'ai chi t'u or "Supreme Principle of Unity," with the complementing and harmonizing forces of yin and yang. This unity is also known in Islamic cosmological language as "al-wahdah fi'l-kathrah wa 'l-kathrah fi 'l-wahdah," "the one in the many and the many in the one."

Unani medicine bases its medical theories of diagnosis and treatment upon a specific understanding of the intradependent relationships that exist among the four humors. This knowledge concerns the subtle (latīf) aspects of human physiology and of philosophical or spiritual psychology.

Teleology and Spiritual Correspondences

The teleological component of Unani medicine's philosophy entails extracting meaning and purpose as it relates to the realities of creation and the human experience. To what end is human life? What is the goal of all of creation? Does a plan of purpose exist that governs the created order and all of its particular elements?

The Qur'ān explicitly states that "inna li 'Llāhi wa inna ilayhi rāji'ûn," "Verily we belong to God and to Him is our return" (2:156). The human race hurtles itself forward in time and space toward an inevitable reunion with its creator. It is this understanding of the nature of things that permits traditional hakīms (or doctors of the Unani medical tradition) to find solace in the fact that every patient is ultimately a patient at the door- steps of the Divine Doctor.

Unani practitioners always insisted on thorough analysis within the larger ethical framework of the Divine's presence in human life. As a prayerful supplication of Ali, the son-in-law of the Prophet Muhammad and the spiritual founder of Shiite Islam, presents this attitude: "Yā man Ismuka dawā', wa dhikruka shifâ," "Oh Ye whose Name is a sacred medicine, and in whom remembrance of Thee (dhikruka) is a healing balm." Avicenna described the contamination of the body by "foreign bodies" prior to infection, and Ibn Khatima also described how "minute bodies" enter the body and cause disease—well in advance of Pasteur's discovery of microbes.

Al Razi was the first to describe smallpox and measles. He was accurate to such a degree that nothing has been added since.

Avicenna described tuberculosis as being a communicable disease.

Avicenna devised the concept of anesthetics. The Arabs developed a "soporific sponge," (impregnated with aromatics and narcotics and held under the patient's nose), which preceded modern anesthesia.

The Arab surgeon, Al Zahrawi was the first to describe hemophilia.

Al Zahrawi was also the first surgeon in history to use cotton, which is an Arabic word, as surgical dressings for the control of hemorrhage.

Avicenna accurately described surgical treatment of cancer, saying that the excision must be radical and remove all diseased tissue, including amputation and the removal of veins running in the direction of the tumor. He also recommended cautery of the area if needed. This observation is relevant even today.

Avicenna, Al Razi, and others formed a medical association for the purpose of holding conferences so that the latest developments and advancements in the field of medicine could be debated and passed on to others.

Benefits

What began as an advanced medical system that set world standards, has now come to be regarded as a system of folk medicine. This decline coincided with the decline of the Islamic Empire and the dissolution of the caliphate (spiritual head of Islam), as these were directly responsible for the direction and impetus of Islamic scientific scholars in all fields.

Unani-tibbi practitioners still treat people with herbal remedies and manipulation, for a variety of illnesses. In the Islamic world, many of the poorer people who cannot afford allopathic medicine still resort to this traditional medicine. There are also people who prefer unani-tibbi to allopathic medicine, as indeed, the traditional unani-tibbi remedies do not bring with them the side effects commonly experienced with allopathic drugs.

Description

Similar to Greek humoral theory, unani-tibbi considers the whole human being, spiritual, emotional, and physical. Basic to the theory is the concept of the "four humors." These are Dum (blood), Bulghum (phlegm), Sufra (yellow bile), and Sauda (black bile). Each is further categorized as being hot and moist (blood), cold and moist (phlegm), hot and dry (yellow bile), and cold and dry (black bile). Every individual has his/her own unique profile of humors, which must be maintained in harmony to preserve health. If the body becomes weak, and this harmony is disrupted, a physician can be called upon to help restore the balance.

This restoration may be done using correct diet and nutrition and/or the unani-tibbi system of botanical therapy, cupping, bleeding, manipulation, and massage, among others, as treatments for all disease and ailments. Herbs or substances used to treat a patient will be matched to his humor type.

Unani-tibbi employs a detailed system of diagnosis, including observation of urine and stools, palpation of the body and pulse, and observation of the skin and eyes.

It also employs a system of prophylactics in order to preserve health and ward off disease. This includes the adherance to strict hygiene rules, protection of air, food and water from contamination or pollution, sufficient rest and exercise, and attention to spiritual needs. Certain herbs are also taken on a prophylactic basis, such as black cumin and sage.

In general, unani-tibbi treatment is not expensive, and it is certainly less expensive than allopathic medicine. However, charges vary according to area and practitioner. Fees should be discussed with a practitioner before treatment begins.

Preparations

Remedies are often provided by the practitioner or are obtained from a specialized herbalist. The ingredients are mainly herbs and honey. It must be noted that the honey used will be raw and unadulterated, rather than the type found in supermarkets, which is usually heat-treated.

A famous and widely used medicinal herb is black cumin (Nigella sativa), also known as Hab Al Baraka in Arabic, which means blessed seed. Black cumin has been cultivated since Assyrian times and it is beneficial for a very long list of ailments. It is widely mixed with other herbs for greater beneficial effect and is said to strengthen the immune system when taken over a period of time. Research has proved that it has the ability to slow the division of cancer cells.

Ingredients of Unani medicine are sourced from Plants Kingdom, Mineral Kingdom as well as Animal Kingdom.

Following is the list of Plant Kingdom

1. Aatrilal (*Ammi majus*)
2. Anjir (*Ficus carica*)
3. Atis (*Aconitum heterophyllum*)
4. Azaraqi (*Strychnos nux-vomica*)
5. Badam shireen (*Prunus amygdalus*)
6. Baqla (*Vicia faba*)
7. Bazrulbanj (*Hyoscyamus niger* L.)
8. Chashmizaj (*Cassia absus*)
9. Chobchini (*Smilax china*)
10. Dudhi (*Lagenaria siceraria*)
11. Dudhi khurd (*Euphorbia thymifolia*)
12. Fifil Siyah (*Piper nigrum*)
13. Gaozaban (*Borago officinalis*)
14. Gulnar Farsi (*Punica granatum*)
15. Habbun Neel (*Ipomoea nil*)
16. Halela Siyah (*Terminalia chebula*)
17. Heel Kalan (*Amomum subulatum*)
18. Heel Khurd (*Elettaria cardamomum*)
19. Heeng (*Ferula foetida*)
20. Hina (*Lawsonia inermis*)
21. Inderjao Shireen (*Wrightia tinctoria*)
22. Inderjeo Talkh (*Holarrhena antidysenterica*)
23. Ispand (*Peganum harmala*)
24. Kaiphal (*Myrica esculenta*)
25. Karanj (*Pongamia pinnata*)
26. Karanjwa (*Caesalpinia crista*)
27. Karnab (*Brassica oleracea*)
28. Kasni (*Cichorium intybus*)
29. Khar-e-Khasak (*Tribulus terrestris*)
30. Khatmi (*Althaea officinalis*)
31. Khella (*Ammi visnaga*)
32. Khulanjan (*Alpinia galanga*)

33. Khurfa (*Portulaca oleracea*)
34. Khurfa Khurd (*Portulaca quadrifida*)
35. Kishneez Khushk (*Coriandrum sativum*)
36. Konch (*Mucuna cochinchinensis*)
37. Kulthi (*Dolichos biflorus*)
38. Madar (*Solanum nigrum*)
39. Mako (*Solanum nigrum*)
40. Mawiz (*Vitis vinifera*)
41. Methi (*Trigonella foenum-graecum*)
42. Mundi (*Sphaeranthus indicus*)
43. Nakhud (*Cicer arietinum*)
44. Narjeel Daryaee (*Lodoicea maldivica*)
45. Nilofar (*Nymphaea alba*)
46. Panwad (*Cassia tora*)
47. Palas (*Butea monosperma*)
48. Sambhalu (*Vitex negundo*)
49. Saranjan (*Colchicum luteum*)
50. Turbud (*Operculina turpethum*)

Following is the list of Mineral Kingdom

1. Adamas
2. Alumen
3. Kaolinum
4. Ammonicum chloride
5. Antimony sulphide
6. Argentum
7. Arsenum
8. Arsenic disulphidum
9. Arsenic trisulphidum
10. Asphaitum
11. Aurum
12. Makaradhwaja
13. Calcium
14. Calcil carbonas
15. Caicii hydras
16. Calcii oxidum
17. Calcii sulphas
18. Carbo ligni

19. Cuprum
20. Cupri sulphas
21. Ferrum
22. Ferroso-ferric oxide
23. Ferri sulphas
24. Ferri sulphuratum
25. Hydrargyrum
26. Plumbum
27. Plumbi carbonas
28. Plumbi oxidum
29. Plumbi oxidum rubrum
30. Plumbi sulphuratum
31. Potassii carbonas impura
32. Potassii nitras
33. Silicum
34. Silicate of Alumina
35. Silicate of Alumina and Oxide of iron
36. Silicate of lime
37. Silicate of magnesia
38. Silicate of magnesia and iron
39. Saline substances
40. Sodii carbonas impure
41. Sodii biboras
42. Sodii chloridum impure
43. Sodii chloridum
44. Stannic sulphidum
45. Stannum
46. Sulphur
47. Talcum purification
48. Zincum
49. Zinci carbonas
50. Zinci oxidum

Following is the list of Animal Kingdom

1. *Acipenser huso* linn
2. Adeps
3. *Ambra grasea*
4. Animal flesh

5. *Bombyx mori*
6. Castoreum
7. Cateria lacca
8. Cera
9. *Cervus dama*
10. *Cervus elephus*
11. Cetaceum
12. Chelonia
13. *Coccus cacti*
14. *Corallium rubrum*
15. *Cypraea moneta*
16. *Callus domesticus*
17. *Turbinella rapa*
18. *Hirudo medicinalis*
19. Lactus
20. Mel
21. *Moschus moschiferus*
22. *M. trianthema*
23. *Mytilus margaritiferus*
24. Ossepiae
25. *Ostrea edulis*
26. Pisces
27. Serpent poison
28. *Serum praeparatum*
29. Snake venom
30. *Spongea officinalis*
31. *Squalus carcharius*
32. Urine
33. *Viverra civetta*

Below is the list of Compound Unani Medicines and their chief ingredients:

Name of Compound	Chief Ingredient
Anooshdaru Sada	Amla (*Emblica officinalis* Gaertin fruit)
Anooshdaru Lului	Amla (*Emblica officinalis* Gaertin fruit)
Arq Brinjasif	Brinjasif (*Achillea millefolium* herb)
Arq Gazer	Gazer (*Daucus carota* root)

Name of Compound	Chief Ingredient
Arq Kasni	Tukhme Kasni (*Cichorium intybus* seed)
Arq Maullehem	
Mako Kasni wala	Gosht Hulwan (Milk feeding goat meat), Mako (*Solanum nigrum* fruit) and Kasni (*Cichorium intybus* herb)
Arq Musaffi	Nim (*Melia azadarachta*)
Arq Sheer Morakkab	Bakri Ka Doodh (Goat Milk)
Dawaul Kurkum Kabir	Zafran (*Crocus sativa gynacium*)
Dawaul Kurkum Sagheer	Zafran (*Crocus sativa gynacium*)
Dawaul Misk Motadil Sada	Musk (*Moschcus moschiferus* secretion)
Gulqand Gulab	Gule Surkh (*Rosa damascena* flower)
Gulqand Mahtabi	Gule Chandni (*Colonyction aculeatum* flower)
Gulqand Sevti	Gule Sevti (*Rosa alba* flower)
Habb Asgand	Asgand (*Wiithania somnifera* Dunal)
Habb Ayarij	Ayarij Feqra (*Aloe vera* dried juice of leaf)
Habb Jadwar	Jadwar (*Delphinium denudatum* root)
Habb Kabid Naushadri	Naushader (Ammonium chloride)
Habb Papita	Papeeta (*Carica papaya* fruit)
Habbe Shifa	Tukhm Datura Safed (*Datura alba* seed)
Habb Sooranjan	Sooranjan (*Colchicum leutium* root)
Habb Tinkar	Tinkar (Borax)
Habb Paan	Sammulfar (Arsenic)
Habb Muqil	Muqil (*Commiphora mukul* gum)
Habb Marvareed	Marvareed (*Mytilus margaritiferus* pearl)
Itrifal Kashneezi	Haleelajat: Haleela Zard (*Terminelia chebula* half-ripe fruit), Haleela Kabuli (*Terminelia chebula* ripe fruit), and Haleela Siyah (*Terminelia chebula* unripe fruit) and Kashneez (*Coriandrum sativum* fruit)
Itrifal Ustokhuddus	Haleela (*Terminelia chebula* fruit), Balela (*Terminelia belerica*), Amla (*Emblica officinalis* fruits) and Ustokhuddus (*Lavandula stoechas* flower)
Itrifal Zamani	Haleelajat (*Haleela Zard* (*Terminelia chebula* half-ripe fruit), Haleela Kabuli (*Terminelia chebula* Khameera Banafshah

Name of Compound	Chief Ingredient
Khameera Gaozaban Ambari	Gaozaban (*Borage officinalis*) and Amber (*Ambra grasea*)
Khameera Gaozaban Ambari Jawaher wala	Gaozaban (*Borage officinalis*), Amber (*Ambra grasea*) and Jawahrat (*Precious Stones*)
Khameera Gaozaban Ambari Jadwar Ood Saleeb wala	Gaozaban (*Borage officinalis*), Jadwar (*Delphinium denudatum* root) and Ood Saleeb (*Paonea officinalis* root)
Khamira Gaozaban Sada	Gaozaban (*Borage officinalis*)
Khameera Khashkhash	Khashkhash (*Papaver somniferum* seed)
Khameera Marvareed	Marvareed (*Mytilus margaritiferus* pearl)
Sandal	Sandal safed (*Santalum album* wood)
Kohl-ul Jawaher	Jawahrat (*Precious stones*)
Kohl Chikni Dawa	Sabun (Soap) ripe fruit), Haleela Siyah (*Terminelia chebula* unripe fruit) and Turbud (*Ipomea turpenthum* root)
Jawarish Amla	Amla (*Emblica officinalis* Gaertin fruit)
Jawarish Anaren	Anar (*Punica granatum* Linn fruit) Sweat and sour
Jawarish Bisbasa	Bisbasa (*Myristica fragrans* fruit coat)
Jawarish Jalinus	Zafran (*Crocus sativa* gynacium), Mastagi (*Pistacia lantiscus* gum)
Jawarish Kamoni Sada	Zeera Siyah (*Carum carvi* seed) Jawarish Mastagi Mastagi (*Pistacia lantiscus* gum) Jawarish Ood Sheereen Ood (*Aquilaria agallocha* fungus)
Jawarish Ood	Tursh Ood (*Aquilaria agallocha* fungus)
Jawarish Tabasheer	Tabasheer (*Bambusa arundinasia* dried exudate on node)
Jawarish Tamer Hindi	Tamar Hindi (*Tamarindus indica* fruit pulp)
Jawarish Zar Ooni	Ambari Amber (*Ambra grasea*)
Jawarish Zar-Ooni	Sada Zafran (*Crocus sativa gynacium*)
Khameera Abresham	Sada Abresham (*Bombax mori* cocoon)
Khameera Abresham	Hakim Arshed wala Abresham (*Bombax mori* cocoon)
Khameera Abresham Ood Mastagi wala	Abresham (*Bombax mori* cocoon), Ood Gharqi (*Aquilaria agallocha* fungus) and Mastagi (*Pistacia lantiscus* gum) Abresham (*Bombax mori* cocoon) and Sheera Unnab (*Zizyfus sativa* fruit ground with water)

Name of Compound	Chief Ingredient
Kohl Byaz	Tobal Nihas Muharraq (Blue dust, Copper sulphate roasted)
Kohl Roshnayee	Tobal Mis (Blue dust, Copper sulphate)
Kushta Abrak Safed	Abrak Safed Mehloob
Kushta Abrak Siyah	(White Talcum)
Abrak Siyah (Black Talcum)	
Kushta Faulad	Faulad (Iron)
Kushta Godanti	Godanti (Arsenic ore)
Kushta Hajralyahud	Hajralyahud
Kushta Marjan	Marjan (*Corallium rubrum*)
Kushta Nuqra	Nuqra (Silver)
Kushta Post	Bezae Murgh Post Bezae Murgh (Hen Eggshell)
Kushta Seemab	Seemab (Mercury)
Kushta Sadaf	Sadaf (*Turbinella rapa* shell)
Kushta Sammulfar	Sammulfar (Arsenic)
Kushta Sang	Jarahat Sang Jarahat (Soapstone)
Kushta Tila	Tila (Gold)
Kushta Zeher mohra	Zeher mohra (Serpentine)
Laboob Kabir (Sparrow brain)	Maghz Sar kanjishk
Laboob Sagheer	Maghz Chilghoza (*Pinus gerardiana* fruit kernel)
Laooq Katan	Tukhme Katan (*Linum usitatissimum* seed)
Laooq Khashkhash	Post Khashkhash (*Papaver somniferum* seed coat)
Laooq Khayar-shamber	Khayarshamber (*Cassia fistula* pulp)
Laooq Nazli	Khashkhash (*Papaver somniferum* seed)
Laooq Nazli Aab Tarbooz wala	Aab Tarbooz (*Citrullus vulgaris* fruit juice)
Laooq Sapistan	Sapistan (*Cordia dichotama, C. latifolia* fruit)
Majun Aarad Khurma	Khurma (*Phoenix dactylifera* fruit)
Majun Azaraqi	Azaraqi (*Strychnos nuxvomica* seed)
Majun Beladur	Beladur (*Semicarpus anacardium* fruit)
Majun Dabeedulvard	Gul surkh (*Rosa damascena* flower).
Majun Falasfa	Maghz Chilghoza (*Pinus gerardiana* kernel)
Majun Fanjnosh	Amla (*Emblica officinalis* Gaertin fruit)
Majun Muravvahul	Arvah Yaqoot (Ruby)

Name of Compound	Chief Ingredient
Majun Salab	Salab (*Orchis latifolia* root)
Majun Sangdana Murgh (Gizzard of cock)	Post Sangdana Murgh
Majun Supari Paak	Supari (*Areca catechu* nut)
Majun Ushba	Ushba (*Sarsaparilla indica* herb)
Marham Dakhilyun	Luabat (Mucilage)
Marham Kafoori	Kafoor (*Cinnamomum camphora* dried extract)
Marham Mazoo	Mazoo (*Quercus infectoria* abnormal growth)
Marham Muqil	Muqil (*Commiphora mukul* gum)
Marham Rusul	Behroza (*Pinus longifolia* latex)
Marham Siyah	Ral (*Pinus longifolia* latex)
Marham Ushq	Ushq (*Dorema ammonicum* latex)
Marham Zangar	Zangar (*Cupric sulphate*)
Mufarreh Aazam	Musk (*Moschcus moschiferus* secretion), Amber (Ambra grasea)
Mufarreh Barid	Marvareed (*Mytilus margaritiferus* pearl)
Mufarreh Sheikhurrais	Daroonaj (*Doronicum hookarii* root)
Mufarreh Yaqooti	Yaqoot (Ruby)
Murabba Amla	Amla (*Emblica officinalis* Gaertin fruit)
Murabba Behi	Behi (*Cydonia quincy* fruit)
Murabba Haleela	Haleela (*Terminelia chebula* fruit)
Murabba Zanjabil	Zanjabil (*Zingiber officinalis* rhizome)
Qurs Kafoor	Kafoor (*Cinnamomum camphora* dried extract)
Qurs Kehruba	Kehruba (*Vateria indica* gum)
Qurs Mullein	Senna (*Cassia senna* leaf)
Qurs Musallas	Afiun (*Papaver somniferum* Linn)
Sherbat Bazoori Braid	Beikh Kasni (*Cichorium intybus* root)
Sherbat Bazoori Motadil	Beikh Kasni (*Cichorium intybus* root)
Sherbat Bazoori Haar	Beikh Kasni (*Cichorium intybus* herb)
Sherbat Deenar	Tukhme Kasoos (*Cuscuta reflexa* seed)
Sherbat Faulad	Burada Faulad (Iron Powder)
Sherbat Fawakiha	Favakihat (Fruits)
Sherbat Vard Mukarrer	

Name of Compound	*Chief Ingredient*
Shyaf Abyaz	Gule Surkh (*Rosa damascena* flower)
Safeda (Plumbi carbonas)	
Shyaf Ahmer	–
Shyaf Asvad	–
Sikanjbeen Bazoori	Tukhme Kasni (*Cichorium intybus* seed)
Sikanjbeen Nanayee	Podina Khushk (*Mentha arvensis* dried herb)
Sikanjbeen Unsuli	Unsul (*Urginea indica* bulb)
Tiryaq Arbaa	Habbul Ghar (*Laurus nobilis* fruit)
Tiryaq Farooq	- Tiryaq Samania -

18

Holistic Medicine

History of Holistic Medicine in Eastern Culture

Vedic Culture

Ayurveda or ayurvedic medicine has more than 6,000 years of history as a system of medicine based on a holism/holistic approach to health that is rooted in Vedic civilization/ Vedic culture. As with traditional Chinese medicine, the concept of it as an alternative form of therapeutic practise is only found in the Western world.

Chinese Culture

Traditional Chinese medicine has more than 5,000 years of history as a system of medicine that is based on a philosophical concept of balance (Yin Yang/Yin and Yang, Qi, Blood, Jing (TCM)/Jing, Bodily fluid/Bodily fluids, the Five Elements, the emotions, and the soul/spirit) approach to health that is rooted in Taoist philosophy and Chinese culture. As such, the concept of it as an alternative form of therapeutic practise is only found in the Western world.

History of Holistic Medicine in Western Culture

Western approaches to alternative medicine have more than 3,000 years of history behind them as systems of medicine based on natural philosophy/natural philosophies that are rooted in all aspects of Western culture. This is a history of how Western natural philosophies developed over the ages.

European History

Throughout Western European history there were two major trends: the professionalism of physicians who belonged to the upper classes and the folk healers who lived among the peasant population. The professionals developed in order to enhance their status in life, while the folk healers developed out of the necessity to survive. Herbalism and the water cure, hydrotherapy, or naturopathy developed slowly over 2,000 years of history. Autocratic traditions developed over time that gave today's European physicians social status and acceptance.

The Greco-Roman Period

In Europe, interest in the hydrotherapy can be traced back to the Roman Baths/ancient Roman spas and the hot mineral springs at Bath, England.

The Dark Ages

In Europe Church played a central role. At first, the Church suppressed all development. Later on, the Church supported the development of professional physicians. Eventually, the power of the Church literally exterminated much of the competition from folk healers during the witch-hunting period which spanned more than four centuries (from the 14th to the 17th century).

Healers throughout the medieval period could come in many varieties. Physicians who studied the works of the Greek masters at Universities, were the elite of the medical profession in the middle ages. However few people other than the well-off or the nobility had regular access to these. Folk Healers passed on their knowledge from master to apprentice, and were more accessible to the peasant or labourer than physicians. Unregulated, but knowledgeable on herbs and folk-remedies, they were gradually excluded from the medical system. Monastic Medicine monasteries played a big part in the provision of medieval medicine. Virtually every monastery had an infirmary for the monks or nuns, and this led to provision being made for the care of secular patients.

The 1200s

From the middle ages through the Reformation personal health in Europe was generally poor. It was a time of plague, pollution and quacksalver mercury poison.

The 1300s

The 1500s—the Renaissance

The 1600s—the Reformation

The 1700s—the Age of Enlightenment

The Age of Heroic Medicine/Heroic Medicine (1780-1850)

1st half of the 19th century—Age of Romanticism

A medical reform movement was started in Europe as a reaction against heroic medicine.

2nd half of the 19th century—The Birth of Modern Science

Germany became the world centre of medical research, training, and pharmaceuticals drawing students from all over the world by the end of the 19th century.

Hygiene and public health became the central focus of emerging urbanization.

The 20th century

In the first half of the 20th century a number of factors including internal conflict and the relative success of conventional medicine led to the decline of alternative medicine in the western world. In the second half Alternative medicine staged something of a recovery as conventional medicine failed to live up to the unrealistic expectations that many people had of it. This combine with the increasing cost of conventional medicine and greater awareness of alternative medicine brought alternative medicine to the position it now has.

American History

Western healing practices developed differently in the New World than they did in the Old World.

In Europe, physicians already had a centuries old monopoly over the right to treat patients. But in United States of America, medical practice was literally open to anyone who called themselves a doctor.

The 1700s—The Colonies

The American public, newly liberated from England, was hostile to professionalism and foreign elitism of any kind. And, the educated physicians who emigrated to the New World from Europe were nothing more than Quacks practicing heroic medicine.

The Popular Health Movement (1830-1850)

In America, the Popular Health Movement played a central role in the development of alternative therapy practices. Herbalism, Homeopathy, Eclecticism (medicine)/Eclecticism and Natural Hygiene developed during the Health Reform Movement.

Only homeopathy, natural hygiene and eclecticism managed to last from the 1830s through the rest of the 19th century.

Antebellum America

Postbellum America

Progressive ERA of Health Care Reform (1890-1920)

Osteopathy, Chiropractic, and Naturopathy developed at the turn of the century.

The 20th century

The high-technology of medicine becomes firmly housed in the hospital. Hospitals are transformed from institutions designed for long-term care of the sick into facilities designed to test, treat and release patients as fast as possible.

In 1970, Modern medicine first appears with the McMaster Medical School in Canada that used a clinical learning strategy that eventually develops into evidence-based medicine (EBM). It further came into vogue in the 1980s at the Harvard University.

In 1977, George L. Engel, MD (1913-1999) first proposed the bio psychosocial model of health, illness and healing.

Holistic Medicine

Holistic medicine refers to alternative health practices which claim to treat "the whole person." To holistic practitioners, a person is not just a body with physical parts and systems, but is a spiritual being as well. The mind and the emotions are believed to be connected to this spirit, as well as to the body. Holistic practitioners are truly alternative in the sense that they often avoid surgery or drugs as treatments, though they are quite fond of meditation, prayer, herbs, vitamins, minerals and exotic diets as treatments for a variety of ailments. Holistic medicine is a system of health care which fosters a cooperative relationship among all those involved, leading towards optimal attainment of the physical, mental, emotional, social and spiritual aspects of health.

It emphasizes the need to look at the whole person, including analysis of physical, nutritional, environmental, emotional, social, spiritual and lifestyle values. It encompasses all stated modalities of diagnosis and treatment including drugs and surgery if no safe alternative exists. Holistic medicine focuses on education and responsibility for personal efforts to achieve balance and well being.

This little scene is illustrative of the basic assumptions of alternative medicine in its various forms. Let us describe four principles:

1. Body and Mind are part of a universal energy field. which carries various names: vital force, bioenergy, Chi, Prana, or simply God. The energy is supposed to circulate in meridians or energetic channels. The Chakras are a Hindu concept of seven energetic centres which control the flow of energy and are supposed to correspond with endocrine glands and major organs.

2. Macrocosm finds its correspondence in microcosm: Thus the constellation of the stars reflects human destiny (astrology), the microcosm of the eyes tells you about the physical condition of the macrocosm of the body (iridology), the microcosm of the feet reflects the disharmony of the organism (foot reflexology), or the ear is a microcosmic representation of the energy flow in the body (ear acupuncture). Bach flower remedies correlate the characteristics of plants with a person's emotional condition (*e.g.*, holly is said to be a remedy for hatred, envy and jealousy).

3. Disease is the result of energetic imbalance or disharmony with universal cosmic energy. This imbalance may be caused by a multitude of causes, such as pathogenic vibrations, energy blockades in the Chakras or in the meridians. Emotional blocks are being seen as a major factor to explain energetic disharmony. Thus homeopathy and flower therapies talk of undesirable emotional conditions in which a person has been estranged from the universal cosmic stream of energy.

4. Healing is the restoration of harmony with cosmic energy. Many alternative therapies try to restore vital energy, to harmonize vibrations or to balance Yin and Yang. Healing in this concept is not just applying a method, rather it is the ritual with deep spiritual overtones. One of the proponents of acupuncture in the United States, Dr. Duke writes, "The acupuncturist sees the lives of his patients as integral parts of the universe. He brings his patients back to health not only for their own sake and happiness, but so that the whole world may function properly. Every needle the acupuncturist twirls between his fingers bears the heavy weight of universal harmony in its slim, pointed end."

Terms Associated with Holistic Medicine

Alternative Medicine is often used by the general public and some healthcare practitioners to refer to medical techniques which are not known or accepted by the majority "conventional" or "allopathic" medical practitioners. Such techniques could include non-invasive, non-pharmaceutical techniques such as Medical Herbalism, Acupuncture, Homeopathy, Reiki, and many others. However, the term Alternative Medicine can also refer to any experimental drug or non-drug technique that is not currently accepted by "conventional" medical practitioners. As non-invasive, non-pharmaceutical techniques become popular and accepted by large number of "conventional" practitioners, these techniques will no longer be considered Alternative Medicine.

Alternative Medicine refers to techniques that are not currently accepted by "conventional" practitioners, but what is currently accepted is quickly changing. Even the definition of "conventional practitioners" is quickly changing. Therefore, techniques that are now considered part of Alternative Medicine may soon be considered part of "conventional" medicine. The terms Holistic Healing and Holistic Medicine are slightly more stable than Alternative Medicine and are therefore preferable.

Complementary Medicine is often used by "conventional" medical practitioners to refer to non-invasive, non-pharmaceutical techniques used as a complement to "conventional" medical treatments such as drugs and surgery. The term implies that "conventional" medicine is used as a primary tool and the non-invasive, non-pharmaceutical techniques are used as a supplement when needed.

In many cases, properly chosen non-invasive and non-pharmaceutical healing techniques plus properly chosen lifestyle changes can completely and safely heal both acute and chronic illnesses. In other cases, "conventional" medicine is only needed in emergencies or when the safer non-invasive, non-pharmaceutical methods fail. In some cases "conventional" medicine will be a major part of a Holistic Healing Plan, but in some cases it is not needed at all.

Natural Healing usually refers to the use of non-invasive and non-pharmaceuticals techniques to help heal the patient. When most people use the term Natural Healing, they are usually referring to physical healing techniques only.

Holistic Medicine has been used by millions of people around the world to:

Heal chronic diseases (including serious chronic illnesses)

Treat psychological disorders

Successfully treat childhood illnesses

Treat acute illnesses

Normalize weight

Prevent disease

Improve overall health and increase energy level

Transform one's outlook upon life

Those who are most successful tend to apply some of the key Holistic Healing Fundamentals as discussed in articles on this web page and receive direct treatments from Holistic Medicine practitioners (or perform self-treatment when possible). While there may be occasional ups and downs (especially when treating chronic illnesses), a successful outcome is more likely in these cases. Practitioners of holistic medicine are doctors who understand the basic principles of life and are there to assist you in understanding and dealing with the fundamental causes of illness so that healing is a complete experience which will last for the rest of your life. Here we will deal with some areas that are commonly in need of attention.

Life Style

This refers to the way you organize your life and focus on your priorities. When you list your priorities, is your health at the top of the list? If it isn't, you are cheating whoever or whatever is at the top of your list because you are going to be less effective in achieving those objectives and/or serving those people. You are likely to die sooner, perhaps before your destiny on this earth is fulfilled. Putting your health at the top and quality time with the people you love and serve is a good beginning to reorganizing your life style.

The life style disease of the modern world is "fragmentation." This involves hurrying from one job or person to the next without ever really "being" anywhere. In a highly competitive society people feel that they must hurry to compete. The idea is to make as much money as possible. People are living out with the aim to acquire more and most bucks. If you continue in the direction you are going, you are likely to get to where you are headed. Is that really the destination for which you were placed on this earth? More bucks? Is your life style consistent with what you know to be so about your life?

Many people are living lives not unlike that lived by a hungry mouse frantically searching for the cheese in a maze. The hypnotic effect of mass media and the frenzied herd instinct generated by the fact that almost everyone else is doing it is almost to powerful to break away from. A good long talk with a holistic health doctor can go a long way toward reorganizing your life style.

Food and Nutrition

Next in importance to the way your prioritize your life is the fuel you present to your body to generate the energy you need to go forward. There was once a time when people were more closely connected to the land and the opportunity to have an unhealthy diet was difficult to locate. You either grew your own or you went to what we would now call organic groceries or farmers' markets. Now that industrialization, capitalism, urbanization, and specialization have been thoroughly integrated, fewer people are connected the farm and more are connected to the pharmacy.

Rather than tell you what a healthy diet looks like (because this is different for each person), let us lay down some principles which will guide you to a healthy diet. Redirecting your life nutritionally means staying away from processed foods. It means finding a source of real food, food which has been grown the way food was once grown, without pesticides, artificial fertilizers, which has not been sprayed with weed killers and which has been grown from rich soil which has been cared for properly.

Here comes organic grocery. Your money is better spent on health now than illness later. You may also find that if you eat organically grown food that you eat less, spend less, and effortlessly lose those pounds you have been trying to lose.

You know how you tell your children to choose their friends carefully, to hang out with the right crowd? Those principles apply just as well in food selection. Stay out of food stores, eat and dine at home, join in group workshops or classes for learning to prepare organic meals.

Once you are choosing only organic foods, you may find that your health is not dramatically better. The fact is people become sensitized, "allergic" to certain foods. Until these foods are identified and eliminated the chances of a super-healthy life are still dim. If you are on what seems to be a good diet and are still ill, we suggest you seek professional help.

In general terms it is best to derive your nutrition from a wide variety of sources. The chemistry of the body is complex and requires many different types of fuels and additives. It is not like your car which takes three things and goes. Your body needs hundreds of things to go.

Equally important to the food you eat is the water you drink. In descending order, from the healthiest to the least health, here is the water list:

> Double steam distilled, carbon filtered
>
> Steam distilled
>
> Reverse osmosis, carbon filtered
>
> Spring water
>
> Tap water
>
> Fizzy bottled water

The purpose of water is to cleanse the body. You get all your nutrients, minerals *etc.* in the food you eat. These minerals are chelated, meaning bound to organic molecules and thus transportable and bio-available. Mineral water contains unchelated minerals, most of which will be a burden on the kidney because they must be excreted. Anything else you hear about it is water company propaganda. The best source of water is not listed (rain water collected in stainless steel containers in non-industrial areas of the world) because it is almost impossible to find. Rain water is distilled water. When water is distilled by man there are still volatile organic substances like chlorine to remove. That is where carbon filtration is important. You may invest in a steam distiller. The pay back in terms of money saved, both on bottled water, and in expenses avoided for illnesses which never happened, will be significant.

Never ignore the importance of water to your health. Next to the air you breathe, it is probably the next most important thing you put in your body.

Well, now you are eating organic food and drinking distilled water. None of this matters much if you are not digesting and absorbing the food you eat. If you are still not feeling well, consult a doctor to investigate your digestion. There are specific tests which can be done to reveal what needs to be done to improve your digestion.

Transformation

Transformation refers to change, but a specific type of change: that which could not have been predicted. Staring at a tadpole or a caterpillar for the first time, one would never guess that in a few days hopping and flying will be happening. One can see transformation by looking into a child's kaleidoscope and turning it slowly. You know change is happening, but it is unpredictable. A healthy life is in a constant state of transformation. Aliveness and transformation are one in the same. When transformation ceases, illness begins.

Education, as it is conceived in our schools, does not take into account certain subjects. For example relationship, good nutrition (in place of real education we get the "food pyramid"), and transformation. A quality education in transformation would allow one to direct the course of his or her life to make a real difference. The one thing that is profoundly missing in the modern world is the experience of making a difference.

Adequate Sleep

A common failure of people leading an otherwise healthy life is going to bed too late and getting up too early. I am myself guilty of this. For most people the ideal amount of sleep is eight hours and many of us can make it on a lot less. Because we can make it does not mean it is the best for us.

When you close your eyes, your pineal gland begins to release melatonin. This marvellous hormone/antioxidant/anti-aging agent puts you to sleep within 30 minutes. The pineal continues to release melatonin through most of the night unless light of significant luminescence strikes the retinae. If you get up to have a snack or use the bathroom and turn on the lights at 3 AM, the pineal shuts down the supply of melatonin and although you may get back to sleep, that sleep with not be as deeply restorative as it would have been had you not turned on the light. If you must get up, grope around in the dark and leave the lights off. If you want to sleep late, sleep in a room without windows to the outside. When the sun comes up, your melatonin level goes down.

Most important, go to bed at a time which allows you eight hours of sleep. You will live longer, feel better and get more accomplished.

Aerobic and Strength Exercise

A healthy life-style requires that you keep your muscles in good strong condition. The only way to do this properly is to exercise those muscles. The heart muscle is made strong through sustained moderate stress. The skeletal muscles are made strong through weight bearing exercise. To do one form of exercise and not the other is to ignore an important aspect of your health. Of course you should be cleared by your doctor to do this type of exercise and if you are not familiar with the parameters of the exercise you contemplate, you should find a trainer to help you get started.

Loving Relationships

The presence of people to relate to who are supportive of your life and with whom you can interact on a regular basis is important to your physical health. This has been proven by scientific study and demonstrates as well as anything the relationship between your spiritual self and your physical self. It is not necessary to be married to have such relationships, so don't rush out and get married because you think that is the only way to live longer. A bad marriage can shorten your life as well as a good one can lengthen it. Relationship is the key and it need not be sexual relationship. The important item is someone with whom to talk and share your inner thoughts and feelings. If you find yourself without such relationships, the responsible and wise course for you would be to establish a relationship with a psychotherapist and explore your life. Insight brings openness and openness brings loving relationship.

Creative Expression

You are on this earth for a purpose. You have a destiny. Life has a plan for you laid out by God himself. These three statements are examples of paradigms or, as we would like to call them, contexts. A context is an idea so deeply believed that reality is created around it. A lot of self exploration is required before you find the presence of one single context.

The joy of creation is something few people experience. It is also something worth working for and finally, it has a definite impact on your health. For example, doctors who cannot transcend the contexts they learned in medical school are destined to be unhealthy specimens and die early, that is to say, they will manage their own health with medical school dogma they learned decades earlier. Most docs are coming along rather slowing in that regard.

Spiritual Life

What kind of spiritual life you should have, only that you need to have one. For the best health possible it needs to be clear to you that you are more than your body. A few of us are lucky and have had near death experiences. We know as a certainty that we are more than this skin bag in which we live. Until you have the direct experience of that fact, take it on faith. You do have a spiritual life, but to be in direct experience of that spiritual life requires creative intention on your part.

Detoxification

While ingestion of pure food and water is commonly recognized as essential to health, except among some allopathic doctors who think drugs produce health, elimination of toxic by products of normal digestion and metabolism, as well as of diseased states of digestion and metabolism, is often overlooked or shoved aside. Without detoxification you would die within a few days. You have large organs in your body charged with almost no other responsibility than detoxification. The colon, liver and kidneys work as a team to get rid of substances which can easily kill you if not excreted. A holistic doctor will take a close look at your state of toxicity. The fact is, all of us are in some state of toxicity. It is only a matter of degree. Relatively low levels of toxicity we call health. For example creatinine is a toxic product of protein breakdown. There are measurable levels of creatinine in your body right now. However, unless these exceed normal levels, we call you healthy, at least from that perspective.

The colon is your most important organ of elimination. If that organ is not working well, your overall health is in jeopardy. You want a doctor who will not overlook, or gloss over, a problem you have with your colon which you may not even be aware of yourself. You want your doctor to look carefully at your kidney and liver function. Doctors who do this and also consider your relationships, spiritual life, creative expression, *etc.* we call holistic.

Each of the aspects of your life reflects your cells, DNAs and all the others. A holistic doctor attempts to see the whole picture, not just stare at your blood pressure readings, or your BMI.

Principles and Practice of Holistic Medicine

Holistic medicine is the art and science of healing that addresses the whole person - body, mind, and spirit. The practice of holistic medicine integrates conventional and alternative therapies to prevent and treat disease, and most importantly, to promote optimal health. This condition of holistic health is defined as the unlimited and unimpeded free flow of life force energy through body, mind, and spirit.

Holistic medicine encompasses all safe and appropriate modalities of diagnosis and treatment. It includes analysis of physical, nutritional, environmental, emotional, spiritual and lifestyle elements. Holistic medicine focuses upon patient education and participation in the healing process.

The Principles of Holistic Medical Practice

Holistic physicians embrace a variety of safe, effective options in the diagnosis and treatment, including education for lifestyle changes and self-care, complementary alternatives, and conventional drugs and surgery.Searching for the underlying causes of disease is preferable to treating symptoms alone. Holistic physicians expend as much effort in establishing what kind of patient has a disease as they do in establishing what kind of disease a patient has. Prevention is preferable to treatment and is usually more cost-effective. The most cost-effective approach evokes the patient's own innate healing capabilities. Illness is viewed as a manifestation of a dysfunction of the whole person, not as an isolated event. A major determinant of healing outcomes is the quality of the relationship established between physician and patient, in which patient autonomy is encouraged. The ideal physician-patient relationship considers the needs, desires, awareness and insight of the patient as well as those of the physician. Physicians significantly influence patients by their example. Illness, pain, and the dying process can be learning opportunities for patients and physicians. Holistic physicians encourage patients to evoke the healing power of love, hope, humour and enthusiasm, and to release the toxic consequences of hostility, shame, greed, depression, and prolonged fear, anger, and grief. Unconditional love is life's most powerful medicine. Physicians strive to adopt an attitude of unconditional love for patients, themselves, and other practitioners. Optimal health is much more than the absence of sickness. It is the conscious pursuit of the highest qualities of the physical, environmental, mental, emotional, spiritual, and social aspects of the human experience.

Holistic medicine has a completely different perspective on health, the body and what is needed for healing. The Western medicine approaches illnesses and diseases as problems that need to be fixed, with the remedy usually in the form of a pill or injection. Holistic medicine believes the body and mind need to be in balance with its surroundings.

19

Homeopathy

Homeopathy is a system of medicine, which was discovered and developed by a German physician, Dr. Samuel Hahnemann, between 1796 and 1842.

In the 18th century, the medical science was still very unscientific. The knowledge about human body, diseases and the modalities of treatment were poor and vague. Methods like blood-letting, leeching, purging were the common treatments for most ailments. Practically the whole of the 18th century in Europe was marked by a plethora of theories and hypothesis concerning the nature of disease and its causation. Consequently methods of therapeutic practice were as numerous and diverse as the theories propounded. The uncertainty and lack of any fixed principle of healing disappointed Dr. Hahnemann.

So Dr. Hahnemann relinquished his medical practice and devoted himself to the translation of great medical classics of his time. In 1790, when Dr. Hahnemann was engaged in translating Cullen's materia medica from English to German, his attention was arrested by the remark of the author that cinchona bark cured malaria because of its bitterness and tonic effects on stomach. This explanation appeared unsatisfactory to him.

To discover its true mode of action, he himself ingested 4 drams of Cinchona juice twice daily for a few days. To his great astonishment, he very soon developed symptoms very similar to ague or malarial fever.

This unexpected result set up in his mind a new train of thought and he conducted similar experiments on himself and other individuals with other medicines whose curative action in certain diseases had been well established. He found that in healthy persons the medicines produce symptoms very similar to what they cure in diseased individuals. So he was led to the inference that medicines cure diseases only because they can produce similar symptoms in healthy individuals. The whole of homeopathy derives from this law. He developed from it the whole system of healing — Homeopathy.

In 1796, after 6 years of Dr. Hahnemann's first experiment, he published an article in Hufeland's Journal volume-II, parts 3 and 4, pages 391-439 and 465-561. "An essay on a new principle for ascertaining the curative powers of drugs and some examinations of the previous principle."

He thus put forward his new doctrine of Similia Similibus Curantur (like cures like) in contrast to the age-old doctrine of Contraria Contraris Curantur (opposite cures opposite).

1796 is considered to be the year of birth of homeopathy. Doctrines of homeopathy were attempted to be formulated, for the first time by Dr. Hahnemann in his article "The Medicine of Experience" published in 1805 till the complete systematization of the principles and practice of the homeopathic art of healing was effected with the publication of Dr. Hahnemann's Organon of Medicine in 1810. After he had laid a solid foundation for reconstructing medicine as a science by the publication of his Organon of Healing Art, and Pure Materia Medica, he issued his valuable Chronic Diseases.

The art of medicine was thus placed on a scientific footing when Dr. Hahnemann discovered the method of testing the positive action of each individual drug and a law guiding the selection of drug to cure diseases.

Every science has certain fundamental principles which guide the whole system. Homeopathy as a science of medical treatment has a philosophy of its own and its therapeutics is based on certain fundamental principles. These are:

1. Law of Similia
2. Law of Simplex
3. Law of Minimum
4. Doctrine of Drug Proving
5. Theory of Chronic Disease
6. Theory of Vital Force
7. Doctrine of Drug-Dynamisation

These fundamental principles are elaborated in the following sections.

Law of Similia

Homeopathy is a system of medicine founded on a definite law 'Similia Similibus Curantur' which means 'like cures like'. The word Homeopathy is a Greek derivation where 'homeos' means 'similar and pathos means 'suffering'. So Homeopathy may be defined as the therapeutic method of symptom-similarity. The recognition of this law was there even before Hahnemann. Paracelus, Hippocrates, and ancient ayurvedic texts have on occasions mentioned this law. But it was Hahnemann who recognized the universality of this law and lifted it from oblivion to make it the basis of a complete system of medicine.

According to this system, the choice of the medicine is fundamentally based on the principle that the medicine must have the capability of producing most similar symptoms of the disease to be cured in healthy persons. In aphorism 26 of 'Organon of Medicine', Hahnemann states this law: "A weaker dynamic affection is permanently extinguished in the living organism by a stronger one, if the latter (whilst differing in kind) is very similar to the former in its manifestations."

Law of Simplex - The Single Remedy

Hahnemann in aphorisms 272-274 of ' Organon of Medicine' states that only one single, simple medicinal substance is to be administered in a given case of time. This is due to the following reasons:

1. The homeopathic remedies were proved singly, and the Materia Medica was built up on the observed effects of drugs given singly, either in planned provings or in accidental provings.

2. Only one remedy can be the most similar at any given time to the condition of any given patient.

3. Moreover, if more than one remedy is used the doctor will never know which element was curative and our source of future guidance is obscured.

4. If more than one drug is given in one prescription the possibility of synergistic action cannot be ruled out, but it cannot be argued that the effect will be the sum total of the effects of the separate drugs. The ingredient drugs may even result in interactions that may have adverse effects in the body. A mixture of more than one remedy in a single dose would constitute a new remedy which would require to be proved as such for a proper estimate of its probable effects.

Law of Minimum

The suitableness of a medicine for any given case does not depend on its accurate homeopathic selection alone, but likewise on the proper size of dose too. Under this principle we give medicine to the patients in very minute doses. The minute dose means that quantity of a medicine which is though smallest in quantity produces the least possible excitation of the vital force and yet sufficient to effect the necessary change in it(§ 246). The quantity is minimum, yet appropriate, for a gentle remedial effect. This concept of minimum dose lead to the discovery of a practical process called potentisation. Administration of the minimum dose has the following advantages:

1. To avoid unwanted aggravation

2. The specific dynamic action which produces the uncommon, characteristic, distinguished symptoms of the drug, is produced by the minimum quantity of drug.

3. The smallness of the dose does not allow the drug to do any organic damage nor there is any risk of drug addiction and drug effects.

4. The concept of minimum dose can be verified by Arndt-Schultz law that small doses stimulate, medium doses paralyze and large doses kill. I other words, the action of small and very large doses of the same substance on living matter is opposite.

5. The Law of Least Action, formulated by Maupertius, the French mathematician, states : "The quantity of action necessary to affect any change in nature is the least possible, the decisive amount is always a minimum, an infinitesimal."

Health is a matter of perfect equilibrium, perfect balance, trifling circumstances may sway it, and so may it be balanced by the least possible in medication.

Doctrine of Drug Proving

In Homeopathy we prescribe only those medicines whose medicinal properties are known through 'drug proving'. Drug proving is a systematic investigation of pathogenic (disease-producing) power of medicine on healthy human being of different ages, both sexes and of various constitutions. These recordings of drug proving give the only reliable knowledge of medicines which is very essential to cure disease homeopathically. Different medicines must be proved thoroughly in order to obtain full details of their curative properties. The drug must be proved on human beings because:

1. Animals do not give subjective or mental symptoms.
2. Effects of the same drug on animals and on human beings are different.
3. We do not get the modalities and finer symptoms in animal provings.

The drug must be proved on the healthy human being because:

1. The symptoms of the drug and the disease will be mixed together.
2. Moreover, the action of the drug on the sick person is different from normal person.

Theory of Chronic Disease

During the early age of homeopathic practice Hahnemann observed that in spite of best homeopathic treatment some cases would return with a recurrence of symptoms at intervals. This failure led him to investigate thoroughly a large number of chronic cases and after 12 years of observations he reached the conclusion that the chronic diseases are caused by chronic miasms. The miasms are Psora, Syphilis and Sycosis.

Psora is the real fundamental cause and producer of innumerable forms of disease. It is the mother of all diseases and at least 7/8th of all the chronic maladies spring from it while the remaining eighth spring from Syphilis and Sycosis. Cure is only possible by proper anti-miasmatic treatment.

Theory of Vital Force

It is Homeopathy which stresses the existence and operation of the vital force in a living organism. The human organism is a triune entity consisting of body, mind, and spirit. This spirit which is responsible for different manifestations of life was termed by Dr. Hahnemann as 'Vital Force'. Hahnemann speaks of the vital force in Aphorism 10 of his Organon of Medicine as : "The material organism without the vital force is capable of no sensation, no function, no self preservation; it derives all sensations, and performs all functions of life solely by means of the immaterial being (the vital force) which animates the material organism in health and disease."

In the healthy condition, it is the vital force which maintains normal functions and sensations of the organism. But when the vital force is primarily dynamically deranged by morbific influence, it causes abnormal sensations and functions which are manifested outwardly through the material body as abnormal signs and symptoms, the totality of which constitutes the disease.

Again if a cure is to be established it is the vital force that must arouse itself or be abided to arise for the recovery. If the vital force is too debilitated and exhausted then no medicinal aid is of help.

Law of Similars, Minimum Dose, Single remedy

The view one holds of symptoms is crucial to one's choice of cure. When symptoms (fever, diarrhea, cough, *etc.*) are viewed as the disease, as in conventional medicine, the principle of cure by opposites prevails and drugs are used to suppress or counter the symptoms, thus, supposedly, curing the disease. When, instead, symptoms are viewed as the body's attempts to heal itself, then the principle of cure by similars prevails and a remedy, with characteristics similar to the disease symptoms, is given that will further stimulate the body to heal itself

Hahnemann used the phrase "Similia Similibus Curentur," like cures like: a disease is cured by a medicine which would create in a healthy person symptoms similar to those the patient is experiencing. In this way, the body's own immune system is activated rather than suppressed, and the patient experiences a true cure rather than a temporary relief from disturbing symptoms.

Two other important principles of homeopathic curing are the use of small doses to stimulate the self-healing process and the use of only one remedy at a time. These principles can be appreciated for their logic. One of the well-known laws of drug reaction, the Arndt-Schultz Law, states that "Minimal doses of a drug stimulate, medium doses inhibit, and large doses destroy cellular activity." While conventional medicine uses large drug doses over long periods of time, often causing severe or fatal side effects, homeopathy's minimal dose, high dilution remedies, often given only once, have rarely been known to cause adverse effects. Likewise for the principle of using only a single remedy at a time. This eliminates any potential for harmful drug interactions, which is often the case with conventional medicine.

Doctrine of Drug Dynamisation

Homeopathic dynamisation is a process by which the medicinal properties which are latent in natural substances while in their crude state, become awakened and developed into activity to an incredible degree.

According to Dr. Stuart Close, "Homeopathic potentisation is a mathematico-mechanical process for the reduction, according to scale, of crude, inert or poisonous medicinal substances to a state of physical solubility, physiological assimilability and therapeutic activity and harmless, for use as homeopathic healing remedies."

Drugs are potentised by two methods:

1. Trituration - in case of insoluble substances.
2. Succussion - in case of soluble substances.

The objectives of potentisation in Homeopathy are:

1. To reduce the medicinal substance which helps to avoid unwanted medicinal aggravations and side effects.

2. Homeopathy believes that vital force is dynamic in nature and that is affected by disease, can only be cured by the dynamic power of serviceable medicine, not by its material quantity.

3. By this process the most virulent and deadly poisons are not only rendered harmless, but are transformed into beneficial healing remedies.

4. Substances which are medicinally inert in their crude natural state are thus rendered active and effective for healing the sick.

5. The medicinal qualities of other drugs which are more or less active in their natural state are enhanced and their sphere of action is broadened by this process.

The action of potentised medicines is deeper, longer and more wide-spread.

Homeopathic physicians treat the patients as whole beings, by respecting the symptoms first and recognizing that there is a reason that they exist in the first place. And then they work reactively - by giving the remedy that in a well person would create the totality of symptoms that the patient is already experiencing. That is the heart of the homeopathic method and practice. It is called similarity. It is the principle of treatment with which Hahnemann revolutionized medicine. But remember that the homeopathic method is a reactive method and it is the job of the remedy to act as a catalyst. As the body's own immune system rises up against the remedy in equal and opposite reaction, the reactive force moves the patient from a state of disease to a state of health. In Homeopathy, we are constantly aware that anything you do to the smallest tissue in the body you do to all the tissues in the body.

There are over two thousand Homeopathic Medicines made from substances in the plant, animal, and mineral kingdoms. The inherent irony of the homeopathic pharmacy and one of the central mysteries of homeopathic pharmacy and one of the central mysteries of homeopathic philosophy has to do with the act of dilution itself. This irony is best expressed in the Arndt-Schultz law, which gives insight into the biological activity of drugs. The law states that small doses of drugs will actually encourage life activity in general. Large material doses of drugs impede life activity in general, and very large doses of drugs will actually destroy life activity.

Hahnemann began to dilute his remedies in water. First on one part of the original substance with nine parts water and then one part substance to ninety-nine parts of water or alcohol to form a "1C" medicine. Not only were the remedies as powerful in their diluted states, but they were more curative. Another method he used called Succussion which is a process of systematically shaking the remedy while it is still in its liquid form. Hahnemann, through his empirical approach to medicine, found again and again that the remedies that had been shaken during their process of dilution were more potent than those that were not.

Homeopathy provides a safe, effective, natural, nontoxic treatment for many acute and chronic illnesses. It is important to remember that Hehnemann tells us that it is quite all right to move on to the second remedy if the first has not worked, but he does not suggest a third or fourth. While allopathic medicine is to always be considered suppressive, in time of true emergency you should consult with your physician. Learn homeopathic philosophy and practice now, so that when an emergency occurs, you will be in the best position to deal with it.

Arnica montana (Arnica) Arnica is also commonly called leopard's bane. The arnica plant has a bright yellow, daisy-like flower that blooms around July. Preparations made from the flowering heads have been used in homeopathic medicine for hundreds of years. It is popular in Germany and over 100 drug preparations are made from the plant. Arnica is a perennial that is protected in parts of Europe. The active components in arnica are sesquiterpene lactones, which are known to reduce inflammation and decrease pain. Other active principals are thymol (an essential oil), flavonoids, inulin, carotenoids and tannins. Arnica works by stimulating the activity of white blood cells that perform much of the digestion of congested blood, and by dispersing trapped, disorganized fluids from bumped and bruised tissue, joints and muscles. Arnica is known to stimulate blood circulation and can raise blood pressure, especially in the coronary arteries. The plant is used externally as an arthritis, burns, ulcers, eczema and acne. It has anti-bacterial and anti-inflammatory qualities that can reduce pain and swelling, improving wound healing.

Arsenicum album is a mineral remedy. The remedy Arsenicum Album was created, not just from arsenic alone, but from a chemical compound of arsenic and oxygen. This homeopathic remedy was created by Hahnemann and was one of his first and most important discoveries. It is a good remedy for respiratory infections of all sorts, cold, flu, fever, chronic fatigue, indigestion, colitis, irritable bowel, chronic diarrhea, insomnia, food allergies, hay fever, seasonal allergies, shingles, pneumonia and influenza.

Acontitum napellus (Aconite) is created from the roots and stems of the plant Aconitum napellus, which is also known as wolfsbane and monkshood. the plant is native to Europe, and must be gathered in the floweingstage in order to me made into the remedy Aconite. This remedy is good for Upper respiratory infections, colds and flu, conjunctivitis, croup, otitis media, pneumonia, tonsillitits, neuralgia, vertigo, angina, arrhythmia, headache, injury and trauma.

Belladonna is taken from a plant that for many years has been used as an herbal medicine. The remedy is created from the whole plant, picked at the flowering stage. This remedy aids fever, digestion migraine, convulsions, hemorrhoids, influenza, low back pain, sinusitis, tonsillitis, vertigo. Also Meniere's disease and otitis media.

Bryonia is created from wild hops. The mother tincture is made from the roots of the plant. It's remedy is good for rheumatoid arthritis, physical trauma, tendonitis, connective tissue disease, low back pain, sciatica, bursitis, headache, migraines, vertigo, nosebleed, toothache, constipation, diarrhea and gastritis.

Causticum, like the Kali family of remedies, is a potassium-based remedy. In this case the remedy is taken from a mixture of lime and bisulfate of potash. This mixture is dissolved in alcohol, from which the mother tincture is then made. This remedy is good for bronchitis, cough and cold, whooping cough, ear ailments, goiter, headaches, hemorrhoids, constipation, rheumatism, arthritis, sciatica, neuralgia, carpal tunnel syndrome, muscle aches and pains, hip pain, back pain, tendon pain, also warts, boils, acne, eczema and herpes.

Chamomilla is taken from the German chamomile. The plant must be picked when in full flower. The entire plant is used in the creation of the remedy. Situations that suggest Chamomilla: coughs, fever, colic, pain, especially associated with teething and toothache. This remedy is also used for hot flashes that accompanymenopause. It is a great remedy for use with those with behavioral problems. It is an excellent remedy for cases involving diarrhea that follow general pain pattern.

China is also known as cinchona. It was the very first remedy in the homeopathic pharmacy. It was created from the bark of the Cinchona officinalis. a bark from which the allopathic medicine quinine is also take. The substance, which was know for centuries for its medicinal qualities, was named Kina-Kina, or "bark of barks," by the natives of Peru. China was created by Samuel Hahnemann. This remedy is used for fevers of all sorts, influenza, and respiratory infections of all sorts, coughs. Also colic: constipation and diarrhea, hemorrhoids, headache, vertigo tinnitus and menieres disease, deafness, rheumatism, hip pain, backache. Also food poisoning and indigestion, suffocation, acute asthma, home recovery from strokes, sleep disorders.

Colocynthis is taken from the bitter cucumber, or Colocynth cucumis, of the Natural Order of Cucuritacae, which is native to Egypt and Turkey. Also known as bitter apple. The tincture from which this remedy is taken is made from the dried fruit of the plant from which the rind and seeds have been taken. This is a remedy of Hahnemann and is used for colic, diarrhea, PMS, rheumatic pains, toothache and especially sciatica.

Eupatorium perfoliatum is taken from the plant known as both "boneset" and thoroughwort, a perennial plant with white flowers and leaves that grow at right angels to those that are growing immediately above or below. It belongs to the Natural Order of Coposite and is native to the US and Canada. This remedy was proven in 1846 by Williamson and Neidhard. This remedy aids influenza, fever, bilious fever, diarrhea, measels, ringworm, rheumatism and hiccup.

Ferrum phosphoricum is taken from the chemical compound of iron and phosphoric acid. Ferrum Phos is very helpful in treating problems with the blood, like anemia, and with the circulatory system as a whole. It aids fevers, colds, coughs, flu, croup, sore throat, vomiting and whooping cough. Also good for injuries, sprains, rheumatism and gastritis, dyspepsia, diarrhea and frostbite. This remedy was first proved as a homeopathic by J.C. Morgan in 1876 but was refined by Schussler.

Gelsemium is taken from the plant yellow jasmine (called Chambeli Hindu). It belongs to the Natural Order Longaniaceae and is native to India, Europe, and North America. It aids influenza, diarrhea, fever, hay fever, laryngitis, upper repiratory troubles, chronic fatigue syndrome, headaches, vertigo, tremor and insomnia.

Kali bichromicum - The Kalis are a large group of remedies, all based in part on the substance potassium. Its chemical formula is $K_2CR_2O_7$. It is a compound of potassium and chromic acid. The remedy was created by Dr. J. Drysdale, a British homeopath. The proving remedy was performed and published by the Austrian Society of Homeopaths. The major action of this remedy is on the sinuses as well as gout,headaches, all physical pains, acute cold, and coughing.

The remedy Lachesis is taken from the poison of the surukuku snake, also called Trigonocephalus lachesis. The snake is native to South America. The remedy was created by homeopath Constantine Hering. This remedy has many acute applications. It aids coagulation of the blood, acute asthma, swollen glands, sore throat, varicose veins, ulcers, warts, tumor, ulcers, whitlows, vertigo and eye disorders.

Lycopodium is taken from the spores of club moss. The spores of the plant are shaped like a wolf's paw, which gives the remedy its name: lyco means "wolf" and podo means "foot". This remedy is helpful for male issues such as impotence, premature ejaculation; digestive disorders, especially bulimia. Also aids kidney and urinary disease, kidney stones, urinary tract infections, cystitis, headache and migraine, allergies, rhinitis, acute asthma, infections of all sorts, colds, flu, bronchitis, otitis media, chronic fatigue. This remedy was created by Samuel Hahnemann.

Mercurius (Merc) - There are several remedies based on the metal mercury. Mercurius Vivus was a remedy devised by Hahnemann himself, although Hering must be given credit for the actual creation of the remedy. Situations that suggest Mercurius are infections of all sorts, sinusitis and tonsillitis, colds, ear infections, upper respiratory infections of all sorts, toothaches, acne, eczema, colitis and gastroenteritis, ulcerative colitis, sexual disorders, vaginitis. Also phobias, anxiety disorders.

Natrim muriaticum (Nat Mur) is made from sodium chloride, or regular table salt. Because this chemical compound is so common to our environment and is indeed a natural part of our own bodies, the remedy taken from it is one of most benign and curative. It is one of Hahnemann's first remedies. Natrum Mur is is one of the best remedies for digestive ailments that can be traced to an overuse of salt, especially chronic migraines and headaches that come from over studying and overwork, especially hyperthyroidism and diabetes.

Nux vomica (Nux vom) is taken from the gray seeds of a tree. This tree is related to the Saint Ignatius bean, making Ignatia and Nux Vomica was created by Samuel Hahnemann. This remedy aids constipation, hemorrhoids, influenza, also diarrhea, colitis, irritable bowel syndrome,hypertension, spasms and twitches of all sorts. Sleep disorders, insomnia, addictions to coffee, tea, drugs, alcohol, tobacco and also coughs and colds, acuteasthma, chronic fatigue, environmental illness and neuralgic pains.

Rhus toxicodendron (Rhus Tox) is taken from poison ivy, and is also known as snow rose. The leaves of the plant are used in the creation of the remedy. This remedy is good for joint and muscle pain, rheumatism and rheumatoid arthritis, low back pain, connective tissue disease, toricolis, tendonitis, sciatica, housemaid's knee, lumbago and sprains.

Sulphur is created from the element of the same name and is also called brimstone. The substance occurs naturally as a brittle crystalline mineral that is found near volcanoes. This substance was used curatively for at least two thousand years. This remedy was first proved as a homeopathic remedy by Hahnemann. It is good for skin conditions of all sorts, allergies, digestive disorders, irritable bowel syndrome, constipation, diarrhea, ulcer, depression, anxiety, chronic fatigue, arthritis, menstrual problems, headaches, migraines, alcohol abuse, especially sleeplessness and vertigo.

Veratrum album (Verat alb) also known as White hellebore treats fainting, restlessness, vomiting and diarrhea.

Case Taking

The Symptoms for Analysis

- ☆ Mind: Place all symptoms of mind under this heading, but be sure and leave space on your sheet for symptoms that you may discover at subsequent sittings.
- ☆ Head: Here will be placed pains, hair symptoms, movements of head, *etc.*
- ☆ Stomach: This group will include pains, food desires and aversions, eating and drinking, appetite, thirst, nausea, vomiting, eructations and sensations.

☆ Abdomen: you may place under this heading symptoms referring to constipation, diarrhoea, sensations as pain, pressure, *etc.*, symptoms of urination, defecation, bladder and male genitals.

☆ Menses: These symptoms are of so much importance to the female case that a separate heading should be made. All symptoms referring to the female generative organs, to child birth, haemorrhage, *etc.*, may be placed in this group.

☆ Chest: Symptoms referring to coughs, pains and sensations, expectoration, breathing, heart, pulse, breasts, *etc.*, may be place under this head.

☆ Back: Sensations, pains, *etc.*

☆ Extremitites: All symptoms as refer to the condition of sleep, dreams, *etc.*

☆ Generalities: Here place all symptoms which refer to conditions of modalities that affect the patient as a whole, not already covered by the mentals.

With the symptoms of your case arranged in this orderly manner, from Mind to Generalities, we have a record to which it is easy to refer and from which it is easy to individualize the record for repertory study.

Analysis of Forty Homoeopathic Remedies

These forty remedies will be far from the number required in all your cases, and the forty we have included in our list will contain, no doubt, some which you will never use in your individual work, while, on the other hand, some will be lacking which you find of daily use. Any list of so small a proportion of our vast materia medica would necessarily be open to such criticism; but we think that by the arrangement of this list of remedies you will acquire - by giving them a few minutes study each day - a working knowledge of the remedies you use. If it is possible for us to enable you to systemize these few remedies then we are sure that you will arrange those which you find most often indicated, but which absent from our list, so that you may then have a working knowledge of the remedies in which you are personally interested.

Consistent use of the repertory leads us to the study of our remedies in a scientific, rational manner, from center to circumference, from the mind to skin, noting the effect of the drug upon the provers, as given in the pathogenesis, in the will, the intellect and responses to every environment, thus learning to observe the disordered patient rather than pathological changes in the organs or parts.

In trying to have an image of a remedy in mind learn to keep an orderly general picture of its action as a whole, following these generalites through the particular manifestations as referred to parts rather than only a few so-called characteristics of the remedy for your daily use. Kent's Materia Medica has the remedies so arranged and their pathogenesis is so graphically portrayed that, after reading over a remedy in this book, a picture of the general action of the drug is left with you.

The way we study a remedy and the kind of picture we try to carry in mind, for daily use, are illustrated by the following short study of one of our familiar remedies, Arnica.

Arnica

The red strand running through this remedy is soreness. A general state of soreness throughout the whole body. The joints become sore, the periosteum is sore, the muscles are sore, and the soreness will continue until stiffness begins and we find the sore, stiff rheumatic pains of the Arnica patient. The soreness is manifest in the skin, so that there are black and blue marks. The soreness is so marked that pressure is painful and the parts lain on are sore, so sore that he wants to move, to change position frequently, for the longer he lies on a part the more sore and sensitive it becomes. He is stiff, so the motion is painful; still the bed is so hard, the parts so sore, that he must move. Therefore, when we see our Arnica patient we must expect to find this soreness; if not, Arnica will not be the remedy.

There is a general relaxation of the blood vessels in our Arnica patients, and this is manifest in the haemorrhages from various organs.

In the subcutaneous tissues this is represented by extravasation of blood under the skin which results in black and blue spots. The Arnica state which is associated with or preceding many acute diseases is manifest by this weakened state of blood vessels, and the patient will wonder how she got so many black and blue marks; even the slightest bruise or pressure will result in this discoloration. Little injuries produce bleeding. On mucous surfaces these result in haemorrhages. Haemorrhages of bright red blood which soon clots. The blood of the Arnica state soon clots, as is manifest by the blood-streaked or blood-flecked sputa which will contain many clots.

Arnica developed in its provers violent chills and fever; the fevers are a low, slow form that is associated with inflammation. From the results of the relaxed condition of the blood vessel all the organs of the body are prone to inflammations and haemorrhages; but with these haemorrhages we will have this general condition of soreness.

With these conditions we have pains, and the general characteristic pains that call for Arnica are, crawling, pricking or paralytic pains as if joints are dislocated. Unsettled pains which shift from one part to another; tingling and tearing pains. With all these conditions are the bruised sore sensations, and a deep, profound disturbance of the economy which is manifest by weakness; great and profound prostration, fatigue and sleepiness.

The countenance in these profound cases will be flushed and dark; there will be a besotted look, as if he was intoxicated, and he speaks and thinks with difficulty.

Many cases of cerebral haemorrhage and the low forms of typhoid will present this typical Arnica picture, and unless these patients receive this remedy they will die. From this you will be led to look for Arnica in your septic condition, and it has many symptoms which correspond to septic processes, such as are associated with typhoid and scarlet fever and other low forms of diseases. In septic diseases of every sort we find our patients running into Arnica conditions.

Surgical septicaemia and blood changes due to surgical shock. Where arnica covers the condition of your patient it will do more to restore the antibacterial power of the blood than any number of vaccines. Arnica represents the surgical septic condition more closely than that of the puerperal type. (This latter condition corresponds more closely to Sulphur.) Wonderful is its action in preventing suppuration. A severe inflammation will be set up in an injury, a severe bruise upon the muscles, there will follow the pain and soreness and induration with final suppuration. A dose of Arnica in the beginning will prevent all this and quickly restore the part to normal.

Bruises

This name at once makes you think of Arnica, and for this condition it has been applied externally by all schools and by all people. The external application is better than nothing, but the administration internally is best of all. It is not the bruise per se, that we can expect to relieve; that has happened and cannot be undone, but it is the resulting effects of the bruise that we wish to prevent and remove, and this came from the center from the internal structure and can best be overcome from the center by the internal action of the remedy.

Injuries to the head, with the resulting nerve and brain symptoms, send the patient into an Arnica state, and they will need this remedy to bring about order no matter how long ago the injury took place. The resulting shock of surgical operations calls for Arnica, and this remedy is given in routine practice by the surgeons of our school. The symptoms following operations which Arnica will remove, are those which are producted by handling and bruising of the soft tissue and no others. That is the reason the results are so often disappointing. Those sharp cutting pains, the results of the needle or the knife, will never be removed by Arnica, but are rapidly dispersed by Staphisagria. Cuts and open wounds never call for Arnica, only as there are shock, bruises and contusions.

Thus we have outlined the general action of our remedy, and these general conditions are always present in a greater or less degree in every case that calls for Arnica. Where there is no soreness never think of Arnica.

The mental symptoms of Arnica are striking, and many of them are symptoms which you would expect to result from shock. Fear, excitement, emotion and horror stand out prominently. The fear that something awful is going to happen, that he is going to die instantly. This is marked and the patient has a horror of death and of the unexpected. In many of the acute conditions we have an obstinate and irritable patient. He will want to fight with you and drive you from the room. This excessive irritability will often be followed by a delirium. Indifference, anxiety and hopelessness run through the mental state. In the low states we find a stupor. He is hard to arouse, and when you do wake him he will be confused and will not know where he is. Mental exertion, motion or physical exertion, all aggravate his condition.

The pains in the head are pressive, cramp-like, darting and tingling, and are made worse by walking, ascending and mental exertion. There is nothing very distinctive about the particular symptoms of the head, but any pains or conditions that arise from injuries will lead one to think of Arnica.

There is a peculiar symptom under this remedy which is associated with the eyes. He must keep his eyes open. They come open spontaneously, he cannot hold them closed himself. As soon as the eyes are closed he gets dizzy, things go round and it makes him sick.

The pains of the nose are sore pains, as if bruised; much nosebleed when first blowing nose in the morning. The coryza of Arnica comes in the evening when going to sleep, but with this will be the general bruised condition, the soreness that will differentiate it from Nux or Pulsatilla.

One of the keynotes of Arnica is manifest in the face; heat and redness of the face with coldness of the body. It seems as if the blood had left the body and gone to the head. The expression of the face is peculiar. We have a deep mahogany redness, with an intoxicated, besotted look; he looks as if his mental condition was benumbed; looks as if he was making an effort to find the right thing to say or do but cannot. He is stupid and looks it. In injuries about the face, especially about the eye and cheek bones, where the periosteum seems to have been injured, we find Arnica will remove the first effects, the superficial soreness, the black and blue condition; but after this has been done away with there will remain a soreness that appears to be in the bone itself. We could give Arnica indefinitely, and these symptoms would not disappear, but Hypericum will follow and remove them speedily.

The general condition of Arnica is exhibited, in the mouth by soreness of the teeth. Soreness at the roots of the teeth, as if they we being pressed out. The gums bleed easily. Haemorrhages from the gums after extraction of the teeth. This is one of our leader in bleeders after teeth extraction. Soreness of the gums after extraction. This remedy will do more to remove the soreness from the gums after extraction that all the mouth washes you ever heard of. (Sepia is another remedy which is useful in this condition, specially in the nervous women who have been made sick by having a few teeth extracted.) The mouth tastes bitter and like rotten eggs. This is from eructations, which are bitter and have the odor of spoiled eggs; this taste remains in the mouth and you can almost smell it on the breath; therefore, the books give "putrid smell from the mouth;" this as well as the eructations are worse in the morning. These eructations burn as they come up and cause a burning from the stomach to the fauces.

With this large amount of gas in the stomach we have a loss of appetite.

A loathing of food; even the sight of food is repulsive and nauseates. Meat, milk and broth are especially repugnant, and even his tobacco nauseates. Aversion to tobacco, to even the smell of tobacco smoke, stands high in this remedy. (What does a peculiar symptom like this mean and what weight shall we place on it. We cannot place all the ladies and others to whom tobacco may be offensive a dose or two of Arnica and make them lovers of the weed, but where a man has become a user of the weed, where the habit has become fixed so that his tobacco is a necessity, and then have some disturbance of his economy so effect him that what he desired and craved he now dislikes, and has such an aversion to it that even the odor is nauseating, we have what we are justified in calling a peculiar condition, and when this arises we will give it a prominent place in our symptom picture.)

The generals are still with us when we study the effects of Arnica on the stomach. The sore, aching extending through to back. The stomach is so sore it feels as if it rubbed the spine, and as if the spine was made sore by this pressure. Pressing pains in the stomach; as it was pressed by the hand. This pressure continues until it seems to rise to the neck; then he feels nauseated and bitter water comes into the mouth. The stomach is so sore that everything seems to press against it as if the xiphoid process was pressed inward; as if a weight was on or in her stomach; as if a stone laid in the stomach. Nausea; retching; ineffectual retching; they retch and retch and try to vomit, and after straining for some time they vomit blood and bloody mucus. The blood will be dark and coagulated. After this the stomach will be more sore and burn.

Inflammation of the liver and spleen often taken on Arnica symptoms. Shooting and stitches in the spleen and pressure as if from a stone in the liver are found under this remedy;

with this condition we have a distended tympanitic abdomen with passage of much foul flatus smelling like rotten eggs. The soreness and bruised sensations are stronger in all the abdominal symptoms.

With a condition in the stomach and bowels which led to the above symptoms you would expect to have trouble with the stools; you would look for a diarrhoea, and under Arnica we find slimy, mucus stools; brown, fermented, like yeast; undigested; bloody; purulent; dark, bloody mucus; large fetid, faecal; yellow, offensive and sour.

A peculiar stool symptom of Arnica is the involuntary stool during sleep. The rumbling and colic in the abdomen are relieved after stool. Another of the peculiar symptoms of this remedy is that the diarrhoea is aggravated, as well as the accompanying bowel symptoms, by lying on the left side. During the stool there is urging, tenesmus, sore bruised pain in abdomen; cutting in intestines; rumbling and pressure in abdomen. Tenesmus in rectum and bladder. After stool they are weak and prostrated and are obliged to lie down.

From the low state that the Arnica patient represents we would look for its counterpart in typhoid, where its general soreness and weakness resemble Baptisia, Pyrogen, and Rhus; but where the general and characteristic symptoms of Arnica are present it will be curative in cases where vaccines and other remedies fail.

The peculiar urine of Arnica is dark brown, with brick dust sediment; the urine is full of urates and uric acid that we find associated with rheumatic cases. From the general relaxed condition of the blood vessels we get bloody urine, haemorrhages from the bladder. "Urination involuntary when running" is peculiar to Arnica.

The symptoms of Arnica referring to the female sexual organs are distinctive, here we find the character of the haemorrhage changed to a bright red flow mixed with clots. The flow feels hot as it passes the vulva. Menses are profuse, especially after a blow, a fall or a shock to the system. The general soreness is marked, and the pelvis is so sore it prevents her from walking erect. The uterus is sensitive, bleeds easily; discharges of blood between perios, with nausea. Bleeding after coition. Arnica is especially useful in nervous women who cannot stand pain.

Not only for the resulting shock and effects of the bruising resulting from labor is Arnica useful, but it has a field of usefulness in changing the character of the labor pains. These pains in your Arnica patient will be too feeble and irregular, resulting from fatigue of the muscular tissue. They do nothing, although so severe that they drive her to distraction. Feels sore and must often change her position. Vagina sore and sensitive so she does not want to be examined. Great soreness of the back during labor. Arnica high will often prevent afterpains. It will contract the blood vessels and prevent post-partum haemorrhages. Used in routine practice it does much to relieve the distressing after symptoms, both mental and physical, of labor.

The cough of Arnica is dry and is caused from tickling in larynx and trachea; the cough is worse in evening until midnight, from motion, warm room and after drinking. The expectoration is scanty, difficult, of glairy mucus mixed with tiny clots of dark blood. The general soreness of the remedy is marked in the chest and is shown in whooping cough where the child will cry before the paroxysm. The coughing causes blood-shot eyes, nosebleed and expectoration of foaming blood. With the cough is a burning rawness of the chest, stitches in left chest, which are worse form motion and pressure.

From the general soreness and bruised sensations in the muscles you would be led to think of your Arnica patient as a rheumatic patient, and such is the case. Arnica is full of bruised, paralytic, sore and stiff rheumatic pains. The joints ache and feel as if they were bruised. The soreness is so marked that the Arnica patient is full of fear; afraid he will be touched; afraid of jars; doesn't want you to come near for fear you will touch and hurt the sore joint or muscle.

In the back we have violent pains in the spine, sore pains; spine feels as if it would not hold the weight of the body. Small of the back feels as if it had been beaten. Pressive pain between the scapulae.

The rheumatic pains in the extremities are associated with heaviness. The legs are so heavy that it seems as if he could not lift them; this heaviness is due to the paralytic pains in the joints, and is constant both when at rest and in motion. Limbs are sensitive to concussions, as the jar of carriage or of walking. In the arms we have violent twitchings going from the shoulder to joints of middle finger. Crackings in wrist joints, worse in right, as if dislocated; drawing pains in wrist relieved by letting hand hang down. Pressing, tearing pains in fingers. Cramps in fingers of left hand. These tearing and drawing pains as if sprained are also found in the lower extremities. The hips feel as if sprained, with a pressive drawing in the left hip, which is worse from extending the thigh when sitting. The tearing pain on right external malleolus and on dorsum of foot with drawing in outer half of foot is peculiar to Arnica. Gout in joint of great toe with redness; pain worse towards evening and from pressure. These pains as if bruised and sprained with discoloration are a picture of sprains and here the remedy administered internally will take the soreness and discoloration from the sprained ankle and remove the first effects of the sprain; those symptoms which remain after Arnica are usually amenable to Ruta and Rhus.

The most severe action of the remedy on the nerves is the paralysis, the prostration, the general weakness and sinking of strength; so weak he can scarcely move a limb. The prostration and general sinking of strength corresponds to the low state found in typhoid and other zymotic fevers.

The Arnica patient has many symptoms during sleep, those symptoms which resemble the stupor of apoplexy and the sleep symptoms of meningitis find their counterpart in Arnica. One of the peculiar sleep symptoms is that the patient will be sleepy all day but cannot sleep at night.

Your Arnica patient is full of chills; chilly, with heat and redness of one cheek; head hot, body cold; internal chill with external heat; thirst during chill (resembling Eupatorium), he will drink and drink, becoming more chilly all the time, and will have the characteristic stomach symptoms, and finally vomit a bitter, sour fluid.

Chilly on only one side of the body, and that of the side lain upon. Many of the intermittent symptoms closely resemble Eupatorium, but the general and stomach symptoms will allow you to differentiate in this disease.

Remember the generals of this remedy and you will find its greatest usefulness after mechanical injuries, no matte what disease name you may give to the condition arising from this source. Arnica will help not only to remove the disease condition, but if given early will prevent many of the resultant symptoms of shock from appearing. Most of the particular symptoms of this remedy can be figured out by applying the general state of the remedy to all organs or parts of the body. Keep these in mind and you will see how often many symptoms or disease conditions can be removed by this remedy alone; given internally and without recourse to any adjuvants. If it has the generals of Arnica it is an Arnica case, and does not require Baptisia, Bryonia, Rhus or anything else to be curative.

Suggestion as to Method of Study and Use of the Following Analysis

Take first the twenty-two rubrics and memorize the group of remedies found under each one, paying attention first to the generals. After you have become familiar with your list of remedies then learn the particular circumstance of the remedy under each rubric. This will give you ground work of these remedies that will be of use to you in the daily work of prescribing for your acute cases. After you have become familiar with the above symptoms you may broaden your knowledge of each remedy by reference to the materia medica. It has been my experience (as well as that of my students) that a few minutes' study each day will soon give you a comprehensive knowledge of the remedies that will be in shape to use at the bedside.

Take, for example, a cold patient, one who is shivering with the cold, and, although covered by blankets, cannot get warm. We find this patient having burning pains; he may be thirsty or not, there may be oedema of mucous membrane with stinging pains. There may be scanty urine or any number of symptoms referring to a particular organ or to disease condition, which might lead you to think of Apis, but the fact that your patient was cold would rule that remedy out and turn your thoughts to a remedy found under the first rubric, Cold and aggravation from cold. Here you would find that one of the twenty-six remedies given would be the one which would be homoeopathic to the patient in hand.

Take another example of a patient with throbbing pains. The first though of the majority of our men when they hear throbbing pains mentioned is Belladonna, but fourteen remedies in our list of forty have throbbing pains, and Aconite, Calcarea carb., Phosphorus, Pulsatilla and Sepia all have this characteristic pain in a higher degree than overworked Belladonna. We will know at least from this analysis that one of our fourteen remedies will be indicated, but must individualize more closely to find the one remedy. If the patient who exhibits the throbbing pains is worse after midnight think of those remedies that have an aggravation after midnight, and we will at once see among these ten we have Bry., Calc. c., Phos., Sulph. and Sil. Here we have five, any one of which may be the remedy to help your patient's throbbing pains. We learn that the patient is chilly, that the pains are worse from warmth, but that she desires very cold drinks. This at once lets us know that Phosphorus alone of the above remedies will be the one which the patient requires. Many other examples could be cited as to the use of the preceding scheme, but to those who will look to this work for assistance it would not be necessary, and the student who begins to get a usable knowledge of our materia medica from this analysis will find that his learning of the remedies by this method will enable him to discriminate, individualize and differentiate his remedy and patient quickly, accurately and with an ease which will astonish him.

Forty Remedies

The following remedies are those we will analyze:

Aconite Arnica Arsenicum Apis Antimonium tart. Belladonna Bryonia Calcarea carb. Carbo veg. Causticum China Chamomilla Colocynth Digitalis Drosera Dulcamara Gelsemium Graphites Hepar Hyoscyamus Ignatia Ipecac Lachesis Lycopodium Mercurius Natrum mur. Nitric ac. Nux vom. Phosphoric ac. Phosphorus Podophyllum Pulsatilla Rhus tox. Sepia Silicea Staphisagria Sulphur Thuja Veratrum alb. Zincum.

In order that we may analyze these remedies I have taken twenty-two rubrics which cover generals as to:

1. Heat and cold;
2. Mental states as related to (a) restlessness, (b) fear, © crossness and irritability, and (d) tearfulness; the modalities as to
3. Motion, and
4. Position when lying;
5. The time of aggravation as to (a) afternoon, (b) after midnight, and (c) after sleep; aggravation and amelioration from
6. Pressure; generals and particulars as related to
7. Thirst; aggravation from
8. Eating and
9. Drinking;
10. The character of the pain as found under (a) burning, (b) cutting, (c) sore, (d) throbbing, (e) cramping, and (f) bursting.

Cold and Aggravation from Cold

This is covered by the following twenty-six of our forty remedies, either in the first or second degree:

Acon.; ARS.; Bell.; Bry.; CALC. C.; CHINA; Carbo veg.; CAUST.; Coloc.; DULC.; GRAPH.; HEP.; Ipec.; Ign.; Lach.; Lyc.; Merc.; NUX V.; Nat. mur.; Nit. ac.; PHOS.; PHOS. AC.; RHUS; SEP.; Sulph.; SIL.

In using this rubric we must distinguish between coldness which is a lack of vital heat, and an aggravation from cold in various forms, or amelioration from heat. These are two distinct phases. A patient who craves warmth and cannot keep warm is cold, but the particular symptoms may be aggravated from warmth and ameliorated from cold. An example is seen in Phosphorus, which is a very cold patient, but his stomach symptoms are better from cold drinks. When he is sick he craves cold drinks, which, however, are vomited as soon as they become warm in the stomach. His head symptoms are also better from cold.

Looking to the particular circumstances under which each of the remedies are affected by cold, your leaders will be:

☆ Arsenicum, when patient is cold and has general aggravation from cold, except the headache, which is relieved by cold.

☆ Calcarea carb, has chilliness with aversion to open air and sensitiveness to cold, damp air, with aggravation of pains from slightest draft.

☆ China, where there is chilliness with coldness of internal parts.

☆ Causticum, where there is coldness that warmth does not relieve. The cough, diarrhoea, and rheumatism are worse from cold; paralysis from cold.

☆ Dulcamara, complaints brought on by cold, damp weather and living in damp places; coryza, cough and neuralgia are worse from cold.

☆ Graphites, predominantly chilly; the coryza bone pains and stomach are worse from warmth.

☆ Hepar is another chilly patient; extremely sensitive to slight draft, is worse from cold wind and cold drinks; aggravation from getting a part cold.

☆ Lycopodium, while a warm remedy, stands high in its particulars, being aggravated by cold, as its stomach, cough, throat and headache.

☆ Nitric acid, where there is icy coldness and aggravation from least exposure; soles of feet cold. The coryza and chilblains worse, but cough is better from cold.

☆ Nux vomica has general chilliness over whole body; sensitive to open air; aversion to uncovering. Cough and headache made worse.

☆ Phosphorus is very cold, with coldness locally in the cerebellum, stomach, hands and feet; neuralgia, rheumatism, cough and diarrhoea are worse from cold, while the stomach and head symptoms are relieved by cold.

☆ Phosphoric acid, where there is sensitiveness to drafts; abdomen and one side of face cold.

☆ Rhus tox, where there is internal chilliness; aggravation from cold, wet, open air, drafts, cold drinks and cold east wind.

☆ Silica, where there is general chilliness, always cold; cold weather, cold water and cold in general aggravate.

If the above do not cover your case, examine the following:

☆ Aconite, is worse from cold, dry winds, complaints from riding in; makes the coryza, conjuntivitis, toothache, croup, cough and rheumatism worse.

☆ Belladonna, where there is aggravation by going from warm to cold; aggravation from drafts and cold wind.

☆ Bryonia, where there is chilliness; complaints from cold drinks in hot weather.

☆ Carbo veg., where there is susceptibility to cold, cold nose, knees, *etc.*

☆ Colocynth, where there is coldness of whole body; aggravation from cold weather; stomach, coryza, gastritis and rheumatism are worse from cold; tearing, stinging pain in face from taking cold.

☆ Ipecac has oversensitiveness to both heat and cold; colic from cold drinks; aggravation in winter.

☆ Ignatia has chill predominating; cold winds and air alike aggravate; washing hands in cold water aggravates pains; nose, feet, and legs up to knees are cold.

☆ Lachesis has a coldness over the whole body; limbs and upper lip cold; throat worse from drafts.

☆ Mercurius, cannot bear cold; extremely sensitive. Coldness in ears, testicles and lower limbs.

☆ Natrum Mur., icy coldness about the heart; coldness of feet, joints, back and stomach.

☆ Sepia has coldness over whole body; sensitive to cold, damp air; the cough, eruptions, toothache and rheumatism are worse from cold.

☆ Sulphur is worse from cold, windy weather; in damp, cold weather; the throat and the diarrhoea are worse from cold.

Warmth and Aggravation from Warmth

Are covered by the following eighteen remedies:

APIS, Ant. t., Bry., Dulc., Dros., Graph., Ipec., Lach., Lyc., Mer., Nat. mur., Phos., PULS., Secale, Sulph., Sepia, Verat. and Zinc.

Your leaders will be:

☆ Apis, where there is general condition of warmth and aggravation from warm room. The chill and headache are worse from warmth.

☆ Pulsatilla is too warm, with great internal heat; aggravation from warm room and warm food; from heat of stove, with general aggravation of all complaints from heat.

☆ Secale, cannot bear heat, will throw off all covering; aversion to heat; internal pains much aggravated by heat. Warm drinks aggravate the coldness of stomach.

Others

☆ Antimonnium tart., the head is worse from warmth; cough is worse from warm drinks; aggravation from getting warm in bed; drowsy from warmth.

☆ Bryonia, head, face and chills are worse. Cough worse from warm air and room.

☆ Drosera, while always chilly, has < of cough; ulcers, and pain in long bones from warmth.

☆ Dulcamara, the cough, nettle rash and sneezing worse from warmth.

☆ Graphites, is worse from dry heat in the evening and night; itching is worse from heat of stove; toothache is worse from warmth.

☆ Ipecac, the heat aggravates the chill; worse from warm, moist, south winds.

☆ Lachesis, worse in warm spring weather (*e.g.,* diarrhoea) and from warmth of bed; diarrhoea aggravated.

☆ Lycopodium has desire for open air; warmth < eruptions; warm room < cough and headache. Aversion to warm food (warm drinks > pain in throat); longs for cold food although it < diarrhoea and cough.

☆ Mercurius, the external pains worse from warmth of bed; extremely sensitive to heat; headache, mumps, toothache, rheumatic pains and itching are worse.

☆ Natrum mur., is worse from heat of sun and in summer; cough and headache worse; toothache aggravated from warm food.

☆ Phosphorus, while cold, cannot tolerate heat near back; warm water causes toothache; warm food causes diarrhoea; warm drinks < cough; stomach is worse from heat; hands, face and arms become red from heat, and itching is worse.

☆ Sepia, general aggravation in warm room, warm climate, from covering; conjunctivitis and headache worse; breathing oppressed from warmth.

☆ Sulphur. Too warm. Throws off covers; < warm room, warmth of bed and heat of sun; headache, burning of feet and itching especially <.

☆ Veratrum has cough worse in warm room; neuralgia worse from warmth of bed; diarrhoea worse in warm weather.

☆ Zincum, complaints from becoming heated and getting cold; rheumatism from overheating; warm room aggravates headache.

Restlessness

The following thirty-two remedies have restlessness, either mental or physical:

ACON., Ant. t., Apis, ARS., BELL., CALC. C., Carbo v., Caust., Cham., China, COLOC., Dulc., DIG., HYOS., Ipec., Ign., Lach., LYC., MERC., Nux v., Nat. mur., Nit. ac., Phos. ac., PULS., Rhus, SECALE, SEP., SIL. STAPH., SULPH., Thuj., VERAT. A.

Your leaders will be:

☆ Aconite changes position constantly; impatient and anxious at night; must walk or move about, although it does not relieve pain. Does everything in great haste.

☆ Arsenicum, mental and physical restlessness; goes from one bed to another.

☆ Belladonna, during colic; with cardiac trouble; striking, biting; wants to fly away from pain.

☆ Calcarea carb., mental anxiety and restlessness; child cross, fretful and restless.

☆ Digitalis, where restlessness is associated with great nervous weakness.

☆ Hyoscyamus, turns from one place to another.

☆ Lycopodium, restless from oversensitiveness to pain; during colic.

☆ Mercurius, mental; desire to flee, with anxiety; everything is done hastily; must constantly change places; uneasiness; restless 8 P. M. until morning.

☆ Pulsatilla, mental restlessness and changeability forces him to get up at night; cannot rest, although motion aggravates.

☆ Rhus tox., cannot remain quiet although it hurts to move; mental restlessness.

☆ Secale, spasmodic twitchings with irregular movements of whole body; arms in constant motion; head jerks about from side to side.

☆ Sepia, throbbing in all the limbs will not permit of quiet.

☆ Silica, fidgety; starts at least noise; internal restlessness and excitement; body restless when sitting long.

☆ Staphisagria, restlessness with lack of inclination to move; hurts to move.

☆ Sulphur, uneasiness and excitation of nervous system; constantly moving feet.

☆ Zincum, feet fidgety; must move them constantly.

The following have restlessness in the second degree:

☆ Apis, is very busy; does nothing right; changes kind of work frequently; uneasiness, mental and physical.

☆ Antimonium tart., anxiety; tossing about; throws arms.

☆ Carbo veg., restless at night, or 4 to 6 P. M.; mental restlessness.

☆ Causticum, restlessness of body, worse evening; wants to run away; obliged to walk about.

☆ Chamomilla, child quiet only when carried; kicks when carried; whining restlessness; tosses about in bed; great restlessness with anxiety and impatience; jerking and twitching in sleep.

☆ China, compelled to jump out of bed.

☆ Colocynth, restlessness with diarrhoea; weak but has to move; finds rest in no position; headache compels him to walk.

☆ Dulcamara, great restlessness; impatience; general uneasiness.

☆ Ignatia, trembling of hands when writing; change of position often relieves pains; jerkings and twitchings in various parts of muscles.

☆ Ipecac, is restless in fevers.

☆ Lachesis, must change position frequently, with pain in back and limbs.

☆ Natrum mur., restless with chill; must move limbs constantly; hastiness.

☆ Nitric acid. Restlessness of limbs in evening; twitchings in upper part of body.

☆ Nux vom., great reflex excitability; convulsive twitchings of single muscles; body tossed to right side and back again; legs drawn up to body with sudden jerk, then forcibly thrust out again.

☆ Phosphoric acid, walking relieves oppression of chest, pain in loins, hip joints, thighs, and pain in the bones.

☆ Thuja, tossing about at night from anxiety; mental restlessness.

☆ Veratrum alb., must walk about; mental restlessness; constant twitches and silly motions; cannot dress herself.

Irritability

The following thirty-four remedies are cross and irritable.

ACON., APIS, Ant. t., Arn., Ars., BELL., BRY., CALC., CARBO VEG., CAUST., CHAM., China, Coloc., Dig., Dulc., Gels., HEP., Lach., LYC., Merc., NAT. MUR., NIT. AC., NUX V., PHOS., PHOS. AC., PULS., RHUS., SULPH., SEP., SIL., Staph., Thuja, Verat. alb., ZINC.

Your leaders will be:

☆ Aconite, pains intolerable, drive him crazy; ailments from anger.

☆ Apis, is hard to please; irritable; ailments from rage and vexation.

☆ Arsenicum, peevish, waspish and quarrelsome.

☆ Belladonna, quarrelsome; violent rage; bites and strikes and screams;

☆ Bryonia, weeping; angry; peevish; wants to be alone.

☆ Calcarea carb., is cross during day; obstinate; vindictive; easily angered.

☆ Carbo veg., is excitable and peevish; strikes, kicks and bites in rage.

☆ Causticum, is peevish, fretful, quarrelsome, disturbed and ill-humored.

☆ Chamomilla, is always out of humor; peevish; quarrelsome; angry.

☆ Hepar, gets angry at least trifle; obstinate; cross; extreme violence; threatens murder and arson; passionate fretfulness.

☆ Lycopodium, is peevish and cross of awaking; cannot endure least opposition; obstinate; defiant, arbitrary; morose, worse before menses.

☆ Natrum mur., ill-humor in the morning; great irritability; cross when spoken to; gets into passion about trifles; bad effects from anger or reserved displeasure.

☆ Nitric acid, is headstrong; trembles while quarreling; fits of rage with cursing; vexed at trifles; sad and obstinate.

☆ Nux vom., is sullen; quarrelsome; oversensitive; scolding; ill-humor; gets so mad he cries; stomach complaints after anger; frightened easily.

☆ Phosphorus, is excitable and easily angered; irritability of mind and body; prostrated from least unpleasant impression.

☆ Phosphoric acid has a condition of silent peevishness and aversion to conversation.

☆ Pulsatilla, is out sorts with everything; fretful, easily enraged; taciturn.

☆ Rhus tox., impatient; vexed at every trifle; depressed and ill-humored.

☆ Sepia, vexed and disposed to scold; fretful about business; irritability alternating with indifference; nervous irritability.

☆ Silicea, headstrong; obstinate and violent.

☆ Sulphur, is obstinate; destructive and easily excited.

☆ Thuja is easily angered about trifles; obstinate and quarrelsome.

☆ Zincum is cross towards evening; irritable; peevish; terrified; fretful; cries when vexed.

The following remedies will be less often in use:

☆ Antimonium tart. is worse after anger; weeps and cries in anger.

☆ Arnica is oversensitive; ailments from anger.

☆ China, taciturn; ill-humor increased by petting and caressing; stubborn and disobedient.

☆ Colocynth, throws things in anger; diarrhoea, vomiting and suppressed menses from anger.

☆ Digitalis, is gloomy and disturbed.

☆ Dulcamara, is easily angered and quarrelsome.

☆ Gelsemium, is gloomy and wants to be left alone.

☆ Lachesis, has a sensitive and jealous disposition.

☆ Mercurius, has desire to kill person contradicting her. Taciturn.

☆ Staphisagria, has ailments from vexation or reserved displeasure; child cries for things, which, when it gets, throws away.

☆ Veratrum alb., curses and howls all night; attacks of rage with swearing.

Fear

ACON., Arn., Ars., Bell., Bry., Calc., Carbo veg., Caust., DIG., Gels., GRAPH., Hep., Hyos., Ign., LYC., Merc., Nat. mur., Nux v., PHOS., Puls., Sulph., Verat. alb.

Among these twenty-two remedies, you will find your leaders to be:

☆ Aconite, has ailments from fright; afraid of crowds; ghosts; death; dark; of falling; to cross a street.

☆ Belladonna, has fear, worse in day time; of ghosts; of water; hides from fear.

☆ Digitalis, is constantly tortured by fear of death; fear of future.

☆ Graphites, is apprehensive; full of fear in the morning.

☆ Ignatia, has a dread of every trifle; terror; fear of thieves.

☆ Lycopodium, is timid; apprehensive; easily frightened even by slight noises.

☆ Phosphorus, has a fear and dread of death; fear during thunder storms; of faces, as if horrible faces were looking out of every corner.

The following remedies may also be found useful:

☆ Arnica, has fear of being struck or even touched; of death.

☆ Arsenicum, has great fear, anxiety with restlessness and prostration. Fear and dread of death; of being left alone.

☆ Bryonia, apprehensive; dread of future; anxiety about and fear of death.

☆ Calcarea carb., fears imaginary things that will happen to her; anxiety about recovery; that she will become insane. Fear of death; of consumption; of being alone (evenings).

☆ Carbo veg., is easily frightened and has nightly fear of ghosts.

☆ Causticum is timorous, is afraid to go to bed alone, full of frightful ideas; that something unpleasant will happen; fear of death.

☆ Gelsemium, has lack of courage; fear of death; bad effects of fright.

☆ Hepar, has violent fright on going to sleep.

☆ Hyoscyamus, stands high in complaints from fright; fear of begin alone, of being injured, and of poison.

☆ Mercurius, is afraid that she will kill herself; of being alone; that he will lose his mind.

☆ Natrum mur., fears that foetus will be marked; that something is going to happen; that she will lose her reason; chorea after fright.

☆ Nux vom., inclined to commit suicide, but is afraid to die; frightened easily; anxious about condition; terrifying illusions.

☆ Pulsatilla, has diarrhoea after fright; dread of people.

☆ Sulphur, has a fear that he will be ruined financially.

☆ Veratrum alb., has a fear that takes breath away; coldness, fainting and involuntary stool after fright; of death; easily frightened.

Tearfulness

Patients that are tearful are covered by the following twenty remedies.

Acon., APIS, Ant. t., Bell., Bry., Calc., Carbo veg., CAUST., Dig., GRAPH., Hep., IGN., LYC., NAT. MUR., Phos., PULS., RHUS., SULPH., SEP., VERAT. ALB.

Your leaders will be:

☆ Apis, when they are discouraged and despondent.

☆ Calcarea carb., when they are easily offended. Despair of life.

☆ Causticum, is hopeless, looks on dark side of everything; weeps during day; whines; least thing makes child cry.

☆ Graphites, has inclination to weep; cries about slightest occurrence; weeps from music.

☆ Ignatia, has inward grief; alternating weeping and laughter; sits alone and weeps.

☆ Lycopodium, cries all day; weeps when thanked; sensitive and melancholy.

☆ Natrum mur., is sad and weeps without cause; when spoken to; concern about future.

☆ Phosphorus, sadness regularly occurring at twilight; prostrated from least unpleasant impression; tearfulness alternating with mirth.

☆ Pulsatilla, cries from sadness or joy; from vexation and mortification; over nothing; when telling her symptoms.

☆ Rhus tox., has weeping with prostration, worse evening; desires solitude; begins to weep without knowing why.

☆ Sepia, has involuntary weeping; great sadness with frequent attacks of weeping; worse walking in open air.

☆ Sulphur, cries from consolation; during day and because she is depressed about illness.

☆ Veratrum alb., cries, howls and curses over fancied misfortunes.

Less often indicated will be:

☆ Aconite, sadness, alternating with laughter.

☆ Antimonium tart., cries from anger; from being touched; during cough; whines.

☆ Belladonna, howls; cries from vexation and hopelessness.

☆ Carbo veg., thinks he has committed some crime, which causes him to weep.

☆ Digitalis, sighing and weeping; worse from music; tearfulness with low spirits.

☆ Hepar, is low spirited and sad, must cry for hours.

Aggravation from Lying

Aggravation from lying is covered by seventeen remedies, as follows:

Acon., APIS, Ant. t., Ars., Bell., CHAM., DROS., DULC., HYOS., Lach., LYC., Nux v., PHOS., Phos. ac., PULS., RHUS., SEP.

☆ Apis, worse from lying on left side; chest, breathing, and cough are worse lying on left side.

☆ Arsenicum, must lie but pains are worse; breathing is worse.

☆ Chamomilla, flickering before the eyes, nausea, vertigo, neuralgia, pain in thighs, and swallowing are worse; aggravation from lying on back.

☆ Drosera, is worse lying in bed; on the sore side; aggravation of cough.

☆ Dulcamara, has headache, cough and rheumatic pains worse when lying.

☆ Hyoscyamus, lies on back, but cough is worse when lying.

☆ Lycopodium, the cough is worse from lying on left, and better on right side; lying on back aggravates breathing; abdomen and cough worse lying on right side [?].

☆ Phosphorus, lying on back relieves pneumonia; on right side relieves diarrhoea, stitches in chest after pneumonia. Lying on left side aggravates heart, cough, rheumatism, and diarrhoea.

☆ Pulsatilla, is worse from lying on back during pains, and from lying on left or painless side. Urging to urinate aggravated lying on back.

☆ Rhus tox., lying aggravates the cough; vertigo; back; rheumatism and trembling.

When the above do not cover the case one of the following may be indicated.

☆ Aconite, lying is unbearable during fever; palpitation worse; chest and cough aggravated from lying on right side. Cheek lain on sweats.

☆ Antimonium tart., is worse from lying on affected side; earache; vomits when lying any way but on right side.

☆ Belladonna, headache and cough are aggravated lying on right side; aggravates pain in liver.

☆ Lachesis, has pain in lungs, left arm, back, in spine, and suffocation, all worse lying; involuntary urination when lying.

☆ Nux vom., cough and pains in chest worse lying on back; cannot lie on right side; asthma; sneezing and headache worse lying.

☆ Phosphoric acid, vertigo and tickling in chest when lying in bed.

☆ Sepia, headache worse lying on back; lying on left side aggravates cough.

Aggravation from Motion

Apis, Arn., Ars., BELL., BRY., Carbo veg., COLOC., Dig., Gels., Hep., Ipec., Lach., MERC., Nit. ac., NUX V., Phos., SULPH., SIL., Verat. alb., Zinc.

Your leaders will be found under:

☆ Belladonna, where they are worse from least jar; aversion to least motion; colic, worse from bending backwards. Staggers when rising from seat; headache; vertigo; pains in face, diarrhoea, metrorrhagia and cough worse from motion; cannot bear to stoop.

☆ Bryonia, has general aggravation from least motion; walking, ascending, rising, stooping and a misstep aggravate conditions.

☆ Colocynth, turning head, stooping and walking aggravates; rheumatism, pain in abdomen, and in eyes, are worse from motion.

☆ Mercurius, pain in spine; joints knee, palpitation, stitches and ulcers are worse.

☆ Nux vom., ascending aggravates cough; on rising from seat vertigo and pain in right kidney are worse; turning in bed and walking aggravates brain and abdomen; staggers when walking.

☆ Silicea, has general aggravation from even the slightest motion; stooping; rising and walking, aggravate complaints.

☆ Sulphur, headache, noise in ears; soreness between thighs, are worse from motion; walking aggravates head, sciatica, legs, burning soles (cramps in soles at every step); stooping makes head worse; ascending and rising from seat aggravate.

The following have Particulars aggravated from motion:

☆ Apis, the headache, chill, stiffness and rheumatism are worse; stooping, walking, and least motions of hands aggravate.

☆ Arnica, headache, chills, chest, stomach, stiffness and soreness are worse.

☆ Arsenicum, headache, ovarian pains, constriction of chest, are worse; raising in bed aggravates headache; walking and ascending aggravate.

☆ Carbo veg., has difficult breathing on slightest motion; turning in bed and walking aggravate.

☆ Digitalis, motion brings on angina pectoris; desire to urinate and defecate. Oppressed breathing and asthma when walking; palpitation and cyanosis from motion; cough worse from moving arms upward. Fears to move lest heart should stop.

☆ Gelsemium, fears heart will stop unless he keeps constantly in motion; headache, eyelids, and cramps in legs, worse from motion.

☆ Hepar, pain in back and limbs from walking up and down stairs; stooping and moving head aggravate headache.

☆ Ipecac, slightest motion causes nausea; griping in intestines; sweat; cramps between scapulae; cutting in intestines, and constriction of throat are worse.

☆ Lachesis, has aversion to every kind of motion; walking aggravates vertigo and dyspnoea; headache, chest and suffocative attacks are worse.

☆ Nitric acid, has vertigo; soreness in anus; stitches in vagina and sudden loss of breath when walking. Dyspnoea and palpitation on ascending; headache, chill and pain in abdomen, worse from motion.

☆ Phosphorus, headache; dyspnoea; weakness in abdomen; exhaustion; pain in heel and staggers when walking; vertigo, cardialgia, palpitation, cough and involuntary stools, all aggravated from motion.

☆ Veratrum alb., least motion aggravates nausea and vomiting. Rising aggravates the cough. Headache, cutting in stomach, debility and dyspnoea are worse.

☆ Zincum., slightest motion causes cutting pain from back into calves and feet; walking aggravates vertigo, headache, flatulent colic, burning anus, involuntary urine, and pain in knees and heel. Nausea, liver, chest and intercostal neuralgia are worse.

Aggravation during Afternoon

Is covered by the following eighteen remedies:

Apis, BELL., Bry., Coloc., Dig., Dulc., Ign., LYC., Merc., Nat. mur., Nit. ac., Phos., PULS., RHUS., SEP., SIL., THUJA, ZINC.

Your leaders will be found under:

☆ Belladonna, when worse from 3 P. M. to midnight.

☆ Lycopodium, 3 or 4 and 4 to 8 P. M.

☆ Pulsatilla, 3 to 6 P. M.; general aggravation in evening.

☆ Rhus tox., fever worse at 2 P. M.; paroxysms appear at 5 P. M. in intermittent fever.

☆ Sepia, has aggravation from 3 to 8 P. M.; fever, vertigo and pains worse.

☆ Thuja, has chill at 5:30 P. M.; mucous stool at 6 P. M.; pressing in vertex worse.

☆ Zincum., chill from 4 to 8 P. M.; cardialgia 3 to 4 P. M.; moroseness; vertigo, burning in eyes, sneezing, thirst, weakness and thoughts of death; sensitiveness to open air in afternoon.

Other Particulars that are aggravated in afternoon are found in the following:

☆ Apis, has chill at 3 to 4 P. M.

☆ Bryonia, headache; frequent urination worse 6 to 7 P. M.; sciatica and many complaints worse afternoon.

☆ Colocynth, has aggravation from 4 to 9 P. M.

☆ Digitalis, has 4 to 6 P. M. aggravation.

☆ Dulcamara, general aggravation toward evening; pressing out headache, worse toward evening, on walking in open air; quarrelsome mood

☆ Ignatia, pains gradually increase afternoon till evening; 4 P. M. aggravation.

☆ Mercurius, chilly 5 to 6 P. M.; coldness in testicles in afternoon.

☆ Natrum mur., has head, chill, and cold feet, in afternoon.

☆ Nitric acid, has cough, chill, vertigo, and incarcerated flatus, worse afternoon.

☆ Phosphorus, has aggravation from 3 to 6 P. M.

Aggravation after Midnight

Is covered by the following thirteen remedies:

ARS., Bry., Calc., DROS., Gels., Merc., NUX V., PHOS., POD., RHUS., Sulph., SIL. THUJ.

Your leaders will be:

☆ Arsenicum, worse from 1 to 2 A. M.; anxiety; restlessness; diarrhoea; heat and coldness.

☆ Drosera, has aggravation of nausea; cough; heat and cutting pains.

☆ Nux vom., is worse 3 to 4 A. M.; cough, renal colic and sweat are worse.

☆ Phosphorus, has aggravation of sweat, coryza and cough.

☆ Podophyllum, has a diarrhoea with pain in abdomen at 3 A. M.; cramps in the intestines from 5 to 9 A. M.

☆ Rhus tox., has general aggravation after midnight; restlessness, cramps and itching are worse.

☆ Silicea, has general aggravation after midnight. Chill 1 to 7 A. M.; wakens at 2 A. M.; sweat at 6 A. M.; diarrhoea from 6 to 8 A. M.

☆ Thuja, has aggravation of chill; headache and rheumatism; pressing in vertex from 3 to 4 A. M.; chill at 3 A. M.

The following also have less marked aggravation after midnight:

☆ Bryonia, < 3 to 6 A. M.

☆ Calcarea carb., worse from 2 to 3 A. M.; sweat and cannot sleep after 3 A. M.

☆ Gelsemium, has dreams; enuresis and leucorrhoea.

☆ Mercurius, has thirst, ptyalism with nausea; heat with violent thirst for cold drinks, worse after midnight.

☆ Sulphur, has aggravation at 4 and 5 A. M.; sweat after waking from 6 to 7 A. M.; cough until 2 A. M.

Aggravation after Sleep

Is found in the following fourteen remedies:

Acon., Apis, Arn., Ars., Carbo veg., Caust., Hep., LACH., Lyc., Phos., Phos. ac., Puls., Rhus., SULPH.

Your leaders under this rubric will be:

☆ Lachesis, where there is general aggravation after sleep and where complaints come on during sleep.

☆ Sulphur, starts and screams after sleep; wakens frightened; diarrhoea after sleep.

The following have aggravation after sleep in the second degree:

☆ Aconite, on going to sleep fever becomes intolerable; starts from nightmare.

☆ Apis, sleeps into <; wakes weary. Starts from sleep suddenly with great anxiety.

☆ Arnica, paralyzed on right side; < after a long sleep; unrefreshed by sleep.

☆ Arsenicum, starts from sleep and is weary after sleep.

☆ Carbo veg., has aggravation of coldness of feet and legs after sleep.

☆ Causticum, is worse on awaking; must sit up; cramps in heels after sleep;

☆ Hepar, fright during and suffocation after sleep.

☆ Lycopodium, is hungry and unrefreshed; cross; kicks and scolds after sleep.

☆ Phosphoric acid, has sad thoughts; dry heat and hunger after sleep.

☆ Phosphorus, is anxious and unrefreshed.

☆ Pulsatilla, has indigestion and is languid and unrefreshed after sleep.

☆ Rhus tox., is anxious, weak, restless, trembling, and it seems as if he had not slept.

Aggravation from Pressure

Is found in the following twelve remedies:

APIS, Ars., Carbo veg., HEP., LACH., LYC., Merc., Nat. mur., Nit. ac., Nux v., SIL., STAPH.

Your leaders will be:

☆ Apis, is sensitive to light touch, cannot bear the sheet to touch skin; every hair is painful; child stiffens when touched.

☆ Hepar, has dread of contact and extreme sensitiveness; scalp, eye, renal region, muscles of neck and external throat are aggravated from pressure.

☆ Lachesis, is worse from slightest touch; pressure produces black and blue marks; pressure on larynx causes cough; throat and abdomen sensitive. (Sometimes firm pressure > when light touch is not tolerated.)

☆ Lycopodium, is sensitive to pressure in all soft parts; tight clothes and weight of clothes aggravate; liver specially sensitive.

☆ Silicea, cannot tolerate pressure below floating ribs; scalp and pit of stomach worse from pressure; parts on which he lies go to sleep. Touch aggravates drawing in head, toothache, eye, liver, vagina, and pain in elbows.

☆ Staphisagria, neuralgia of scalp, ovary and ulcers, are worse from pressure; touch aggravates drawing in head, toothache, ulcers and knee-joint.

Particulars under following are aggravated from pressure in second degree:

☆ Arsenicum, has scalp, stomach and abdominal symptoms aggravated from pressure.

☆ Carbo veg., the scalp, liver and perineum are aggravated.

☆ Mercurius, has aggravation of head, teeth, gums, stomach, liver, bladder, spine, ulcers and bone pains.

☆ Natrum mur., must loosen clothing; touching hair causes it to fall out; nose, jaw, teeth, epigastrium and spine are aggravated.

☆ Nitric acid, condylomata bleed when touched; eruption, iritis, teeth, abdomen, anus and ulcer are worse from touch.

☆ Nux vom., tight clothing aggravates soreness over liver; touching with the hand brings on spasm; stomach, liver, scalp and abdomen are aggravated by pressure.

Relief from Pressure

Is found in the following ten remedies:

Apis, BRY., CHINA, COLOC., DROS., Dulc., Graph., PULS., Rhus., SIL.

Your leaders for this amelioration will be:

☆ Bryonia, has general relief from pressure.

☆ China, has a drawing headache and pressure from middle of sternum, which is relieved; pressure in region of liver relieved by bending body forward.

☆ Colocynth, is relieved by firm, hard, pressure.

☆ Drosera, hold chest firmly when coughing or sneezing; pain in face, stomach, and stitches in chest relieved by pressure.

☆ Pulsatilla, hard rubbing relieves; headache, left chest, pains in arm and throbbing in arteries, relieved by pressure.

☆ Silicea, while many of the pains are worse from touch and pressure the headache is relieved by hard pressure or by tying the head tightly.

The following particulars are relieved by pressure:

☆ Apis, has a headache relieved by pressure while all other symptoms are worse.

☆ Dulcamara, the pains in chest and stitches in back are relieved.

☆ Graphites, has a colic relieved by pressure, although the liver and abdomen are worse from tight clothing and pressure.

☆ Rhus tox., has a sciatica relieved by rubbing; pain in back, right nates, crest of left ilium, hip and legs are relieved.

Thirst

The following twenty-one remedies have thirst in the first or second degree:

ACON., Arn., ARS., Bell., BRY., CALC., CHAM., CHINA, DIG., Hyos., Lach., MERC., NAT. MUR., Nit. ac., Nux v., PHOS., Pod., RHUS., SULPH., SIL., VERAT. ALB.

This rubric is common to many disease conditions and to many remedies. If there is nothing to account for the thirst it is an important symptom, but if the patient is running a high temperature, or is working in the heat, or has a disease like diabetes it would be a common thing for him to be thirsty, and under such circumstances your symptoms of thirst would have no place in your symptom picture.

Your leaders for general and particular thirst symptoms will be:

☆ Aconite, has a burning, unquenchable thirst and desires bitter drinks, wine, brandy and beer.

- ✰ Arsenicum, wants cold water a little and often; burning, unquenchable, thirst during sweat; desires acids, coffee, milk, wine, beer and brandy.
- ✰ Bryonia, has a great thirst with internal heat; wants large drinks at long intervals; warm drinks relieve.
- ✰ Calcarea carb., has a thirst which drinking does not relieve, worse at night; desires cold drinks and acids.
- ✰ Chamomilla, has thirst for cold water and weakness and nausea after drinking coffee; toothache relieved by hot water; desires acids.
- ✰ China, has thirst before or after chill and during sweat; wants to drink little and often.
- ✰ Digitalis, has a continuous thirst with dry lips; desires sour and bitter drinks.
- ✰ Mercurius, has a moist tongue with burning thirst for cold drinks.
- ✰ Natrum mur., has a constant thirst without desire to drink, worse in the evening; longing for bitter, sour things and for milk, with aversion to coffee.
- ✰ Phosphorus, wants very cold drinks; his stomach is relieved by them until they become warm, when they are vomited. Desire for refreshing drinks, with aversion to boiled milk, coffee and tea.
- ✰ Rhus tox., has a dry throat at night and wants only cold drinks.
- ✰ Silicea, has want of appetite but excessive thirst; desires cold drinks.
- ✰ Sulphur, drinks much and eats little; violent thirst for ale and beer.
- ✰ Veratrum alb., wants everything ice cold, little and often; desires cold drinks.

The following remedies will be of use when their particular thirst is present.

- ✰ Arnica, has a thirst for cold water without fever; constant desire for vinegar.
- ✰ Belladonna, great thirst, but drinking suffocates; desires lemonade.
- ✰ Hyoscyamus, has a dread of water; unquenchable thirst with inability to swallow.
- ✰ Lachesis, constant thirst, but is afraid to drink; disgust for drink.
- ✰ Nitric acid, violent thirst in the morning.
- ✰ Nux vom., has thirst during chill; in morning; desire for beer and brandy.
- ✰ Podophyllum, great thirst for large quantities of cold water.
- ✰ Desires sour things.

Aggravation from Eating and after Eating

Is found in the following twenty-seven remedies, either in the first or second degree:

Ant. t., ARS., Bell., BRY., CALC., Carbo veg., CAUST., Cham., China, COLOC., Graph., Hyos., LACH., LYC., NAT. MUR., Nit. ac., NUX V., PHOS., PHOS. AC., Pod., PULS., Rhus., SULPH., SEP., SIL., Thuja, ZINC.

- ✰ Arsenicum, feels better on an empty stomach; bitter taste, nausea, painless stools and chill are worse after eating.

☆ Bryonia, has many symptoms directly after dinner; weight and pressure in stomach after eating; complaints from eating oysters, old sausage, old cheese, salads, cabbage and potatoes, fresh, green vegetables. Pertussis worse after eating.

☆ Calcarea carb., nausea and pressure in stomach after eating.

☆ Toothache, cough, heart symptoms, stool and heat worse from eating.

☆ Causticum, complaints from eating bread, fat and fresh meat.

☆ Colocynth, has diarrhoea from least food or drink; colic from potatoes; griping and flatulency after eating; pains worse from eating or drinking.

☆ Lachesis, has vertigo; languor; drowsiness; dyspnoea; flashes of heat; pressing in stomach; diarrhoea after eating or made worse by eating.

☆ Lycopodium, fills up after a few mouthfuls; drowsiness; pressure in stomach and liver; spitting up food after eating; bad effects from onions, oysters and rye bread.

☆ Natrum mur., always feets better on empty stomach; sweat on face, while eating; nausea, palpitation and acidity after eating.

☆ Nux vom., is so sleepy after eating; must loosen clothing after; hypochondrical mood, sour taste, pressure and pyrosis, after eating; also cough is worse.

☆ Phosphorus, has pains which begin while eating and last until he stops; desires cold food and drink; nausea, belching and fullness of stomach after eating.

☆ Pulsatilla, is useful in bad effects from pastry, rich foods, fats, onions and buckwheat.

☆ Sulphur, drinks much and eats little; complaints aggravated from eating even a little; milk disagrees.

☆ Sepia, has pains aggravated immediately after eating; aggravation from bread, milk, fats and acids.

☆ Silicea, has chilliness on back and icy cold feet after eating in evening; sour eructations, fullness in stomach; waterbrash and vomiting large amounts of water after eating. Aversion to mother's milk; vomiting whenever taking it.

☆ Zincum., has heartburn from eating sugar; worse from wine and milk.

Worse after eating is given in the second degree in the following remedies:

☆ Antimonium tart., has somewhat of relief of pressure in stomach after eating; still eating sour food brings on attack of asthma.

☆ Belladonna, has pressure in stomach and putrid taste in mouth after eating.

☆ Carbo veg., dreads to eat because of pain; headache, acid mouth, heaviness, fullness, hot eructations, and burning in stomach, after eating; feets as if abdomen would burst after meals; butter, fats, fish and pastry disagree.

☆ Chamomilla, heat and sweat of face during and after; vertigo, nausea and abdomen puffed up after eating.

☆ China, is drowsy, and uneasy after eating; headache and fullness in stomach after.

☆ Graphites, has disgust for and nausea from sweet things; hot things disagree.

☆ Hyoscyamus, has hiccough when spasms and rumbling after eating.

☆ Nitric acid, has bitter taste; heavy weight in stomach, debility, heat and palpitation after eating; food causes acidity; fat food causes nausea and acidity.

☆ Phosphoric acid, has pressing in stomach and bitter eructations after eating; diarrhoea from acids and sour foods.

☆ Podophyllum, has a craving appetite after eating; nausea and vomiting of food one hour after eating; diarrhoea and sour hot eructations after eating.

☆ Rhus tox., sleepiness, fullness in stomach and giddiness after eating.

☆ Thuja, for the bad effects of beer, fat, acid, sweets, tobacco, tea, wine and onions.

The character of the pain is a symptom always brought out by the patient; under Burning Pains.

We find the following twenty-eight remedies:

ACON., Apis, Arn., ARS., BELL., BRY., Carbo veg., Caust., China, Coloc., Dros., Dulc., Graph., Ign., Lach., Lyc., MERC., NAT. MUR., NIT. AC., NUX V., PHOS., PHOS. AC., PULS., RHUS., SULPH., SEP., SIL., Zinc.

Your leaders for this rubric will be:

☆ Aconite, where there is burning in internal parts; of the lips and tongue.

☆ Arsenicum, has burning pains relieved by heat; through the veins; head, eyes, nose, ulcers, mucous membrane, liver, ovaries, back, spine and joints burn.

☆ Belladonna, has burning eyes, nose, stomach, throat, chest and ovary.

☆ Bryonia, the head, eyes, ribs, liver, abdomen, stool, urine and chest have burning.

☆ Graphites, has old scars that burn; spot on vertex, eyes, tongue, stomach, left hypochondrium, through abdomen, vagina, soles of feet and hands, burn or have pains burning in character.

☆ Mercurius, has general stinging and burning pains relieved by heat; burning internally; burning after scratching.

☆ Natrum mur., has burning pains aggravated by heat of sun and of stove; relieved by washing in cold water and by open air; burning pains in vertex, eyes, ears, nose, throat, stomach, bowels, urethra, vagina, hands and feet.

☆ Nitric acid, general burning, stinging and sticking pains.

☆ Nux vom., has internal burning; burning pains in head; throat, stomach, abdomen, anus, back, bladder and chest.

☆ Phosphorus, has general burning pains in head, brain, chest and under sternum in particular.

☆ Phosphoric acid, burning pains worse lower half of the body; general burning, liver, throat and chest in particular.

☆ Pulsatilla, has burning in eyes, throat, bladder, urethra, feet, chest and heart.

☆ Rhus tox., has burning, stinging and drawing pains worse on left side.

☆ Sepia, has internal burning with relief in open air; feet and palms burn. Hands hot and feet cold or vice versa.

☆ Silicea, has general burning, stinging pains; burning in soles of feet and in ulcers.

☆ Sulphur, has burning in general, with burning heat, burning in skin of whole body and in parts on which he lies; burning pains in vertex, forehead, palms, eyes, lids, nostrils, face, throat, of eczema, fauces, pharynx, stomach, abdomen, urethra, anus, in haemorrhoids, between scapulae, hands, balls and tips of fingers, knee, feet (particularly at night), soles, corns and chilblains.

☆ Zincum, has burning pains in back, whole length of spine, left arm, right wrist and ball of hand, left hip, soles, skin and ulcers.

Those burning pains not covered by the above list will be found under:

☆ Apis, has general burning, stinging pains.

☆ Arnica, has burning pains in brain, eyes, lips, throat, stomach, chest, heart and feet.

☆ Carbo veg., general burning as from coals of fire, without thirst, and better from cold.

☆ Causticum, general burning pains; burning in spots as from ball of fire.

☆ China, has burning of one hand while the other is icy cold; burning of the skin, and in ulcers.

☆ Colocynth, has burning in right side of forehead; eyelids, face, tongue, back, anus, urethra (during stool), right ovary and sciatic nerve.

☆ Drosera, burning deep in throat and center of chest.

☆ Dulcamara, burning in forehead, epigastrium, anus, rectum, meatus, feet, gums and back.

☆ Ignatia, has burning redness of one ear and cheek; burning heat in vagina and feet; pain in head, eyes, epigastrium, stomach, urethra and heels.

☆ Lachesis, has burning, stinging pains in top of head, eyes, mouth, rectum, ovary, wrists, stomach, from hip to foot, throat, hands and soles.

☆ Lycopodium, has one foot burning hot, and the other cold; burning in blisters on tongue; thumb and third finger of left hand; pain in stomach, rectum, lower limbs and ankles; and of wounds.

Cutting Pains

Are covered by the following seventeen remedies:

Arn., BELL., CALC., China, COLOC., DROS., HYOS., LYC., MERC., NAT. MUR., NUX V., PULS., SULPH., SIL., Staph., VERAT. ALB., ZINC.

Your leaders will be found in:

☆ Belladonna, where the cutting pains are in head (right side), face, stomach, abdomen, uterus and in the muscles.

☆ Calcarea carb., where there are cutting pains from within outward; pains in chest; stomach, back and liver.

☆ Colocynth, cutting as from knives in bowels; pain in forehead, left temple, eyes, ears, stomach, abdomen and chest.

☆ Drosera, cutting pains mostly in right side; in calves of legs.

☆ Hyoscyamus, cutting in abdomen, chest and joints.

☆ Lycopodium, has cutting in bladder, rectum, abdomen, liver, chest, scalp and penis.

☆ Mercurius, has dull, cutting, pressive and stitching pains; cutting from stomach to genitals; pains in eyes, abdomen and intestines.

☆ Natrum mur., has pains in head, abdomen, urethra, chest and back.

☆ Nux vom., has shooting, cutting pains about navel.

☆ Pulsatilla, cutting in bowels, throat, abdomen, limbs, liver, chest, back and in abscesses.

☆ Silicea, cutting pains in nerves; in right lung, testes, breast, shoulders, knee, stomach, rectum and about navel.

☆ Sulphur, has cutting, burning pains in eyelids and urethra; cutting in abdomen, loins and sacrum, vesical region, chest, about heart and in great toe.

☆ Veratrum alb., cutting, griping colic; pain in left chest.

☆ Zincum., in small of back during menses; cross umbilical region; pain in right eye and ear, nose, rectum, anus, kidney and urethra.

Cutting pains are also found in:

☆ Arnica, has cutting like knives in kidney; pain in teeth, epigastrium and liver.

☆ China, has cutting pains which shoot through abdomen in all directions before the passage of flatus; cutting in spleen as if it was hardened.

☆ Staphisagria, for injuries caused by sharp cutting instruments; pain over crural nerve; teeth and abdomen; pains in stitches after operations.

Sore Pains

ARN., BELL., CHINA, DROS., HEP., NAT. MUR., NUX V., Phos., RHUS., SULPH., SIL., ZINC.

Your leaders will be found under:

☆ Arnica, for bad effects of bruises and sprains; pain is sore as if bruised in head; brain, throat and stomach; general character of pains sore.

☆ Belladonna, has soreness and rawness; pains in eyelids, throat to ears, abdomen and back.

☆ China, has sore pains worse from light touch but relieved by hard pressure; sore all over in the joints, bones, periosteum, as if they had been sprained.

☆ Drosera, soreness in temples and in skin of right temple; bruised feeling in the larynx, back and ankle.

☆ Hepar, soreness in urethra, in genitals, scrotum, in folds between scrotum and thighs, chest and in all the limbs; bruised feeling in anterior muscles of thighs.

☆ Natrum mur., soreness left side of nose; nostrils; upper arm; epigastrium; chest; tarsal joints; liver; vulva; vagina; larynx and trachea and between the toes.

☆ Nux vom., has soreness all over; great tenderness of abdomen; soreness in liver, stomach, abdomen, across pubis, chest and shoulder-joint; bruised sensation of brain, in small of back, neck of uterus, low down in abdomen; in back and in limbs.

☆ Phosphorus, bruised feeling in bones; soreness and rawness; nose, mouth, chest, lungs, larynx and bronchi are sore.

☆ Rhus tox., has soreness and stiffness; soreness in head, nostrils, tongue, abdomen, of navel, in muscles of abdomen, back, vagina, chest and left side of lumbar region; bruised feeling in head, throat and limbs.

☆ Silicea, the eyeballs are stiff and sore; internal soreness; sore pain in bones, chest, lungs, and head.

☆ Sulphur, sore pain in left eye, in oral commissures, and in whole abdomen, bruised feeling and pain in abdomen, back, coccyx, left shoulder, left hip, thighs, in sciatic region and lower extremities.

☆ Zincum., has soreness in head, vertex, scalp and hair; pterygium; right upper lid; outer canthus; in nose, teeth, tongue, upper chest and left hypochondrium; rectum, anus, left kidney, urethra; as if beaten in the pectoral muscles; chest; outer muscles of thigh and in pimples.

Throbbing Pains

Are covered by the following fourteen remedies:

ACON., Bell., Bry., CALC., Cham., Ign., Nit. ac., PHOS., Puls., Rhus., Sulph., SEP., Sil., Staph.

Your leaders will be found under:

☆ Aconite, where there is throbbing in temples and left side of head.

☆ Calcarea carb., throbbing in ulcers; pain in vertex and forehead, worse from motion.

☆ Phosphorus, throbbing forehead, temples, teeth, heart, extending to throat, back and neck.

☆ Pulsatilla, throbbing in brain, head, forehead, teeth, ear and soles of feet.

☆ Sepia, has throbbing in temple, forehead, cerebellum and teeth.

When the above do not cover your case look to the following:

☆ Belladonna, has throbbing in carotids, in brain, teeth, stomach, ovary and breasts. While this remedy is given in routine practice for throbbing pains, it does not have this symptom in as marked degree as the remedies given above. It will only cure throbbing pains when the rest of the symptoms agree.

☆ Bryonia, has throbbing throughout the body; pain in vertex.

☆ Chamomilla, has a throbbing in one-half of the brain and in the back part of throat.

- ☆ Ignatia, has throbbing pain in left side of head; ears, nape of neck, small of back, teeth and stomach.
- ☆ Nitric acid, has throbbing pain in left side of head; ears, nape of neck, small of back, teeth and stomach.
- ☆ Rhus tox., throbbing in pit of stomach; in temples and from jaws and teeth into temples; in left shoulder and forehead.
- ☆ Silicea, has throbbing pain in forehead and up into head; in eyes, in teeth and in limbs; sacral region; in forehead and vertex with chilliness.
- ☆ Staphisagria, has throbbing in temples and from tooth to eye.
- ☆ Sulphur, throbbing pain in left side of occiput; in hand, teeth, gums, rectum and anus.

Cramping Pains

Are covered by the following ten remedies:

Bell., Calc., Dig., Nat. mur., Phos., PHOS. AC., PULS., Sulph., Staph., Zinc.

Your leaders will be:

- ☆ Phosphoric acid, where there are cramps in joints; upper arm; wrist; chest; stomach; diaphragm and abdomen.
- ☆ Pulsatilla, cramping pain in stomach, through chest; in right leg from knee to groin; in legs, abdomen, and in pit of stomach.
- ☆ The following have crampy pains in second degree.
- ☆ Belladonna, has cramps in jaws; the cramping pain in abdomen and stomach is relieved by lying at an angle of 45 degrees, and is aggravated by bending back; cramps in uterus and muscles are found under this remedy.
- ☆ Calcarea carb., has cramps in the hands and forearms, feet and legs, crampy pains in hypochondria and in stomach, with palpitation.
- ☆ Digitalis, has cramps in chest, abdomen and bladder.
- ☆ Natrum mur., has cramping pains in abdomen at menses; crampy colic pains that resemble labor pains, aggravated after stool and relieved by passing flatus; pains in arms, hands, fingers, thumbs, legs, calves and feet.
- ☆ Phosphorus, has crampy pains in testes, stomach, rectum, calves, between scapulae, and in left side of head.
- ☆ Staphisagria, has crampy pains in abdomen, right knee joints and first joint of fingers.
- ☆ Sulphur, has crampy pains in stomach, chest; cramps in hip joints, middle finger, legs, thighs, calves, soles and toes.
- ☆ Zincum., has crampy pains in epigastrium, hepatic region, sides of abdomen and umbilical region; pit of throat, bladder, in chest to stomach, in heart and lungs; cramps in legs, calves, left foot and muscles.

Bursting Pains

Are covered by the following nine remedies:

BELL., BRY., CALC., CAUST., IGN., NIT. AC., NUX V., SEP., SIL.

Your leaders under this rubric will be:

- ☆ Belladonna, in hemicrania; bursting pains in right temple, above nose, in occiput; in brain towards temples; in eyeballs, over right eye, in chest, stomach, abdomen and hypochondria.
- ☆ Bryonia, has bursting pain in forehead, eyeballs, throat, stomach, right hypochondrium, above left eye and from within outward in head.
- ☆ Calcarea carb., has bursting headache and bursting sensation in the stomach.
- ☆ Causticum, has bursting pain in forehead, small of back, rectum, coccyx, stomach and ears.
- ☆ Ignatia, has bursting above root of nose, in spleen, stomach and rectum.
- ☆ Nitric ac., bursting pain in middle of brain, in forehead, eyes, throat, stomach, rectum and small of back.
- ☆ Nux vom., bursting pain in forehead and vertex, in eyes, stomach, liver towards chest and head, in bladder, anus and pit of stomach.
- ☆ Sepia, bursting pain in forehead, liver, stomach and chest, Bursting sensation from ebullition of blood, which is worse at night.
- ☆ Silicea, has bursting pain in forehead and occiput, relieved by pressure; bursting pain in eyes, stomach and chest.

20

Hypnosis and Hypnotherapy

Hypnotists (professionals including doctors, psychiatrists, psychologists, therapists, and laymen like "mentalists" and entertainers) use a staggering variety of methods to induce hypnosis as well as to make suggestions to the subject, and here we'll discuss several of them.

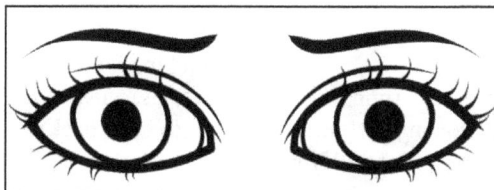

It is probably helpful to state right at the beginning that there is no consensus even among the medical community on what, exactly, hypnosis "is." It has not been studied enough in terms of its medical and/or physiological mechanisms for anything more than conjecture and educated guessing. What the body of literature resulting from the study of hypnosis does indicate is that it can have profound results, and that it is, medically, a waking state in which the subject is tremendously relaxed, so relaxed as to allow an intensity of focus unavailable otherwise, and in which the subject is very suggestible. The suggestions made by hypnotists can have effects long after the hypnosis session is over, and can help people to recover from traumas, stop or modify negative behaviours, and be useful in many more positive applications.

Mesmerism is often referred to be a form of hypnosis, although it has certain similarities to 'traditional hypnosis' techniques, but is not hypnosis in the true sense of the word. Originally developed by Franz Anton Mesmer in the late 1700's, Mesmerism is about the transfer of energy between the Mesmerist and subject to instate a special trance state. In doing this it is possible for the Mesmerist to use energy as a means of healing or to reconcile issues within the client. Mesmer referred to this as 'Animal Magnetism' and that it happened through the transference of 'ethereal fluid'. Today these same principles are applied to a multitude of other disciplines based on Mesmer's methods including energy healing, reiki and even kinesiology. The basic difference between Mesmerism and Modern Hypnosis is shown below.

Mesmerism	Modern Hypnosis
1. Uses little or no words to promote the trance state	1. Relies almost entirely on words and sound to instate trance
2. Uses 'passes (stroking the energy field) to bring change patterns in the client	2. Uses spoken direct suggestion aimed at the subconscious to change patterns
3. Has its roots in 'traditional hypnosis' as practiced by the ancient cultures such as Egypt	3. Derived from Mesmerism but evolved into Ericksonian hypnosis and NLP based speech therapies, dis-empowering the discipline

With that in mind, the varying methods used to induce the hypnotic state are not so much the quackery of pseudoscience (as some cynics might imagine) but simply individual approaches to the same, still-only-partially-understood goal. It does not matter how one separates one's felines from their pelts, then, as long as at the end of the day, the job is done. Let's begin our discussion of hypnosis techniques with the first step in the process: induction. But if you have just begun on the path to becoming a hypnotist, or if you are simply interested in the topic in general, you might do well to check out some online courses in hypnotism that will give you a solid background to draw from.

Hypnosis Tools and Eye Fixation

Hypnotists are generally portrayed using eye fixation when inducing the hypnotic state. How it works, pretty simple, looking for something at or just below eye level for an extended period of time helps to tire the eyes, causing the subject to close his or her eyes to rest them. Following the path of a swinging object back and forth also tires the eyes. And that's all there is to the watch. Hypnotists who practice eye fixation could use anything, really, that was eye catching. A penlight or any attractive object (especially a faceted gem which captures the eye) will suffice for eye fixation.

Some hypnotists simply use their own eyes for eye fixation. In the eye fixation induction technique you "tell" the subject, "look deeply into my eyes…your eyes are becoming heavier, you need relaxation, you are feeling sleepy….". How it works, simply being compelled to make prolonged eye contact with another person tires the eyes and encourages the subject to shut and rest them. There are several methods to energize your mind and prepare your eyes, meditation, self realization, several techniques, which include looking constantly to a burning candle in a dark room or a hypnotism wheel or picture. Eye fixation is, simply put, the technique whereby a hypnotist gets the subject to close his or her eyes. When the eyes are closed, visual sensory information stops flowing into the subject's brain, enabling relaxation and internal visualization, removing roadblocks on the path to hypnosis.

In the 21st Century, most hypnotists, however, don't use eye fixation, finding it corny and more importantly, because other methods are more effective.

Overt and Covert Hypnosis Induction Techniques

There are two ways to hypnotize a subject: with his or her knowledge (overtly), or without his or her knowledge (covertly). While the use of covert hypnosis induction techniques by anyone other than psychiatrists with willing but blocked subjects creates something of an ethical quagmire, we'll content ourselves with discussing the methods and not the morality of using them.

Progressive Relaxation

The vast majority of covert hypnosis induction techniques involve a willing subject who is guided through a progressive series of relaxation states, usually verbally. A serene and non-threatening location is required. The hypnotist calmly explains to the subject exactly what will happen: that he or she will progress through deeper and deeper states of relaxation until arriving at the hypnotic state.

Once the subject is reclining comfortably, the hypnotist guides the subject through a process of relaxing each part or section of the body, generally while imagining that he or she is in a "safe, happy place." Once the body is relaxed, the visualization shifts to a fantasy location, like a cloud-filled sky. The subject imagines him- or herself flying through clouds, with stress or worry vanishing into the clouds.

Then, the hypnotist directs the subject to focus on his or her breathing, as in meditation, and keying the directions to the tempo of those breaths, tells the subject that with each word or phrase, he or she will relax more and more, descending deeper and deeper towards the hypnotic state. Eventually, in this way, little by little, the subject is hypnotized.

Covert Hypnosis Induction Techniques

As for the covert methods of hypnosis induction (and we must say "don't try this at home"), they are also known by the names "conversational hypnosis" and (our favourite) "sleight of mouth." That last name should give you an indication of why this technique is somewhat dangerous and places you on ethically shaky ground.

Generally, conversational hypnosis techniques are derived from renowned psychiatrist Milton Erickson's "Indirect Hypnosis" method, in which the use of certain patterns of speech, tones of voice, eye contact, and key hypnotic words help to create a state of "semi-hypnosis" in which the subject is susceptible to suggestion, usually in the form of a metaphor.

Self Hypnosis Induction Techniques

Many people pursue self-hypnosis, or autohypnosis, as a means to change patterns in their lives. You might want to quit smoking, or relieve stress, or let go of old grudges, work with more focus and initiative towards a goal, or even learn how to manage pain effectively. Self-hypnosis can accomplish all of those things, to an extent. While it is not a magic bullet, it is worth thinking about trying self-hypnosis.

Autohypnosis works in much the same way as conventional hypnosis, except that you act as your own guide, and you do not need to speak out loud. Choose a comfortable and quiet room with dim or mellow lighting. Wear comfortable clothing, and turn off your cell phone. Be certain that you choose a time when you will be undisturbed and alone for at least thirty to forty minutes.

Recline or lay down, and close your eyes. Make a conscious effort to let go of other matters and all stress. Visualize each part of your body, and relax each in turn, letting go of tension in all your muscles. Breathe regularly, slowly and deeply. Some find sound helpful, especially music designed to work with meditation. In general, any droning, continuous, benign sound will work.

Breathing slowly and regularly, guide yourself through the same stages that a hypnotist would guide you through, mentally directing yourself to do all the same things a hypnotist would, such as, imagine yourself, "I am now floating in the clouds. All my tension is drifting away into the clouds…" and so on.

Once you have achieved the hypnotic state, you can focus on what you really need to work on, whatever that may be. You can make the suggestions to yourself by preparing statements ahead of time, like "I don't need cigarettes," or "I can turn off my pain any time I want to," and so on.

It's a good idea to set a timer or choose music that will change after a certain amount of time to make sure that you don't linger too long in the hypnotic state and miss your appointments. You can, of course, end the hypnotic state yourself, and most have no problems, but it pays to be safe, especially if your kids are going to be getting home from school in an hour.

Suggestion Techniques

Suggestion is the tricky part of hypnosis, isn't it? We're going to assume that you're not looking to get anyone to take off their clothes or act like a chicken (not that there's anything wrong with acting like a chicken if that's what you want to do) and go straight to the basic methods of suggestion.

Basically, there are direct and indirect methods. Direct suggestion consists of simple statements intended to reinforce beliefs or behaviours the subject wants to reinforce. For example, a hypnotist might make a direct suggestion to someone with anger management issues like this: "You are a calm, kind, caring individual. You always consider situations from other people's points of view. You think before you speak, and when you do speak, it is from a place of empathy…" and so on. This method works well, and is used widely. Hypnosis can be of tremendous benefit to the subject, whether it is another person, or yourself. Lives are changed for the better every day thanks to hypnosis, putting to rest any suspicions that it is a pseudoscience or not.

Hypnosis is not magic and neither is it mind control. A hypnotist cannot force you to enter a trance state without your full supports.

It can help you too for :

- ☆ Relief from addiction of T.V., Video games, Mobile phones etc; especially for children.
- ☆ Stress Reduction.
- ☆ Release fears, phobias and bad habits (addiction), can help Stop Smoking.
- ☆ Child Birth and Speed up physical healing.
- ☆ Improve Health physically and emotionally and Eliminate pain.
- ☆ Increase confidence and creativity.
- ☆ Ease depression, anxiety, insomnia and stress reduction.
- ☆ Improve school, athletic and professional performance.
- ☆ Treat many more physical, emotional and neurological issues.

☆ Personal development and rehearsal for success.

☆ Negative habitual behaviour and Weight control.

☆ Separation and loss.

☆ Age regression, Past life regression and much more.

Past Life Regression (PLR) is a method of accessing the memories of Past Lives in order to understand certain physical or psychological ailments, which are otherwise inexplicable. This technique is also used to understand relationships in our life, which affect us deeply, but for no logical reason. PLR has its roots in the belief of reincarnation – Law of Karma.

We unconsciously carry forward experiences, attitudes, patterns and relationships from our past lives into our current life. Most often Past Life memories and experiences are beneficial in living our day to day life. And sometimes certain current events triggers off unwanted feelings and emotions in the current life. This Past Life traumatic experiences, *i.e.* like a violent death or loss of a loved one, are left unresolved, relationships are left unhealed, or attitudes, patterns and decisions may be carried forward from a past life that are detrimental to our current life. That is where Past Life Regression therapy has been used to resolve those traumas and other issues that are blocking our progress and happiness now.

The memories of all of your experiences are stored in your unconscious mind since you became a soul, with the awareness of your individuality. Past Life Regression is reaching into those memory banks to recapture the events of past lifetimes. It is not very different from trying to remember events that took place during your early childhood. At first the Past Life memories may be dim and few, but each event remembered sparks another memory and another, until it becomes easy.

The main purpose of PLR Therapy is one of personal, emotional and spiritual growth. However, it can also help overcome fears, phobias and other physical and emotional challenges in the current life. Some of the lessons that we may bring back with us are forgiveness, love, generosity, patience, social skills and emotional intelligence. This helps us to grow as human and spiritual beings and to improve our current lives as we lose many of our fears and become wiser.

A PLR session usually begins with hypnosis and guided visualization, which helps the person to achieve an altered state of consciousness. From there, they experience themselves going into another time and place. Once conscious of another personality, the person will be led through the various key incidents of that lifetime, focusing on unresolved emotions, problems and conflicts. That process takes up about half the session: the remainder is devoted to healing the effects of the negative thought-patterns, emotions and physical traumas that may have arisen in that life. Some people are able to obtain insight and wisdom from inner guides or their Higher Self.

How to Hypnotize

Follow the step by step process below:

1. Have the person sit in front of you – face on face. You should sit yourself, too. Have the person put their hand on top yours – palm to palm.
2. Tell him (her): "Look at my eyes and continue looking until I say something".

3. Tell him (her): "In a moment I am going to count to 3. Press down on my hand and I'll be pressing up against your energy and simply follow my instructions instantly."

4. Say "1...2...3....push, push, push". If he (she) is pushing gently, tell him (her) to push harder.

5. Now with your other hand, put it on top of his eyes, like you are shading them, and slowly caressing down.

6. Then say: "now as you continue to press down on my hand I want you to develop a feeling in your eyes like your up much too late at night watching an old black and white movie, you should go to bed but you're just so tired. You feel your eyes so droopy ...and closing....and drowsy....and....SLEEP !"

The moment you begin to say "sleep" you have to quickly slip your hand away from his (her). This is very important moment and must be done accurately.

Practice this technique on your friends, family members and you will see the results. This is a step by step technique, so everyone who follows it properly can hypnotize people easily.

Clinical Hypnosis

Instant hypnosis is very useful tool for hypnotists, because it creates a powerful effect in a short time.

Now, follow the step by step process below to hypnotize anyone in some seconds.

Firstly, you must relax the person whom you are going to hypnotize. If your subject is not in a relaxed state or if he or she has fear of something, then it will be difficult to hypnotize her. So your first task is to calm the hypnosis subject. When the person is in relaxed state, do the following actions.

1. Take one of her arms and ask her to look at your eyes.

2. At the same time, Look at her eyes deeply while putting your arm on his neck at least 6 seconds.

3. Suddenly cry "SLEEP" and move his head towards you. And you are done.

The moment you say "sleep", you have to quickly slip your hand away from hers. This is very important action and must be done accurately.

Confidence serves as the primary key to success, when performing any act of hypnotism, even one that involves a self-hypnosis. However, it is also necessary to have a cooperative subject. Your subject must be ready and willing to be hypnotized. By the same token, you and your subject should be placed in a comfortable and relaxing setting.

The hypnotic process should not be viewed as something that can be rushed. If you try to hurry it along, it will never succeed. Your subject to be hypnotized should be asked to close his/her eyes. Then that same individual must be placed in a trance.

In your hypnosis experience, you may need to enter the subject in a long trance.

Long Lasting Trance

Long lasting trances are mainly used in a clinical hypnosis. Here is how it works. First, you need to enter your subject in a deep trance. After being sure your subject is in a deep trance, make him suggestion that he sleeps hard and he will sleep too hard to wake up. If you want your subject not to become hungry or drink anything, then suggest him that he is ok with that (use your own words, but don't use negative suggestion).

You should not enter someone in a long trance if you don't know how to hypnotize people. It can be difficult to wake up your subject from long lasting trance. So newbie hypnotists can get excited during the process. This may make it more difficult to wake up the subject.

How to Wake Up Someone from Trance

You must know relevant hypnosis techniques to wake up someone from trance. So remember these rules and strictly adhere to the guidelines below.

In most cases, you need only to make sudden clap and your subject will wake up. You should say with an imperious tone: "Wake up, wake up. Wake up immediately." You can also hit his hands several times and say the suggestion until your subject wakes up. After waking a person, watch him few minutes to make sure that he does not sleep again. Sometimes a person goes out of the trance himself, but it's better if you wake him up. Before waking a person you should tell him impressively that after waking up, he will feel great. Repeat these suggestions several times.

Special Techniques for Those who are Difficult to Wake Up

You must use only words for subjects who are difficult to wake up. You should not use hypnotic ball or any other shiny objects for difficult subjects.

So here is what you should do:

Sit next to your subject (not touching him) and tell him that he must sleep. Repeat the suggestion above for a few minutes. Then say: "Now you will do all I say. As soon as I tell you that you must immediately wake up, you will wake up." Then repeat the suggestion several times. Follow this method and you can easily wake your subject.

What to do if the Person does Not Wake Up

If the person does not wake up, tell him firmly, "I want you to wake up. When do you want to wake up? Will you wake up when I count to seven?" If he does not answer, repeat the same questions until he promises you that he will wake up as soon as you count to seven. If you get that promise, tell him: "Done. I'm starting to count: 1 … 2 … 3 … 4 … 5 … 6 … 7" Saying word "Seven", strongly hit his hands and shout: "Wake up! Enough to sleep! Wake up!"

If that does not work (extremely rare) say with an impressive tone that he should wake up, it is pointless to sleep longer and that you will not let him fool around and lie pretending that he cannot wake up.

After saying this, ask him to promise you that he will immediately wake up. If he does not wake up, ask him if he wakes up or not in 5 or 10 minutes. Then leave the room. After that time return and ask him: "Time has passed, do you agree now to stand up?" Make him talk to you and respond you.

Tell him: "Everything is ready. Time to wake up. When I count to three, you will wake up and be quite cheerful." Count to three, hit his hands and continue to provide relevant suggestions.

Do not rush the person and give him time to wake up. Newbie hypnotists often try to wake up their subjects fast which is not right. When a person sleeps, he can not react this command and overcome sleep all the times.

Never lose your confidence about being able to wake up your subject. If you become excited, nervous and confused, it will affect your subject accordingly and he will never be able to wake up. If a person notices disbelief in you, then it will be equally difficult to wake him up (and hypnotize obviously).

How to Prevent your Subject from Sleeping Again

If the subject is going to sleep again, you should provide the following suggestion: "I'll wake you up again now, but remember that this time you will be awake. You will stay awake…"

Repeat this suggestion several times with confident, decisive to awake. Then wake him up. If it is difficult for the person to wake up, allow him to stay in the room in a calm and cool position because nervous people may react to your own excitement and sleeping which will make it difficult to wake up.Each time that you use this method for easy relaxation, you relax more easily, more quickly, and more deeply. Relaxation is a skill that you are easily developing with trance.

Now, I'm going to count from one to five, and then I'll say, "Fully aware." At the count of five, your eyes are open, and you are then fully aware, feeling calm, rested, refreshed, relaxed.

All right. One: slowly, calmly, easily you're returning to your full awareness once again.

Two: each muscle and nerve in your body is loose and limp and relaxed, and you feel wonderfully good.

Three: from head to toe, you are feeling perfect in every way. Physically perfect, mentally perfect, emotionally calm and serene.

On the number four, your eyes begin to feel sparkling clear. On the next number I count, eyelids open, fully aware, feeling calm, rested, refreshed, relaxed, invigorated, full of energy.

Number five: You're fully aware now. Eyelids open. Take a good, deep breath, fill up your lungs, and stretch. One of the problems that many individuals have where sexual performance is concerned is that they oftentimes worry about whether or not they have pleased their partner, yet they are afraid to discuss the issue with them. Others feel that it is way too personal of a subject to be discussing with other individuals outside the home. This is especially true when your love life and relationship appears to be going downhill. However, hypnosis therapy can be used to help you overcome your fears of inadequacy.

Practice Covert Hypnosis

Another one of the benefits of hypnosis to consider is the fact that you can also learn how to hypnotize someone without their consent or knowledge. This is extremely useful in treating cases of drug or alcohol abuse or behavioural problems in teens, incl., concentration problem or lack of interest in studies. Although appears a difficult process, it really isn't. Our brains are extremely susceptible to hypnotic suggestion. In fact, we go in and out of hypnotic-like trances every day of the week. What is even more remarkable about this is that we usually have no idea that this is happening to us. But if you learn how to control this, you can use it to your advantage.

The first step when learning how to hypnotize someone is to establish a rapport with the individual or subject you are dealing in. That rapport is a deeply connected mutual understanding that enables that enables the establishment of trust between any two individuals.

Once the connection has been established between the two of you, begin using what are referred to as embedded commands or direct instructions to their subconscious. Remember, you cannot do this until you have established a connection with the subject.

Embedded commands are those that are easily hidden during a conversation so that the when you are hypnotizing someone, she or he is not aware of those commands and your intentions. Additionally, these commands help you to strengthen the connection you have established with the individual. As a result, when you are talking to a subject like you normally would and he or she is listening to you intently and will not be able to take their eyes off you.

For every individual who uses this for positive reasons and enjoyable consequences, there are ways to do it without the other person being aware that they are being coerced into the situation. But you want to use it wisely and not too frequently as it can have negative consequences as well as positive ones.

The Effectiveness of Hypnotherapy in Stopping Smoking Addiction

Hypnotherapy is considered one of the most promising fields of medicine when it comes to solving patients bad habit like smoking and gambling. In this type of therapy, the patient is put on trance so that the therapist can create certain contact with his subconscious mind of his patient. Once there is an established connection, the patient becomes more aware to what the therapist say because his mind would hear nothing but the voice of the hypnotherapist, making himself believe that he have to quit smoking or stop gambling.

This type of treatment is known in the field of mind control as the smoking cessation hypnosis which goal is to make people to totally quit their nicotine habits. However, not all people who are capable of hypnotism can do this kind of treatment because this method can only be done by trained professionals and should be administered to patients that agree to follow the intricate process of hypnotherapy. If treatment is done by people who do not know the complete treatment, the patient with problem will not be treated at all or worse may alter part of his behaviour in some ways.

When it comes to smoking cessation, the first step in treating nicotine addiction is for the hypnotherapist to explain to the patient before the treatment about the risks of extreme smoking. He will explain how does person can become addicted to smoking and what factors can tempt the person to start smoking which lead to addiction to nicotine The aim of this discussion is to make the patient understand the risks and the danger he will be exposed to.

The next stage already involves the actual hypnotherapy process. In this phase, the patient will be hypnotized according to his will. He will be put into a drowsy state and then into a trance. When the patient seems to be in a dream like state, the hypnotherapist will ask the patient what lead him to his addiction, who influenced him to smoke and what makes him to smoke even more. When the patient has already brought this out from his subconscious and tell it the hypnotherapist, the hypnotherapist will then tell the patient to liken the cigarette stick into something that can harm the patient and what effect can cigarettes do to his body. The patient will then be given an option how he can stop his addiction and the hypnotherapist will agree and at the same time tell his patient about other option he can take. While the patient and the hypnotherapist discuss these things slowly, the hypnotherapist will inject into his patients mind things that he must do and avoid until the patient finally gives in and agree to stop smoking altogether.

The hypnotherapist will also try to break the myths that the patients believe and negate it with positive ideals. In summary about this treatment, when the conscious mind and the subconscious agreed on the same level, the physical body will follow. When the patient will be awakened, the only things he will remember are those positive ideals that have been injected into his subconscious mind. This kind of treatment is now being recognized by some hospitals because the result for such treatment is promising. At present, this is one of the most popular branches of mental therapies because it caters to a wide variety of patients mental problems like anxieties, phobia and other self-inflicted mental disorders.

Hypnotherapy (Bio Magnetism) and Modern Science

Electricity and Magnetism are very closely related in the material world. They are two branches of the same natural force. They are inseparable and going side by side. But how electrical energy generates in human body and the body become magnetized. Here it is the point of discussion. In our day to day life, we come in contact with the game of Magnetism and Electricity. Astronomical observations indicate the mysterious works in the Universe – Movement of Planets around the stars and the distance maintained among the Stars and Planets. All are nothing just the magnetic field created by astral bodies. The earth, sun, moon and other planets in our own galaxy transmit their own magnetic emanations which greatly influence our life. The whole earth is a huge natural magnet. In magnetic parlance our bodies have considered to have magnetic sides. Considering human body vertically, the head and upper part of the body represent to north pole and their opposite the feet and lower part of the body represents to south pole.

Abiding by natural laws and forces any force, any action or deed done in the natural direction affects peace and pacification and cause the least possible discomfort than when done otherwise. Accordingly, if you will sleep or stand on opposite directions, it will provide more discomfort to us. Sun and Moon provides their Electro Magnetic Effects on earth. the whole creation depend upon fire which we receive from Sunrays and moon controls the water heat, light and moistures for all kinds of growths and developments. They are closely related for the production of green fields and the maintenance of life in earth.

Statistics have proved that body fluids flow more freely at the time of full moon. The custom of fasting in new moon and full moon have been considered highly scientific, since it helps in reducing body fluids and in maintaining proper equilibrium in the body. Tide in the sea is nothing just the magnetics effects of moon. Some diseases like Asthma, Filarisis, Skin eruptions aggravates on the full moon and new moon it is well known to the medical professionals. Now it is obvious that the effects of magnets on human body is from the beginning of the creation. The total life cycle depends upon both magnetism and electricity.

Human Body – Automatice Machinery with Magnetic Properties

The human body is a very complicated and wonderful automatic machinery. Its internal functioning is like that of an electric machine. The brain is the controlling switch board for the whole body mechanism. The nervous system as well as the other system working in the body are regulated through different controlling centres in the brain. In the circulatory system, the heart works as an electrical generator supplying energy to the entire body through circulation of blood. Thus the brain and the heart are the most important parts of the human machinery.

Electromagnetic Effects Conducted in Human Heart

Now let us study the make up of the heart. The heart is made up of thousands of muscles which may be taken as the compositing elements. It consists of two complex system of cells : one constituting the atrium and the other ventricle – which are again divided into two parts each. The recording of the electrical functioning of the heart is technically called Electocardiography(ECG). In taking an electrocardiogram, each chamber is considered separately. Each mechanical contraction – auricular or ventricular, is associated with two electrical processes. The first is depolarisation during which process the electrical charge goes up and the second is repolarisation where the electrical charge comes back and are at resting state. Depolarisation is a rapid process where as repolarisation is slow. The muscular activity of the heart is associated with electrical activity. It is the electrical phenomenon of the heart muscles which produces the electrocardiogram, with its associated change in the membrane's permeability. The heart generates electrical energy and breathing has a close connections with the beating of the heart.

The blood circulatory system originating from the heart is also a very complicated mechanism. The heart is a strong and tough muscular compact, located at the mid centre in the body. It may, however, have a break down like all other machinery. There may be many reasons for the breakdown of the mechanism of the heart or its constituents. When any part of the heart machinery is impaired, the part of the body affected does not receive blood supply it needs and it is damaged. The damage may occur in the heart itself or in any other related part of the machinery, namely brain, lungs, kidneys, limbs or skin.

The chemicals which the body contains, namely, carbon nitrogen, oxygen, phosphorus, *etc.* combine to form a perfect electric battery and the food we eat enables it to charge itself. The body too, therefore, exhibits its electrical responses. The electrical potential of the stomach varies which it goes empty or becomes full, in sickness and in health. It has been observed that when blood sugar turns low, electrical changes take place in the brain. Normally recording of the waves reads 8 to 10 cycles a second but when the concentration of sugar in the blood is lowered, this rate drops to 5 cycles per second.

The electrical activity functioning in the system of a living human being has the capacity to generate electricity within itself for its full requirement. The human body may, therefore, be taken as an electrical battery. It is capable of emanating electromagnetic waves at the rate of 8 million cycles per second, which is beyond the perception of our visual capacity. Every human body is constantly discharging static emanations which may be taken deponent to either electricity or magnetism. To quote an instance, every human being can act as an aerial to receive powerful wireless signals by putting one's hand on the aerial socket. It has also been reported that keeping the hand on a transistor and placing it in North-South direction gives clearer and better sound. A battery needs chemicals for its composition as we need food and drink to work as a battery. By means of some mysterious alchemy, these chemicals generate the electric current that replenishes the battery.

Usually, ill health and sometimes even death is caused by the demolition and disintegration of the cells consequent upon faulty energy. When faulty energy is put into action, it turns things wrong towards evil thing, opposing to good thoughts. There is a common belief that electricity is the cause of life. Innumerable experiments have been conducted and electrical fields have been found to exist in the most elementary from the embryo. It has accordingly been proved that the body of every human being contains some element of electricity and some properties of associated magnetism right from the beginning to the end of life.

Use of Magnets in the Medical Field

Dr. Petit in his learned work on metallic therapeutics shows that in ancient times metals were often applied to the body for curative purposes. Hindu Astrology too advices the wearing of different kinds of Metals as also roots of different trees to counter acts the malicious influences of different stars.

Application of Mr. Borggs Metalic Plates acquired such a reputation in France that he was rewarded for his discoveries by the gift of Berber's Prize by the French Academy of Medicine and the Society of Biology. Mr. Borgg applied metallic plates to parts of the body affected by contraction, cramp of anaesthesia, and thus in the habit of administering as all of the same metal for internal use. He attributed the effecting produced to an unknown force, which appeared to him to semblances to magnetism and electricity.

Professor Charnot suggested it was due to the electric action produced by the contact of metal with the moist surface of skin, followed by Mr. Reynard and later by Leonable who showed the application of magnets on the human body for healing purposes in 1954.

After many subsequent resources a satisfactory demenstration of this great truth was given by Dr. Chazarain who gained honour in French Academy of Medicine and who was a Physician of Mary Bathrust Hospital and Mr. Decle, a Member of the Association for advancement of Science in good, many pamplets theyhave explained the polarity as:

1. The action of the positive pole is not the same as that of negetive the voltaic needle applied to any p0int of a nerve excites it immediately along its entire length and by reflex or derived action the excitement is continued throughout the whole organism.

2. Positive (N) Pole of the Magnet applied to the outer side of the timbs or to the left side of the head or trunk producess contractions of the neighbouring muscles.

3. The negative South Pole of the magenet applied on these sides relaxes the contracted muscles.

4. Negative south pole of the Magnet applied to the inner side of the limbs or to the right side of head or trunk produces contractions of the neighbouring muscles.

   The positive pole applied on these parts relaxes the contracted muscles.

5. Similar effects result from contact with various parts of another body. For instance the little finger which is positive applied on the external side of a limb or on the left

side of head or trunk will cause a construction while the thumb which is negative will relax the contracted muscles. This is the scientific basis of mesmerism – to give passes. It shows that the nervous system and muscles exercise regular and powerful electrification.

Rosenthel mentions that every nerve and every part of a muscle when quiescent is positive in its longitudinal section and negative on its cross section.

Magnetic Fields in Human Body

Today's scientists and medical specialists have proved that the human body is a source of magnetic fields. Attempts have been made to measure the magnetic fields produced by different organs of the human body, namely, heart, brain, nerves, muscles and other tissues as well as the frequency of magnetic fields produced by all the organs in the body. Astoundingly, the peak value of the fluctuating magnetic field produced by the heart is greater than 106 gauss. Similarly, some muscles when flexed produce high frequency magnetic fields whose peak value is 207 gauss. It has been found that the strongest magnetic field from the nerve tissue is from the brain which produces its largest fields during sleep. It has an amplitude of 3 X 108 guess. In certain diseases like epilepsy larger fielder can be produced. The source of magnetic fields in the human body has been ascribed either to the presence of sodium, potassium or chlorine ions that the nerves and muscles generate in the process of contraction or signal transmission. This has led to the magneto cardiogram and magneto encephalogram for measuring the ion currents of the magnetism in the body in the presence of magnetic materials.

Some of the advantages of the of the measurement of the magnetic fields of heart, brain, lungs and other organs may be to evaluate the condition of an organ and to know its impending malfunction or defect. For instance the magneto cardiogram may be used to investigate the heart when it is injured by the curtailment of blood supply, Some of the conditions like ischemia, associated with this type of trouble, can be detected in time and further complications like angina attack or infraction heart could be avoided.

The exchange of ions across the cell wall and transmissive action of the nerves:

As has been explained earlier when electricity flows into the body from an electro stimulator, an impulse (or action potential) is evoked in the nerve cells, which is transmitted to the spinal cord and brain, causing a sensation or perception. Even though electricity is not sent by the device, there are many special nerve cells(receptors) which can sense pain, heat, light, sound, taste and so on existing on the surface of body in the eyes, ears, tongue, nose, skin and other parts. When each receptor gets various kinds of simulation, the electric impulse is evoked and transmitted to the sensory nerves, reaching the brain through the spinal cord which causes a sensation or perception. On the other hand, from the brain and spinal cord, impulses are transmitted retrogressively through another kind of nerve (called motor nerve) to each part of the body's muscles so that arms, legs, eyes and every organ work well. These impulses of the nerves are generated through the instantaneous action of the ions moving between the inside and outside the nerve cells; the nerve fibres then transmit this action. When a nerve is not active (a state of repose) there are many kalium ions (K^+) distributed inside it and the nerve fibre. Many natrum ions (Na^+) are distributed outside the cells. This is the same as the above mentioned pattern for cells in general when various kinds of stimulations (for example, pain, heat, light, sound or taste) are given to the nerves from outside the body. The nerves react to them, the character of the walls of nerve cells changes and instantaneously numerous natrum ions rush into the nerve cells. In the meantime kalium ions are going out.

At this very moment the body loses its electrical balance and the voltage(impulse, action potential) rises instantly, which is transmitted in a wave-form to the spinal cord through the nerve fibres then to the brain where it courses various perceptions(like pain, heat, *etc.*) Like this, by being given some kind of stimulation the character of walls of nerve cells changes with outer and inner ions exchanging places. Thus arises the action potential. We call this phenomenon the "excitement" of a nerve. After one excitement is finished, natrum comes back in its position. The movement of the ions in the retroactive stage is caused by the nerve cells having spent their natural energy in taking in and pushing out these ions by force. This action is called "active transport". Thus the nerve cells generate electricity by the movement of ions between nerve fibres; this movement is what is transmitted.

Now let us took more closely at the process of how nerve transmission takes place when your finger touches a hot pot. By reflex, you pull your finger away from it instantly. In the first place the moment your finger touches the hot pot, the heat is transmitted to the heat receptor nerves on the skin of the finger. These nerve cells get excited and generate their impulses which are conveyed to the nerve cells on the dorsal of the spinal cord through sensory nerve fibre. Some parts of this impulse is also transferred to the nerve cells on the frontal side of the spinal cord (this is called the transmission of nerves by "reflection"). At the same time, the other part of the electric impulse which has been transmitted to the nerve cells on the dorsal of the spinal cord from the tip of the finger is transmitted to the sensory nerves and goes up to the brain, then we get the perception of "heat". This perception is memorised in our mind so that we become careful before touching a hot pot again. We can say that reflection is an unconscious protective function and perception is a learning function for future use. Electricity in the living body is involved directly with these functions. So people should know that electricity and magnetism go hand in hand.

From the above preceding, the following has been laid down for the practical use of the law of polarity:

(a) The left half of head, face, chest neck and back is positive and the upper side of the arms and legs is also positive.

(b) The right half of head, face, neck, chest and back is negative and inter side of the arms and legs is also negative.

(c) If a line is drawn along the centre of the human body parts on it are alternately positive and negative.

We know very well energy neither created nor destroyed. It only changes its position. The electrical energy generated inside the body and changes its surface unitedly from cellular bed to organic stage. Such electrically charged organ possesses a magnetic field, on which field the maintenance of its polarity started from the cellular bed. Effects of magnetic energy on the body seen on two opposite directions :

1. The contractile effect and

2. The relapsing effect.

Such contraction and relapse begins from cellular bed and the two principal works conducted inside the body, the work of absorption and excretion are directly depend upon it. Muscular contraction in an organ helps to drain the unabsorbed or waste particles from the body surface through its natural drainage system and relapse receives the arterial flow which has been empowered by the heart to enriched by nutrients. It is a continuous phenomenon occurring inside the body for its growth and maintenance. It is the point to mention here that some morbid affections of the positive parts can be removed by the external application of negative electricities and of the negative parts by positive electricities.

21

Inner Healing

Some of the most powerful and important techniques for healing the body, mind, and spirit are detailed below. For people who are trying to heal a debilitating chronic illness, the use and eventual practice of one or more of these Inner Healing techniques can be a very important part of a healing program. Diet, stress reduction, avoidance of toxic substances, *etc.* are extremely useful and can provide enormous healing benefits.

Introduction

Like many powerful healing and transformational tools, body-oriented techniques focus on reconnecting, embracing and integrating all aspects of ourselves. These aspects include physical experiences (*e.g.*, chronic disease), psychological/emotional experiences (some of which may not be fully resolved), and possibly energetic or spiritual experiences. Such a process, while difficult on occassion, can be tremendously healing.

Body-oriented healing and transformation techniques primarily use experiences in the body as a way to begin the reconnecting, embracing, and integration process. The movement of energy and the experience of emotions can play a very important role, but experiences in the body, and its structure and movement are a common starting point for the processes.

Physical and psychological healing benefits can be significant. This is nothing like looking for a supplement or herb to heal a particular ailment, but rather a process that can help promote deep "to the core" healing (especially when used in conjunction with other holistic healing techniques. But like most things in the world, such a process is not for everybody. It is a process, and as such, there will be times of significant change as well as some times of uneven change/healing or frustration. Persons with a deep desire to heal or transform and with at least a small sense of adventure are best suited to this process.

Working with a practitioner/teacher on a semi-regular basis and attending workshops whenever possible is the best way to pursue these healing and transformational processes (if onsite, intensive programs are not possible). While individual sessions can be very healing, emerging oneself regularly in the processes can help keep the healing and transformation process moving.

Holotropic Breathwork

Holotropic Breathwork helps the breather go into a nonordinary state of consciousness. The use of nonordinary states of consciousness are seen in almost every culture. In such states, healing and reconnecting with various aspects of life are possible. One might experience a release of physical or emotional tension, deep relaxation, joy, or other experiences including transpersonal experiences.

According to experts, there are fourteen aspects of a Holotropic Breathwork session which work together to help create a successful session:

Accelerated breathing

Evocative music

Trained and experienced facilitation

A safe setting or container for the work

Partnership of participants

Preparatory information

Focused energy release work

Fasting at least four hours before the session

Open body position of the breather

Guided relaxation

Mandala drawing

Group sharing

Community support

Silence during the workshop

The training for the breathwork facilitators is fairly extensive. It requires hours of preliminary work as a participant of Holotropic Breathwork sessions (breather and sitter). Residential theoretical and experiential modules and private consultation are also required. Finally, a two-week training seminar is required for certification. This makes me feel confident about the dedication and training that the Holotropic Breathwork facilitators have.

Those participated in such programs have found the focused energy release work to be extremely helpful when it feels like the energy/feelings are "stuck" in a certain area of the body and the breathing and music cannot seem to change this situation. Such bodywork has, at times, helped me feel much better and more relaxed after the session.

It is advisable to do the process in a small group setting. Openly sharing your own process with others in a very non judgemental and caring atmosphere can be very healing. It can also be a very good way to meet others who are using holistic healing techniques.

The evocative music(especially Eastern Classical either vocal or instrumental) is a big help for going into a non ordinary state of consciousness. We trust this makes it possible to more easily get in touch with other aspects of our story such as the birth process and transpersonal experiences.

It can be slightly easier to find a breathwork facilitator with Holotropic Breathwork because it is relatively popular (as breathwork techniques go), there are a lot of trained faclitators, and can be easily located in your area.

Integrative Breathwork

Integrative Breathwork is very much like Holotropic Breathwork in that the breather goes into a non ordinary state of consciousness where some aspect of healing and transformation can occur. It seems that the main difference is that other therapies and holistic healing techniques(*e.g.*, yoga, bodywork, meditation, traditional psychology techniques, nutrition counseling, shadow work, Twelve Step meetings, *etc.*) are used along with the breathwork to help people integrate their breathwork experiences.

Radiance Breathwork

Radiance Breathwork is another healing and transformational process in which all of the tools and techniques used work together to promote healing and deep transformation.

The program includes Breathwork and a short healing breathwork/presencing program, Movement Therapy, and Body-Centered Psychotherapy. One may find it difficult to understand how a process which includes "psychotherapy" would be useful for a person with a chronic physical disease such as asthma or arthritis. These techniques are not used to treat those diseases, but it has my experience and the experience of many others that the deep healing which can come from such a program can have a significant positive impact on chronic physical diseases.

The Healing Journey is into the Body, Not Out of It

In our years as body-oriented psychotherapists, we have been privileged to be with thousands of courageous people as they have journeyed through deep and life-changing transformations. Although these changes have been infinitely and intricately different, they have had one element in common: They were accomplished by journeying into the body, not out of it. Each person tuned in to his or her bodily experience, amplified it through breath and movement, and flowed with the emerging waves of energy. The results have often been miraculous, and always wondrous to behold.

Vivation

Vivation is different than the other breathwork techniques since it is not a "cathartic breathwork" meaning that there is generally not a huge release of emotions. It tends to be gentler than the other breathworks. There may be crying at times, and in some cases physical vocalization, but it is not as common as in the other breathwork techniques.

Instead of a release of emotional energy, Vivation Professionals helps the client focus on inner feelings using some very simple, yet effective techniques and then lovingly accepting and embracing these feelings. Often what is felt emotionally physically are surpressed feelings which cause energy blockages within the body. (Such energy blockages can contribute significantly to the developement of a number of chronic and some acute diseases.) By using the Vivation process to embrace and integrating the feelings, the energy is released and changed. As the session continues, more and more feelings are felt and integrated leading to the deeper healing of the body, mind, and spirit.

The focus is not on "getting rid" of the feeling or "bad" emotion, but on deeply feeling, embracing, and integrating it, thereby transforming the energy around that supressed feeling. This type of process can lead to a more permanent healing and transformation as the emphasis is put on activating the feelings, feeling them deeply, and then significant emphasis is put on integrating the feelings.

Vivation is more gentle than the other breathwork techniques (usually) and can therefore be used by nearly everyone.

There is a emphasis on the eventual learning of the Vivation process on your own so that you can exchange sessions with other experienced persons. There is also a goal to use the Vivation process to your benefit at times during your daily activities. This is similar to the way practitioners of mindfulness meditation sometimes practice during daily life to enhance their wellbeing and spiritual growth. The facilitators are clear that it is very important to get a significant number of sessions to practice feeling, embracing and integrating before trying to do the process on your own.

22

Iridology

Iridology and Iris Diagnostic

The first recollections about iridology can be found in the book of the Old Testament. It states, "The eyes are the projector of the body. If your eyes are bright and shiny and healthy, the whole body will be full of light and health" Matthew, chapter 6, verse 22.

Iridology has its place in this final category of alternative methods of diagnostics and treatment – a method of identification of abnormal processes or physical changes in different organs and tissues of the body that was invented entirely by Ignatz Peczely, a Hungarian medical doctor who first issued a scientific commentary about his ideas in 1893. According to the old story, Peczely as a youngster found an owl with a fractured leg. When he took the owl home he observed a noticeable black line in the iris of the owl's eye on the same side as broken leg. Over the next few months he nursed the bird back to health. Before releasing the bird into the wild Peczely examined the bird for the last time. It was then that Peczely realized that the mark in the iris had disappeared, and was replaced by shabby white elements. From this distinct incident Peczely discovered the concept of iridology.

Peczely's hypothesis was that the iris relates to the rest of the body in some way, and therefore the marks of color in the iris reflect the condition of overall wellbeing of the different organs of the body. This practice of diagnosis and treatment is named the homunculus methodology and it's based on the idea that one organ of the body maps the other parts of the same body, including organs and systems.

Ignatz Peczely after his initial observations, he opined that there is connection between the iris and different organs and parts of the body. Years after, anatomists would revealed the essential mechanism of these links – a vast system of interconnections between the iris, the body and different organs. Additional exploration developed by several iris practitioners and researchers discovered more techniques to utilize this interesting aspect of this holistic science. The current reputation of iridology, particularly in the United States is related to

Bernard Jensen, a chiropractor, who used iris diagnostics techniques in his everyday practice to better pinpoint the most affected organs or parts of the body. He wrote the manuscript The Science and Practice of Iridology in 1952. Iridology, or iris diagnostics, is practiced by holistic medicine practitioners, Acupuncturists, Reiki healers, Reflexologists, Traditional Chinese Medicine practitioners, Chiropractors and Naturopaths.

Application of Iridology

Often as a result of the iris evaluation the practitioner will be able to prescribe the appropriate dietary supplement and herbal medications.

So, how does the iris disclose altering conditions of the parts and organs of the body?

As it was mentioned above, every organ of the body is mapped on the iris. Various spots, dots, and discoloration in the specific areas of the iris are signs of pathological processes that happened in particular organs and or parts of the body. Some of these signs could be normal, so it's important not only to "photograph" the iris, but also to talk to the patient and discuss his/her condition. For example, bright red dot located on the iris in the area of uterus may symbolize cancer of this organ, but it may also be present during a short time period which may mean a minor disease. There is little difference in the nature of this dot but an experienced iridologist will probably see it. When the bright red dot represents normal menstruation it has very neat edge, however if this edge looks broken – it is usually a sign of a atypical condition.

Based on principals of iridology, a practitioner can discover congenital and developing health problems and diseases, in present health state, general health, and the condition of every specific organ.

Because there are many nerve endings in the eye, all linked to the body's organs, iridologists assert that examination of the eye, especially the iris, its colour and markings, can indicate problems in other parts of the body and reveal weaknesses and strengths in general health, such as a tendency towards a particular disease.

Iridology is not a restorative therapy; it doesn't cure and it doesn't pinpoint particular ailments Iridologist identifies weaknesses or over activity in the body so that future problems can be avoided. For example, if the pancreas is seen to be underactive, it could indicate a condition like diabetes. Iridology also recognises where toxins and inflammation could be, as well as certain chemical deficiencies. to determine your general state of health and constitution.

For example, the area immediately around the pupil relates to the stomach and the area around that represents the intestines. It is said that the pattern of the iris is like a genetic map of the body, showing inherited traits and pinpointing weaknesses. Changes in this pattern can reflect changes in the body.

Sclerology is a non-invasive alternative medicine practice in which the sclera of the eye is examined for information about a patient's systemic health.

As an alternative health modality we typically combine this revealing technique with Iridology Assessments. Sclerology can be traced back over 1000 years.

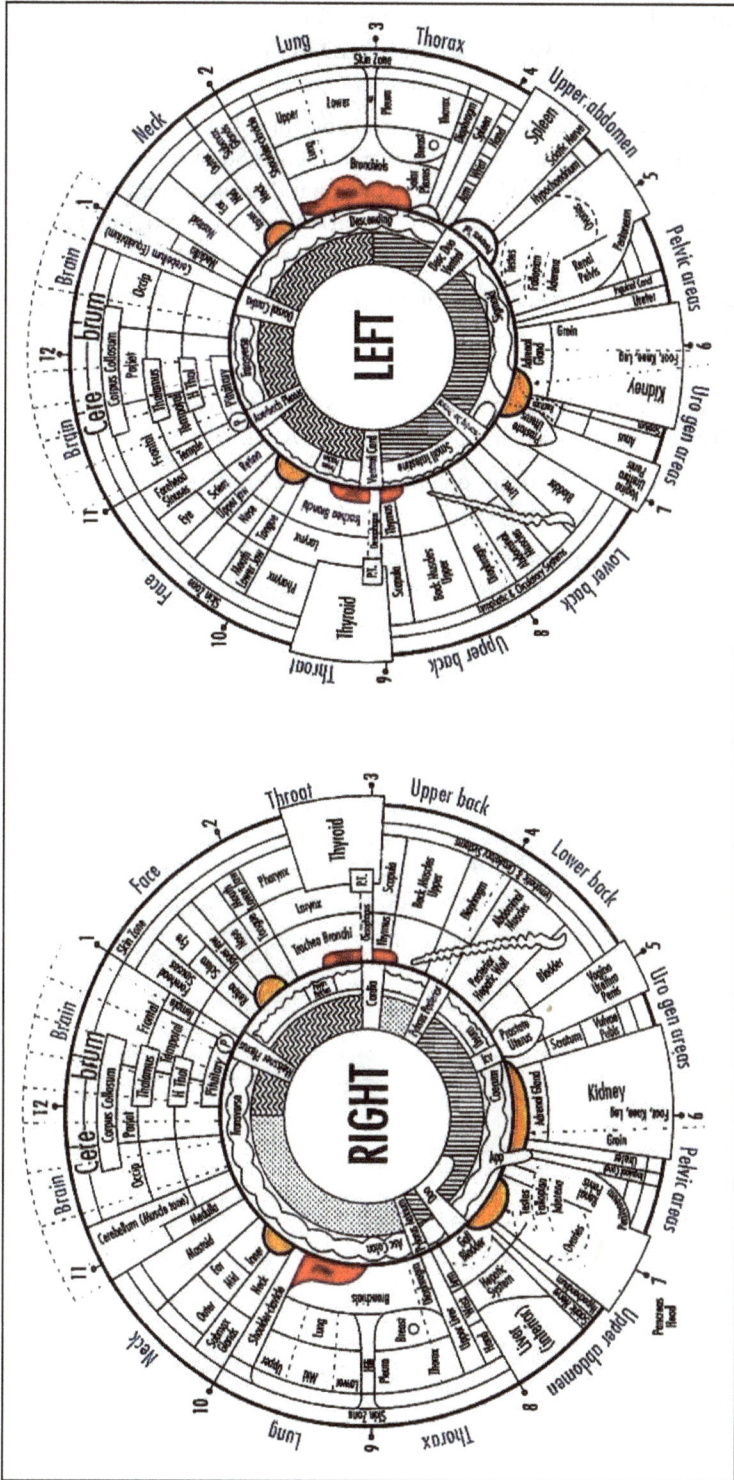

The Chinese have used the sclera markings for at least 1000 years. Chinese medical texts, Secrets of the Bronze Man, written 1046 AD in the Song Dynasty dating from 1046 AD show drawings of the sclera with some sclera markings.

The Native Americans also have a long history of using Sclerology though they did not keep or there no longer remains any written evidence. Sclerology may be defined as the science of the markings in the whites of the eyes that describe conditions of health. To be more precise, Sclerology is the study of those markings and colorings in and on the sclera layers vs. the signs within the bulbar sclera conjunctiva membrane.

The red lines and other markings and colorings in the whites of the eyes correspond to real conditions in the body. Their size and shape plus various other aspects and shadings describe health qualities in the body.

It shows where the highest stress and congestion are located, the originating causes, how much is emotionally based and how much is physical.

Sclerology is therefore a means of looking into a person's eyes to see what is happening with his/her health, to get a sense of what is most important.

Alternative health sciences like Sclerology are attracting increased scientific study and some have been studied by western researchers for over 150 years.

|23|
Meditation

Meditation - Mind Body Therapy

More and more doctors are prescribing meditation as a way to lower blood pressure, improve exercise performance in people with angina, help people with asthma breathe easier, relieve insomnia and generally relax the everyday stresses of life. Meditation is a safe and simple way to balance a person's physical, emotional, and mental states. It is simple; but can benefit everybody.

Meditation is not just for yoga masters sitting cross-legged on mountaintops in the Himalayas. It's a flexible approach to coping with stress, anxiety, many medical conditions and the day-to-day "static" that robs us of inner peace. Today, the Pittsburgh International Airport boasts a large meditation room featuring a quiet ambiance, comfortable furniture and paintings of clouds. What better place than one of the nation's largest, busiest airports for a refuge from all the hustle and bustle?

The Taoist sage Chuang-tzu referred to meditation, which the Chinese simply call 'sitting still, doing nothing', as 'mental fasting'. Just as physical fasting purifies the essences of the body by withdrawing all external input of food, so the 'mental fasting' of meditation purifies the mind and restores the spirit's primal powers by withdrawing all distracting thoughts and disturbing emotions from the mind. In both physical and mental fasting, the cleansing and purifying processes are natural and automatic, but the precondition for triggering this process of self-rejuvenation is emptying body and mind of all input for a fixed number of minutes or days. Taoists believe that only by 'sitting still, doing nothing' can we muster sufficient mental clarity to focus fully on the difficult task of taming and training the two aspects of temporal mind that govern our lives - the mind of emotion and the mind of intent.

Introduction

The use of Meditation for healing is not new. Meditative techniques are the product of diverse cultures and peoples around the world. It has been rooted in the traditions of the world's great religions. In fact, practically all religious groups practice meditation in one form or another. The value of Meditation to alleviate suffering and promote healing has been known and practiced for thousands of years.

Of the religions that use meditation, perhaps Buddhism, practiced widely in eastern and central Asia, is the best known. To Buddhists, the practice of meditation is essential for the cultivation of wisdom and compassion and for understanding reality. Buddhists believe that our ordinary consciousness is both limited and limiting. Meditation makes it possible to live life to the full spectrum of our conscious and unconscious possibilities.

In spite of its rich history and traditions, it is only during the past three decades that scientific study has focused on the clinical effects of meditation on health. During the 1960s, reports reached the West of yogis and meditation masters in India who could perform extraordinary feats of bodily control and altered states of consciousness. These reports captured the interest of Western researchers studying self-regulation and the possibility of voluntary control over the autonomic nervous system. At the same time, new refinements in scientific instrumentation made it possible to duplicate and substantiate some of these reports at medical research institutes. Health care professionals who were often dissatisfied with the side effects of drug treatments for stress-related disorders embraced meditation as a valuable tool for stress reduction, and today both patients and physicians enjoy the health benefits of regular meditation practice.

Healing Power of Meditation

Research has shown that Meditation can contribute to an individual's psychological and physiological well-being. This is accomplished as Meditation brings the brainwave pattern into an alpha state, which is a level of consciousness that promotes the healing state.

As discussed in the section "How Meditation Work?", there is scientific evidence that Meditation can reduce blood pressure and relieve pain and stress. When used in combination with biofeedback, Meditation enhances the effectiveness of biofeedback.

Benefits of Meditation

Physical Benefits

Deep rest-as measured by decreased metabolic rate, lower heart rate, and reduced work load of the heart.

Lowered levels of cortisol and lactate-two chemicals associated with stress.

Reduction of free radicals- unstable oxygen molecules that can cause tissue damage. They are now thought to be a major factor in aging and in many diseases.

Decreased high blood pressure.

Higher skin resistance. Low skin resistance is correlated with higher stress and anxiety levels.

Drop in cholesterol levels. High cholesterol is associated with cardiovascular disease.

Improved flow of air to the lungs resulting in easier breathing. This has been very helpful to asthma patients.

Younger biological age. On standard measures of aging, long-term Transcendental Meditation (TM) practitioners (more than five years) measured 12 years younger than their chronological age.

Higher levels of DHEAS in the elderly. An additional sign of youthfulness through Transcendental Meditation (TM); lower levels of DHEAS are associated with aging.

Psychological Benefits

Increased brain wave coherence. Harmony of brain wave activity in different parts of the brain is associated with greater creativity, improved moral reasoning, and higher IQ.

Decreased anxiety.

Decreased depression.

Decreased irritability and moodiness.

Improved learning ability and memory.

Increased self-actualization.

Increased feelings of vitality and rejuvenation.

Increased happiness.

Increased emotional stability.

How Meditation Works

Studies have shown that meditation (in particular, research on Transcendental Meditation, a popular form of meditation practiced in the West for the past thirty years), can bring about a healthy state of relaxation by causing a generalized reduction in multiple physiological and biochemical markers, such as decreased heart rate, decreased respiration rate, decreased plasma cortisol (a major stress hormone), decreased pulse rate, and increased EEG (electroencephalogram) alpha, a brain wave associated with relaxation. Research conducted by R. Keith Wallace at U.C.L.A. on Transcendental Meditation, revealed that during meditation, the body gains a state of profound rest. At the same time, the brain and mind become more alert, indicating a state of restful alertness. Studies show that after TM, reactions are faster, creativity greater, and comprehension broader.

A laboratory study of practitioners of Maharishi Mahesh Yogi's transcendental meditation (TM), carried out by Benson and Wallace at Harvard Medical School towards the end of the 1960s, provided the first detailed knowledge of the many physiological changes that go with meditation.

Some of the meditators, whose ages ranged from seventeen to forty-one, had been meditating only a few weeks, others for several years. All recorded changes associated with deep relaxation.

The fall in metabolic rate was the most striking discovery. This was indicated by a dramatic drop in oxygen consumption within a few minutes of starting meditation. Consumption fell by up to twenty per cent below the normal level; below that experienced even in deep sleep. Meditators took on average two breaths less and one litre less air per minute. The meditators' heart rate was several beats less per minute.

During meditation, blood pressure stayed at 'low levels', but fell markedly in persons starting meditation with abnormally high levels.

The meditators' skin resistance to an electrical current was measured. A fall in skin resistance is characteristic of anxiety and tension states; a rise indicates increased muscle relaxation. The finding was that though meditation is primarily a mental technique, it soon brings significantly improved muscle relaxation.

Meditation reduces activity in the nervous system. The parasympathetic branch of the autonomic or involuntary nervous system predominates. This is the branch responsible for calming us.

During anxiety and tension states there is a rise in the level of lactate in the blood. Lactate is a substance produced by metabolism in the skeletal muscles. During meditation blood lactate levels decreased at a rate four times faster than the rate of decrease in non-meditators resting lying on their backs or in the meditators themselves in pre-meditation resting.

The likely reason for the dramatic reduction in lactate production by meditators was indicated when further studies of meditators showed an increased blood flow during. Benson and Wallace found that there was a thirty-two per cent increase in forearm blood flow. Lactate production in the body is mainly in skeletal muscle tissue; during meditation the faster circulation brings a faster delivery of oxygen to the muscles and less lactate is produced.

The two investigators summed up the state produced by their meditating subjects as wakeful and hypometabolic. The physiological changes were different in many ways from those found in sleeping people or those in hypnotic trance states. Meditation, they said, produces 'a complex of responses that marks a highly relaxed state'. Moreover, the pattern of changes they observed in meditators suggested an integrated response, mediated by the central nervous system.

"Through meditation we can learn to access the relaxation response (the physiological response elicited by meditation) and to be aware of the mind and the way our attitudes produce stress," says Dr. Borysenko, author of 'Minding the Body, Mending the Mind". "In addition, by quieting the mind, meditation can also put one in touch with the inner physician, allowing the body's own inner wisdom to be heard."

Taoists believe that the mind of emotions is governed by the Fire energy of the heart. When your emotions are not controlled, the fire energy of the heart flares upwards, wastefully burning up energy and clouding the mind. The mind of intent, or willpower, is controlled by the Water energy of the kidneys. When unattended, the water energy flows down and out through the sexual organs, depleting essence and energy and weakening the spirit. Taoists believe that when you are 'sitting still, doing nothing', as in meditation, the flow of Fire and Water are reversed: Water energy from the kidneys and sacrum is drawn up to the head via the Central and Governing channels, while emotional Fire energy from the heart is drawn down into the Lower Elixir Field in the abdomen, where it is refined and transformed and enters general circulation through the energy channels. On the spiritual/mental level, this internal energy alchemy enables the mind of intent (Water) to exert a calming, cooling, controlling influence over the mind of emotion (Fire).

Taoist classical texts often deal with symbolisms and abstractions. Microcosmic Orbit Meditation works by visualizing the energy channeling through different parts of the human anatomy. Hence it is of interest to find out the anatomical description of the human body from the Taoist perspective as it relates the channeling of energy or vital essence. This description is from a translation of an eleventh century manuscript written by Chang Po-tuan by Thomas Cleary.

Taoist texts speak of the 'medicinal elements' of internal alchemy, using metaphors such as 'red lead', 'black mercury', 'cinnabar', 'white snow', 'green dragon', 'white tiger', 'sun rays', and 'moon beams'. These metaphors refer to various aspects and elements of the Three Treasures - essence, energy, and spirit - which are the only real elements of internal alchemy.

The vital junctions, or 'passes', used in circulating internal energy are also given mysterious names, such as the 'Yellow Chamber', 'Red Cauldron', 'Mysterious Pass', 'Lead Furnace', 'Flower Pond', 'Dragon Lair', and 'Vermilion Palace'. These colorful names denote the invisible but highly functional points inside and along the surface of the body, where energy collects, transforms, and enters various channels for circulation. Taoist internal alchemy is actually a highly scientific method of harnessing, controlling, conserving, converting, and circulating essence and energy under the guidance of spirit in order to replace depletion with accretion, reverse disintegration with integration, and counteract degeneration with regeneration.

A term of particular importance in Taoist internal alchemy is the 'firing process', which has nothing at all to do with fire. The firing process refers to breathing, which acts as a bellows to gently fan the 'fire' of energy in the 'cauldrons' of the Elixir Fields.

Taoists believe that the human body has three posterior passes and three anterior passes. The three posterior passes are in the coccyx, at the base of the spine; in the mid-spine, where the ribs join the spine; and at the back of the brain.

The pass in the coccyx, at the bottom of the spine, connects with the channels of the genital organs. From this pass ascends the spinal cord, which is called the Zen Valley, or the Yellow River, or the Waterwheel Course, or the Mountain Range up to the Court of Heaven, or the Ladder up to Heaven.

This is the road by which positive energy ascends; it goes right up the point opposite the center of the chest, the pass of the enclosed spine, where the ribs join in back, then it goes straight up to the back of the brain, which is called the Pass of the Jade Pillow.

The three anterior passes are called the Nirvana Center, the Earth Pot, and the Ocean of Energy. The Nirvana Center is the so-called upper elixir field. It is a spherical opening 1.3 inches in diameter and is the repository of the spirit. That opening is three inches behind the center of the eyebrows, right in the middle.

The space between the eyebrows is called the Celestial Eye. The space one inch inward is called the Bright Hall. The space one inch farther in is called the Hidden Chamber. One inch farther in from that is the Nirvana Center.

The windpipe has twelve sections and is called the Multi-storied Tower; it goes to the openings in the lungs, and reaches the heart. Below the heart is an opening called the Crimson

Chamber, where the dragon and tiger mate. Another 3.6 inches directly below that is what is called the Earth Pot, which is the Yellow Court, the middle elixir field.

The umbilical opening is called the Door of Life. It has seven channels connecting with the genitals. The leaking of sexual energy takes place through these channels. Behind the navel and in front of the kidneys, right in the middle, is the place called the Crescent Moon jar, or the Ocean of Energy. And 1.3 inches below that is what is called the Flower Pond, or the lower elixir field. This is where vitality is stored, and it is the place where the medicine is gathered.

Stay Healthy - Learn To Meditate

Wondering how people who live to be 100 with a great quality of life do it? researcher from Harvard George Vaillant found out just what centurions do. They cultivate a sense of peace, well-being and maintain a positive attitude. How? Here's one of their biggest secrets: meditation. What's ironic about meditation is, it has just become known in the West as a healing technique, but it has been practiced for ages in the East. So in my quest to give you easy sensible ways to purify your energy, let's begin by defining meditation, then you will realise how easy it is to apply to your daily routine.

Meditation Defined

Meditational exercises primarily use the experience of the body and thought as a means to reconnect with the environment and its healing power. Meditation, when practiced frequently, has been proven to promote inner peace and wellness. Meditation is also a mental practice in which the mind is directed to one area, often the breath. It draws its energy from the human connection to nature and creates a sense of unity or one-ness with it.

This unity has been shown to increase communication with the spirit of the body. It has also been known to allow positive thoughts in and to stimulate positive physiological and psychological effects. Meditation techniques are easy to learn and can easily be incorporated into any lifestyle. If practiced regularly, meditation will bring balance to your body and mind.

General benefits of meditation and breathing exercises include:

- ☆ Deep inner peace
- ☆ Improved self-esteem
- ☆ Increased creativity
- ☆ Physical health/healing
- ☆ Reduced medical care
- ☆ Slowing/reversal of aging
- ☆ Reversing of heart disease
- ☆ Stimulation of the body's immune system
- ☆ Reduced stress

Here are a couple of meditational exercises :

Meditative Grounding Exercise

☆ Sit with your legs crossed in a comfortable (Indian-style) position with your hands relaxed on you lap. Close your eyes and imagine a beam of light dropping from the base of your spine through the earth and connecting you to its center.

☆ Allow this beam of light to expand in width until it is wider than your own body and envelopes it. This is your personal space.

This exercise places you totally in your body and reminds you that you are anchored to the earth. Remember, the more grounded you are, the more aware you are. Sense the presence of your higher self: listen to its voice.

Energy Cleaning Exercise

Now that you are grounded, it's important that you cleanse this personal space. Often we collect other people's energies and are not aware of it. We do this both through interaction with others and basic activities of daily living.

☆ To remove all foreign energies from your space, imagine holding a brush and sweeping away the debris.

☆ Allow the debris to fall to the ground and become washed away. Let the light from the previous exercise envelop your body and spread its healing energy to the edge of your space, forming a protective force field around you.

Cleaning out the area surrounding your body will keep you grounded, define your personal boundaries and declare your space. Then choose who and what you wish to enter you space, keeping disease and illness out.

Breathing Exercise

☆ Follow your breath as you slowly inhale through your nose and exhale through your mouth. Count with each exhale until you reach 10 then begin again at one.

☆ If you find yourself past 10, acknowledge this and begin again at one from wherever you are.

☆ Imagine your body's cells being replaced with fresh, pure oxygen and positive healing energy from this power source. Picture yourself exhaling old cells, stress, illness and worries.

☆ Let your thoughts pass through your mind like drifting clouds. Let them in and gently let them pass through. If the mind should harbor a negative thought, refocus on the breath.

☆ Thank any persistent negative thoughts for coming into your mind then gently let them go.

☆ Listen only for the positive, pronounced voice; the voice of your body.

Retrieve Your Energy Exercise

Since foreign energy often resides in your space, let's also assume that you leave energy in other places. After completing the preceding exercises it is necessary to re-energize and call energy back.

☆ Imagine you have an energy magnet used to attract your energy back to you. Visualize energy flowing back to you, filling your body with light, health and empowerment.

☆ Allow a few minutes for the process.

After a few sessions you will feel energized from inside also that your body and mind are becoming calmer and clearer.

24

Magnetotherapy

Magnetotherapy is the art of healing through application of magnets. The magnet relieves distressing symptoms of many diseases and cures them without medicine. It is a forgotten ancient system of treatment which is being revived now in India as well as in other countries.

Main Principles of Magnetotherapy

1. A disease is the results of imbalance between the various electromagnetic forces acting inside the body. Balance can be restored with creation of artificial magnetic fields in and around the body.

2. Properties of water change when it is brought in contact with magnetic fields. Blood is also similarly influenced and gets ionised. It flows more easily. Improving circulation and avoiding clotting. Magnetic emanations exercise a positive effect on the haemoglobin of blood and remove excess of Calcium, Cholesterol, *etc.*

3. Magnetic emanations pass through the tissues and secondary currents are induced. These currents produce impacting heat which reduces pains and swellings.

4. Magnetic treatment works by reviving, reforming and promoting growth of calls, rejuvenating tissues and increasing the number of sound blood corpuscles.

5. Functions of Autonomic nerves are normalised and the internal organs controlled by them regain their proper functions. It is also beneficial for mental retardation and weak memory.

6. It helps maintain homeostasis of the body. One feels in full vigour and vitality after the treatment.

Mode of Application of Magnets

Each magnet has two poles, North and South. North Pole retards the growth of germs and bacteria and is applied on boils, eczema or infection, *etc.* South pole removes pain and swellings, *etc.* Only one pole is applied on the diseased part of painful spot if the disease is localized in small portion. It extends to greater parts of whole body, both the pollen are applied in the following manner.

1. For ailments of the upper half of the body, magnets are used under both the palms and for the ailments of the lower half of the body, magnets are applied under both the soles.

2. If the magnets are applied on right and left sides. North pole on the right side an south pole on left side. If on upper and lower sides. North pole on upper side and South on lower side, if on front and back sides. North pole on front side South pole on backside.

North pole is marked 'N' and South pole 'S'.

The sides marked 'N' and 'S' should remain in constant touch with the body where they are applied, for 10-15 minutes in one sitting. Generally, one silting in the morning is enough but another sitting may be taken in the evening, if necessary. But remember that during the application or high, medium and ceramic magnets, the face should preferably remain towards the west.

These are general instruction. Special cases may require special treatment.

Dos and Don'ts

1. Do not take cold eatable, drinks and bath for one hour after using strong-magnets.
2. Do not let magnets touch your watch.
3. Do not apply strong magnets for one hour after full meals.
4. Do not touch walls or ground when using magnets.
5. Do not apply strong magnet to head, heart and to pregnant women.
6. Keep a wooden plank below feet when applying magnets.
7. Do not clap strong magnets face to face.
8. Keep east-alloy magnets joined with a 'keeper' when not in use.

Results of Magnetotherapy

Very satisfactory results have been noticed in the following diseases:

Arthritis, asthma, backache, bed-wetting, boils, cervical spondylitis, diseases of bowels and uterus, eczema, flatulence, headache, injuries, pains of the body, rheumatism, scanty menses, toothache, insomnta, sprain, stiffness, swellings *etc.*

Magnetic Instruments

There are different specialized magnetic instruments. These are mentioned below with their common utility.

Backache and Belly Reducing Belt

The flat elastic belt is the latest achievement in the field of magnetic treatment and is applied successfully for Backache. Belly reduction and Sciatica. In its application on must remember that it should be applied on empty stomach or when there has been an interval of atleast two hours in the food already taken by the patient. Its application should be form 1 to 2 hours at one time and stopped when the complaint is over. But if the complaint is still persisting, it can be applied twice a day, morning and evening till there is complete relief. The application in different ailments, should be as follows:

(i) *Backache and Sciatica*: The belt should be wrapped around the waist.

(ii) *Belly Reduction*: The belt should be wrapped opposite to that of the method suggested for Backache.

Head Belt

Headache: The belt is very useful in all kinds of pains, like pain in any part of the head, one sided headaches, migraine, *etc.*

That belt has also been found useful in other conditions like insomnia, neurasthenia, mental fatigue and weakness, *etc.*

Spondylitis Belt

The flat rigid magnetic belt is successfully applied in the treatment of Spondylitis and Stiff Neck. This belt is studded with three sets of the magnetic poles at three different places, *i.e,* 2 sets about the requisite application. The curved portion of the belt should always remain on the throat below the lower jaw.

Knee Caps

In the treatment of knees, the knee-caps are successfully used in certain diseases like:

Gout, Rheumatism or Rheumatic pains especially in the lower parts of the legs, Arthritis, Osteoarthritis or Rheumatoid arthritis. Synoitis and other inflammatory conditions.

Wrist Band

A very useful magnetic ornament to maintain blood pressure. It normalizes both High Blood pressure (Hypertension) and the Low Blood Pressure (Hypotension). In case of former, the band should be worn in the right wrist. In case of later, it should be worn on the left wrist.

Tonsil Belt

This belt is studded with two sets of tonsil-like curved-shaped magnets, so much so that in case of tonsillitis, it is wrapped around the neck the first set being on one and the second set on the other tonsil.

Magnetic Necklace

This magnetic necklace is worn as general necklaces around the neck which helps in getting rid of:

1. Pains in any part of the neck and chest.

2. Stiffness of neck.

3. Respiratory difficulties.

4. Pain the gall-bladder.

5. Gall-stone

6. Palpitation of heart

This necklace, in addition, can help in relieving the pain of Cervical Spondylities and beauty maintenance in ladies.

Ceramic Magnets

These are curved low power magnets which are suitable for application to the delicate organs like eyes, nose, tonsils where curvature of the magnets is necessary. These types of magnets can be used in case of infants as well as diseases like insomnia tonsillitis, opthalmia, conjunctivitis, *etc.*

Eye Belt

The eye-belt has proved very useful in all types of eye-troubles, like dimness of vision, painful eyes, opthalmia, conjunctivitis, *etc.* The belt is appended with two separate sets of magnets according to the distance of the eyes.

Medium Power Magnets

The medium power magnets can be used for children as well as general application to the palms of hands and soles of feet, in case of children and for diseases like earache, toothache, *etc.* They are also useful for adults in certain affections of more tender parts of vital organs of the body.

High Power Magnets

These strong magnets are suitable for all adults for general application to the palms or soles. They are successfully applied in various acute and chronic diseases like paralysis, polio, spondylosis, sciatic, lumbago, gout, dysmennorrhoea and eczema. These magnets are also used for preparing the healing or magnetized water.

President Magnets

These strong magnets are suitable for adults, normally used in chronic diseases and for preparing the healing or the magnetized water in greater quality.

Premier Magnet

This magnet has both the poles, North pole on one side and South pollen on other side. This magnet has very deep penetrating power and give much better results when used with Electromagnet.

Electromagnet

It works on A.C. mains (230 volts). The combined use of premier magnet and electromagnet accelerates the curative effect and reduces the period of treatment of about one-fourth.

Magnetised Water

It can be prepared at home, keep two glass bottles, full of boiled and cooled water on the president or high power magnets for 12 hours, then mix the waters of the bottles and use as a medicine 3-4 times a day in quality of two ounces (50 ml.) at a time of adults and one ounce (25 ml.) for children. This water is found highly effective in all diseases whether relating to the nervous system, circulatory system, alimentary system, urinary system or genitor-urinary system. It greatly helps in throwing out the waste matters from the body.

25

Massage Therapy

Theory

Scientific massage, based upon the Swedish method, moves the blood towards the heart, thus helping in circulation. Basically, the blood leaves the heart to bring nutrition in various parts of the body. At first, there is much pressure, especially when its first leaves the heart. As the blood travels in the arteries throughout the body (arteries are the blood vessels that carry the blood from the heart to various parts of the body whereas the veins return the blood to the heart), the blood gradually enters into tiny blood vessels called capillaries, reaching tiny spaces in the body in need of nourishment. The blood then moves into the veinous system and travels through the veins the back to the heart.

By the time the blood gets into the veins, on the return trip, there is very little blood pressure left, as the pressure is reduced after passing through the tiny capillaries. The veinous blood is assisted in this movement back to the heart in two ways: first, by physical movements of the body muscles which squeeze the blood vessels and second, by Swedish massage, which helps to move the blood through the muscles and other tissues of the body.

In some individuals, physical movement may be limited due either to illness and weakness or to a weak heart. For example, an individual who has undergone major surgery would be quite weak for a considerable time after such surgery and daily Swedish massages would be of great help in returning the body to normalcy. Insurance companies who have to pay for each day of disability would be wise to encourage their insured customers with disability to obtain the services of a good technician skilled in Swedish massage to hasten recovery.

Sometimes with the blood moves too slowly, blood clots may form that may move to the heart or to the brain, block blood passages and cause death. Frequent Swedish-type massages would help prevent this from happening.

Massage Techniques

Massage starts with the right hand, then to the wrist, arm and shoulder, then to the back of the neck, the left shoulder, left arm, left wrist and left hand. It is then followed by massage of the left foot, left leg, right foot, right leg, chest, abdomen and back. For a one-hour massage, use 3 strokes for each movement, 1 stroke for each movement for a one-half hour massage. Be sure to practice all movements with someone until they can all be done automatically without need to refer to study material.

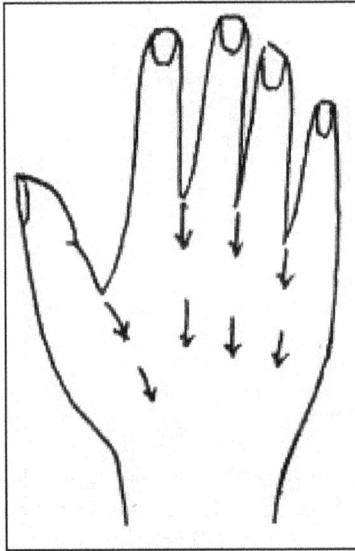

Hand Massage

The Hand

1. Warm up. General rubbing of hand and fingers. Until skilled, use a nourishing cream, oil or powder for lubrication.
2. Knead tips of thumb and fingers with small circles, moving to base of fingers.
3. Stroke from tip to base.
4. Repeat on each thumb and fingers.
5. Stroke with your thumbs from the base of fingers to the wrist, on back of hand.
6. Stroke with heel of hand.
7. Repeat (5) over palm of hand.
8. Stroke with heel of hand over palm.

The Wrist

After massaging the fingers, the back of the hand and the palm of the hand, the next step is to treat the wrist. If there are remains of an oily substance on the hand, it would be necessary to remove it by applying or rubbing some alcohol on the hand and then wiping everything off with a towel.

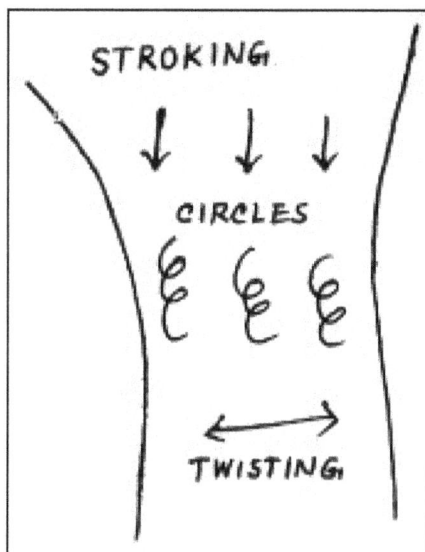

Wrist Massage

1. The first step in working on the wrist is to apply oil or lotion to the wrist. The thumbs are then used on the palm side of the wrist from the hand towards the arm in stroking movements for a distance of about three inches.

2. The thumbs are then used in little circular movements over the entire wrist area. This is followed by stroking again.

3. The fist of one hand is then wrapped around the wrist and a twisting movement applied. This is followed again by a stroking movement.

4. This is repealed on the opposite side of the wrist, the stroking, circles and twisting.

The Arm

Have someone with whom to practice. The ideal situation is for an experienced massage instructor to work on the student and then for the student to the work on the instructor to be checked.

1. In massaging the arm, first the oil or lotion to applied over the entire arm.

2. The patient is told to hold the arm straight up. The operator then grasps the wrists with both hands and strokes downwards to the shoulder. Bring the hands up and repeat the stroke several times. When bringing the hands up, do not squeeze but maintain light contact.

3. Twist the hands from the wrist down to the shoulder and twist back up. Repeat three times. Repeal the stroking.

4. Place the arm alongside the patients side. With the tips of your lingers, use small circles from the wrist to the shoulder and return. Do it 3 times then repeat the stroking.

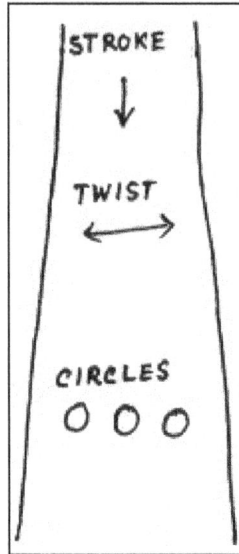

Arm Massage

5. Remove the oil by applying alcohol and then wiping it off. In some cases, the lotion may be of such a quality as to make it unnecessary to remove it, as when it is of the type that dries almost completely, leaving only a thin film is really good for the skin.

The Shoulder

1. Apply lubricant over the shoulder area.
2. Using fingertips, knead area with small circles.
3. Stroke from arm to neck over top and both sides.
4. Using both hands moving in opposite directions, twist over the shoulder area from arm to neck.
5. Stroke from arm to neck over top and both sides.

Back to Neck

1. Apply lubricant in back of neck area. This may be done while patient is on his or her back. There is usually sufficient space to work, due to the head lying on a pillow and there is therefore, a spare between the neck and the table or bed.
2. Stroke from hairline on back of neck with fingers of both hands towards the hack as far as space will permit.
3. Use small circles with fingers of both hands over entire area of back of neck.
4. Repeat stroking action.

Left Shoulder

Same instructions as for right shoulder.

Left Arm

Same instructions as for right arm.

Left Wrist

Same instructions as for right wrist.

Left Hand

Same instruction as for right hand.

Left Foot

1. Apply lubricant over entire foot.
2. Using both hands, circle ankles, heel and back of foot.
3. Stroke from toes, towards leg and form soles of feet towards leg.
4. Knead tip of big toe with small circles and work towards base of toe.
5. Stroke from tip of toe with fingers towards base of toe.
6. Repeat with other 4 toes, first kneading then stroking each.
7. Take both thumbs and stroke between all those to the ankle.
8. Take heel of hand and stroke over the area.
9. Small circles over the top of the foot.
10. Stroke with heel of hand.

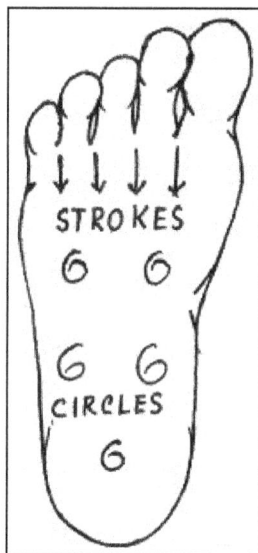

Feet Massage

11. Take both thumbs and stroke on sole of foot from toes (between toes) to middle of foot.

12. Stroke sole with heel of hand.

13. Small circles on sole of foot.

14. Stroke with heel of hand.

The Left Leg

The left knee should be raised from the table and the following steps taken:

1. Apply lubricant over both the upper and lower leg.

2. Stroke from the ankle to knee and form the knee to the hip.

3. Hold the knee with one hand and with the other hand, make small circles from the ankle to the knee, behind the knee as well as on top and continue with small circular movements over the entire upper leg.

4. Stroke from the ankle to knee and from the knee to the hip.

5. Staring to the ankle, use both hands, moving in opposite directions in twisting motions and move slowly to the knee, over the knee and then over the upper leg to the thigh and hip. Some therapists will sit gently on the toes of the foot to hold the leg.

6. Repeat stroking motion as previously. Stroking should be over all areas of the legs, at the sides and bottom as well as at the top.

7. Lower the leg to tin table.

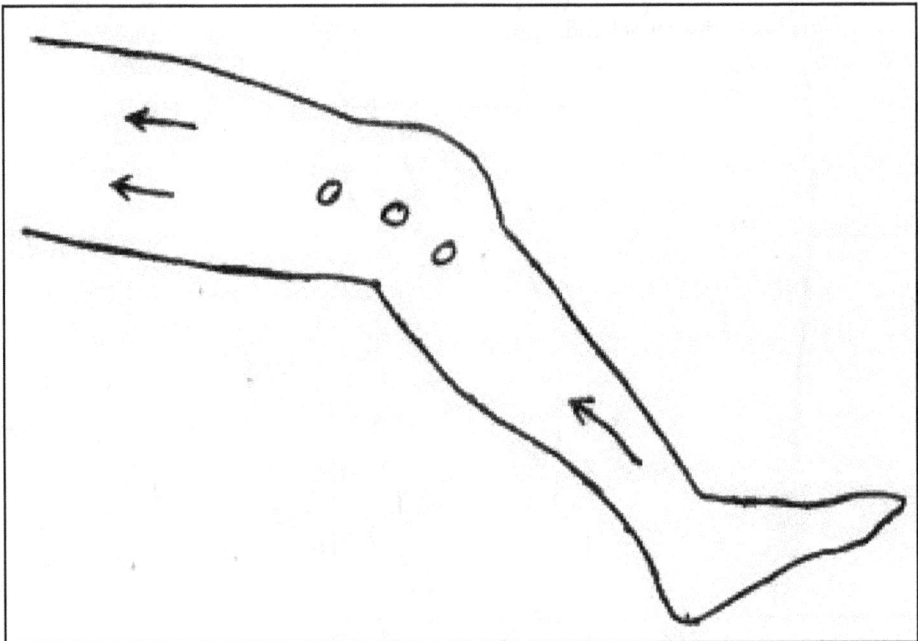

Leg Massage

The Right Foot

Repeat same instruction as for the left foot.

The Right Leg

Repeat same instructions as for the left leg.

The Chest

1. Apply lubricant over entire chest area.
2. Stroke, starting with both hands at the middle of the chest near the abdomen, moving upwards toward the neck, fingers pointed towards the neck, then both hands separate towards the sides, down the sides towards the abdomen and move fingers and hands back towards the middle. Repeat several times, covering the entire chest area.
3. Using the finger tips, use a circular movement around the entire chest area following the same path as during the stroking movement.
4. Repeat the stroking movement.

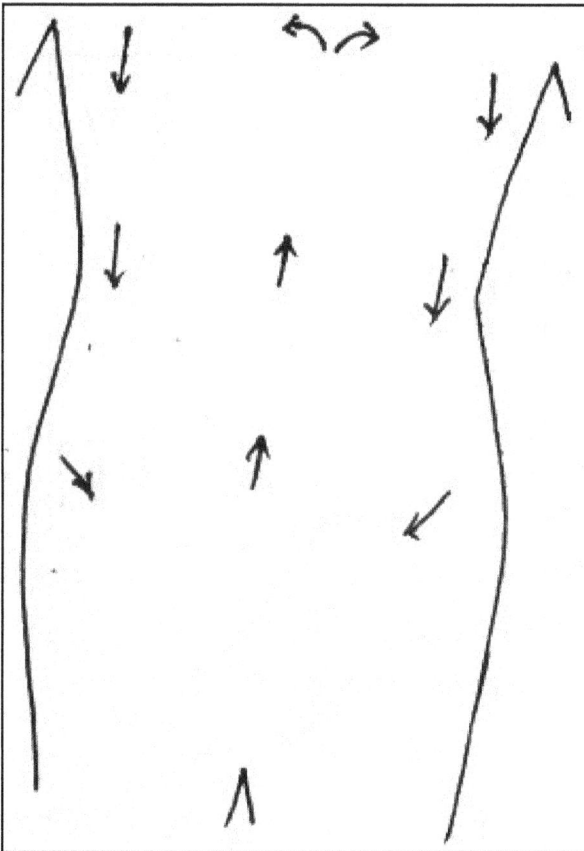

Chest Massage

Note: In female patients, if during massage, it should be noticed that any part of the breast gland has any lump or hardness, this should be brought lo the attention of the patient so that there may be a more careful medical examination.

The Abdomen

1. Apply lubricant over entire abdominal area.
2. Patient should raise both knees to relax the muscles of the abdomen.
3. Stroke in small circles. Start in the area of the navel with small circles around the navel then increase the size of the circles gradually until the entire abdominal area is being stroked – in clockwise direction.
4. Make circles about 4 inches apart over the same path as in the previous stroking.
5. Repeat circular stroking as in 3.
6. Place hands flat on abdomen and move than back and forth in opposite directions, going from pelvic area to chest and return, (Up and down).
7. Repeal circular stroking as in 3 above.
8. Repeat (6) above, going from side to side, with fingers and heel of hand lifting at start of each stroke.

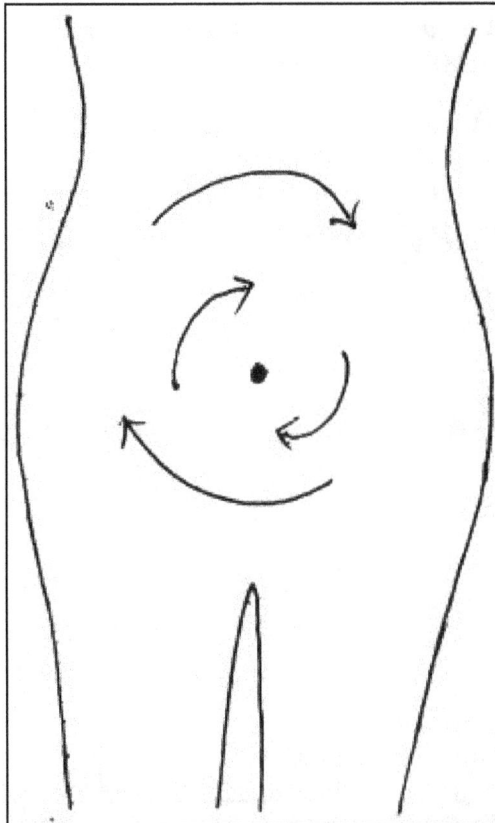

Abdomen Massage

9. Repeat circular stroking as in 3 above.

Note: Avoid abdominal massage where there is any abdominal pain or where a woman is pregnant.

The Back

1. Apply lubricant over entire back including buttocks and upper thighs.
2. Stroke upwards, starting from end of spine to the neck then from right thigh to right shoulder, then left thigh to left shoulder.
3. Small circles throughout following path of previous stroking as in 2 above.
4. Stroking repeated as in 2.
5. Place both hands below neck with fingers pointed towards head. Stroke sideways to sides of body, then stroke to original position.

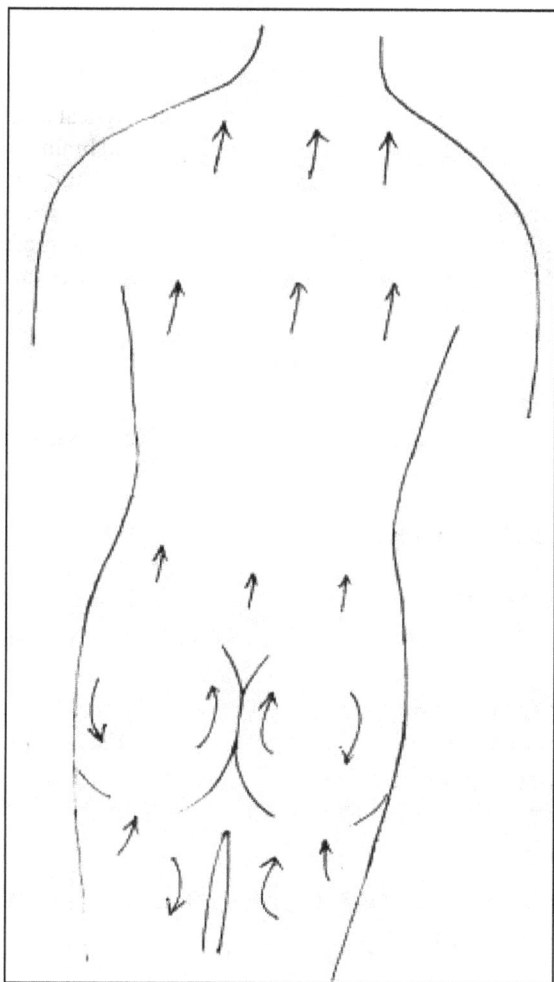

Massage of Back

6. Move hands down about 8 inches and repeat.

7. Move down 8 inches more and keep moving down and repeating until the entire back is stroked. When returning to the centre of the back, press down on the skin.

8. Clap hands over the shoulders.

9. Stroke buttocks, using heels of hands, stroking upwards.

10. Do small circles over buttocks and upper thighs.

11. Stroke buttocks as in 9.

12. Light clapping of the buttocks.

13. Stroke buttocks as in 9.

14. Light punching of the buttocks.

15. Stroking movement over entire back area.

Massage Review

Massage Therapy

"Massage has often being neglected in favour of other physical measures which can be used more easily. Massage requires skilled use of the hands and brain for its curative effects producing or regaining elasticity of tissues, stimulating blood supply, giving the patient confidence and at the same time giving him or her encouragement and psychological stimulation to use the part that is disabled - and no machine can substitute." (Dr. Paul B. Magnuson, M.D., Founder, Rehabilitation Institute of Chicago, Formerly Chief Medical Director, Veterans Administration and formerly professor of medicine, Northwestern University, Chicago).

Ethics, Cleanliness and Hygiene

A professional attitude must always be maintained. The massage expert is often doing work that doctors would be doing if they had the time for It. Unfortunately, there has long been a shortage of doctors and since massage treatments are lengthy and should be taken frequently, medical doctors do not have the time to give hand massages. The massage expert should have poise and dignity and speech and manner must always be courteous. Conversation should be very limited, as usually the person needing massage needs rest and relaxation.

Immaculate personal cleanliness is essential because of the close contact with the person getting the massage. Any body odour is usually indicative of either ill health or careless personal hygiene and is offensive.

It is Important that the uniform not only be neat and clean but that it allows free movements of the arms and shoulders. The shoes should be comfortable and the hair should be neatly groomed. No jewellery should be worn.

The hands should be soft and warm and free from any rough parts such as nails. The hands should be washed either in the presence of the person being massaged or with the knowledge of the person. For instance, after the person about to be massaged has been shown to the massage room and told to disrobe and lie down on the table and to cover himself or herself with a sheet or towel, whichever is being used, the massage expert may say: "Excuse me, while I go wash my hand."

Names of Movements and Manipulations

Stroking is called effleurage. Squeezing and pinching is called petrissage. Kneading is called friction. Clapping, cupping and hacking is called tapotement. A trembling pressure is called vibration. It is used during stroking of the spine and when applying circular movements over the abdomen.

Pressure

There is strong' medical opinion that when the muscles are relaxed, the pressure applied on the surface will be transmitted freely to all structures under the hand, and empty the veins and the lymphatic spaces. It is therefore not necessary to expend and exert any great muscle energy or vigour. Thus, persons need not be of large bone structure or endowed with large muscles to do massage work. Great physical labour is not required.

Names to Use

The most popular name to use is "client". To call the person receiving the massage a patient may not be true, if the client is not ill.

Place for Massage

The ideal place is on a massage table. However. If this is not available, a couch or bed may be used.

Clothing

Most efficient results are obtained when all clothes are removed. Usually, only the part of the body being massaged is uncovered while a sheet or towel is used to cover the other parts.

A Trained Nurse

A trained nurse is not necessarily a registered nurse, qualified to assist a surgeon at an operation or even to give injections of drugs. A trained nurse may not necessarily fall in the category of a practical nurse, that is, a nurse who has taken certain training given by hospitals or nursing schools to qualify men and women for general nursing duty, usually assisting the registered nurses or assigned by doctors or nurses agencies to care for sick persons. A trained nurse may be one who is speedily trained to perform either limited or general functions in connection with care of the sick.

The Sick

It is difficult to define a sick person. One may have a headache and this headache may prevent the person from performing his or her duties or even carrying on with normal functions such as preparing food. A sick person may be one with a touch of arthritis, with aching arms, legs or back, who may need a back rub or a general massage of arms, legs and back. A nurse who can give a good massage may know more than the registered nurse or the practical nurse as few nurses know the methods and movements of the Swedish massage.

Making the Bed

A family may engage a trained nurse to care for a sick member of the family who is not ill enough to be in a hospital. In making a bed, the trained nurse must always be sure that her hands are clean before she touches anything the patient will use. Before making a bed, the trained nurse should first always wash her hands. It is always best to use a waterproof cover on the bed so the mattress will always be kept clean and dry. Also, it is well to place a quilted cotton pad between the bottom sheet and the waitress cover. So the patient will he much more comfortable.

Skin and Muscle Care

When the muscles are not used regularly, they become weak. When a patient is confined to bed, it is therefore for hit or her benefit that the patient have a general Swedish massage every day or even twice a day. This cleanse and refreshes the skin as well as exercises the muscles. Mineral oil may be used for massage or even olive oil or cocoa butter. When oils are used, it is desirable to use some rubbing alcohol for the purpose of removing the oil. Some nurses know how to give a massage with the bare hands without using any oils. This takes practice and must be done carefully so as not to hurt the patient. In some cases a chalk-like powder can be obtained to massage.

Waste Removal

The bedpan is used by patients of both sexes for bowel movements. It is also used by woman for urinating. Male patients usually are provided with a urinal for urination. When nurse helps in waste removal, she should be prepared to do this without embarrassment. She will make her patient uncomfortable if she shows any signs of embarrassment herself. The nurse should give the patient soap, water and a towel alter any removal of body waste.

Care of Teeth

The trained nurse should provide the patient with some tooth paste, a tooth brush and a small basin for brushing the teeth after every meal.

Washing

If the patient is unable to wash himself, use a face cloth and make kind of a mitten of it using it to wash around the eyes, nostrils and face.

Bed Bath

When a patient cannot bathe himself or herself, the nurse removes the top covers and replaces with a bath blanket or large towel. The patient's clothing is then removed. First the face and hands are washed, then the arms, and legs, abdomen, chest and back. Bedridden patients may result in sores or ulcers if the weight of the body block up the blood from circulating. Frequent massages are therefore necessary for bedridden patients.

Blood Pressure

The Heart

We must remember that under normal conditions the heart beats 70 to 80 times a minute and under stress it will beat 180 times a minute or more. Under excitement, adrenalin inlands pour substances into the blood stream that causes the heart to beat faster. The blood, off course, brings nourishing substances to various body parts so that they can perform their functions. Normally, five quarts of blood are pumped, through the circulatory system every minute. Under pressure, the heart may pump the blood through the system perhaps seven or eight times a minute, moving about 40 quarts of blood during the period of only one minute.

Pressure

Two important nerves, one in the aorta and the other in the neck, regulate the blood pressure so that it will not get too high or too low. When there is excessive pressure, the heart slows down and the blood vessels enlarge or dilate.

Diastole Phase

Remember that the arteries carry blood from the heart to the various parts of the body and that the veins carry the blood from the various parts of the body of the heart, from where it is pumped into the lungs to be refreshed and then returned to the heart and then sent out again to the arteries to distribute nourishment throughout the body. The expansion of the arteries due to the dilation of the heart is part of the diastole phase of blood pressure. The large veins of the legs and abdomen are equipped with valves that open only towards the heart. Thus when the blood is pumped upwards, against gravity, the valves prevent the blood from flowing backwards during the momentary period between contractions of the heart. The diastole phase is opposed to the systole phase.

Systole Phase

The contraction of the heart and arteries that Impels the blood outwards into various parts of the body is called' systole phase.

Sphygmo Manometer (Pronounced stigmo-ma-meter)

This is an instrument for measuring the pressure of the blood in the arteries. The procedure is to first wrap the hollow bag around the upper right arm of the person whose blood pressure is being taken. A small bulb is then continually squeezed to cause air to enter into the hollow bag. As the air enters it causes the bag to swell and tighten around the upper arm. As the pressure in the bag increases, the mercury reading gets higher and higher and the blood circulation is shut off. There is a valve opening that allows the air to escape. To take a reading of systolic pressure, which is the highest pressure that is exerted by the heart in sending blood through the arteries, the systolic being the pressure of the blood against the artery walls, observe the following: The person taking the reading uses a stethoscope, applying it to the inside of the elbow. By listening carefully and first elevating the blood pressure high enough on the sphygmo manometer so as to be above the point where any heart beat may be heard, the air release valve is slowly opened so that the blood pressure reading slowly drops. At the point where the heart beat is first heard, the reading will show the systolic pressure. Keep allowing the reading to drop slowly and continue to listen to the heart beat. When the reading drops so low that the heart pulsations can no longer be heard with the stethoscope, it is at that reading that there is a minimum pressure against the artery walls and the reading at that point is the diastolic reading.

26

Medical Gymnastics

Definition

Medical gymnastics refers to a system of exercise wherein the actual exercising is done under the direction of a skilled therapist, usually a massage therapist trained in medical gymnastics. The movements described increase blood supply and nutrition to the joints and muscles and stretches shortened ligaments and contracted muscles. The exercises described also assist in removing lymph, exudations, transudations and infiltrations and also loosen adhesions and increase flexibility of muscles and joints- The oxidizing powers of the blood is increased and the nervous system is simulated, !n addition, the blood pressure and body temperature is raised.

The massage therapist may combine these gymnastic movements while doing massage therapy over the various parts of the body or may commence the medical gymnastics following the massage therapy.

1. The Wrist: Hold wrist with one hand and the hand with other hand, with palm held downward. Lift hand up and down several times, to limit. Also, bend wrist from side to side and then rotate the hand in circular fashion.

2. Arms: Hold back of wrist with one hand and place other hand on upper arm. Flex forearm against upper arm. At this point, twist palms up and down by turning the wrist.

3. Neck: Flex (bend) head forward and then backward. Bend the head to the right shoulder, then to the left shoulder. Bend the head forward, then the to right, then backward, then left, the reverse the movement.

4. Shoulder: with patent lying on side, move top arm as far forward as possible, then as far backward as possible. With patient on back, move arm from side as far towards head as possible. Hold shoulder blade with one hand and with the other

lift arm straight up and as far back as possible. Also, while on side, move arm in circles in one direction, then in opposite direction while on back, lift patient's arms together as far back as they can go. Repeat several times.

5. Spine: Patient lies face down, folding arms under forehead. The therapist assumes position astride patient. With one hand under patient's arms and the other hand on patient's back, therapist lifts forehead as far as possible. This is repeated several times. Patient then bends knees while on table or bed, resting head on table with both arms alongside head to support weight of body. Therapist, astride but behind patient, places both hands on shoulders of patient and flexes spine several times, pulling shoulders towards therapist. In the same position, therapist may turn face and trunk several times to the right and then to the left. While in this position, some massage strokes over the back will tend to loosen and case the muscles.

6. Spines and hip exercise: With the patient lying on the back, one hand is placed on the ankle and the other hand on the knees. The knee is bent and the thigh pushed back to the abdomen. The leg is extended by placing one hand at the buttock and the other hand pushing against the calves of the lower leg. This is done alternately with one leg and then the other and then with both legs together. Keeping legs extended and knees straight, abduct the thighs by separating them, then adduct by crossing them. (Abduct means to separate and adduct is to bring together). With the legs abducted try moving the thighs back against the abdomen as far as they will go. The feet may be placed on the shoulders of the therapist during this exercise. Hold this position for several minutes to allow the muscles and tendons to stretch Relax and repeat this exercise several times. This gymnastic exercise may be compared to standing and reaching down to touch the toes. A minimum of clothing should be worn for these exercises to avoid interference with the free muscular movements.

27
Music Therapy

Music Therapy is one of the alternative form of treatment, it is the planned and creative use of music to attain and maintain health, wellbeing. Individual of any age and ability may benefit from it, regardless of musical skill and background. Music Therapy may address physical, psychological, emotional, cognitive and social needs with therapeutic relationships. It focuses on meeting therapeutic aims, which distinguishes it from musical entertainment or musical education. Music therapy is the therapeutic application of music with proper methodologies and procedure to restore, maintain or enhance the cognitive, socio emotional and physical functioning of the receiving persons.

Long before acoustics came to be understood in Europe as a subject of study, the ancient Arab, Greek and Indian civilizations were already familiar with the therapeutic role of sounds and vibrations and the later day concepts pertaining to them. While music as a whole is well recognized for its entertainment value, the Indian civilization had gone a step forward to attribute the curative aspect to music.

The ancient system of Nada Yoga, which dates back to the time of Tantras, has fully acknowledged the impact of music on body and mind and put into practice the vibrations emanating from sounds to uplift one's level of consciousness. It is the Indian genius that recognized that ragas are not just mere commodities of entertainment but the vibrations in their resonance could synchronize with one's moods and health. By stimulating the moods and controlling the brain wave patterns, ragas could work as a complementary medicine.

What is a Raga?

Raga, we all know is the sequence of selected notes (swaras) that lend appropriate 'mood' or emotion in a selective combination. Depending on their nature, a raga could induce or intensify joy or sorrow, violence or peace and it is this quality which forms the basis for musical application. Thus, a whole range of emotions and their nuances could be captured and communicated within certain rhythms and melodies. Playing, performing and even listening to appropriate ragas can work as a medicine (Bagchi, 2003). Various ragas have since been recognized to have definite impact on certain ailments (Sairam, 2004b).

Historic References on Raga Chikitsa

The ancient Hindus had relied on music for its curative role: the chanting and toning involved in Veda mantras in praise of God have been used from time immemorial as a cure for several disharmonies in the individual as well as his environment. Several sects of 'bhakti' such as Chaitanya sampradaya, Vallabha sampradaya have all accorded priority to music. Historical records too indicate that one Haridas Swami who was the guru of the famous musician in Akbar's time, Tan Sen is credited with the recovery of one of the queens of the Emperor with a selected raga.

The great composers of classical music in India called the 'Musical Trinity', - who were curiously the contemporaries of the 'Trinity of Western Classical Music, Bach, Beethoven and Mozart– were quite sensitive to the acoustical energies. Legend has it that Saint Thyagaraja brought a dead person back to life with his Bilahari composition Naa Jiva Dhaara. Muthuswamy Dikshitar's Navagriha kriti is believed to cure stomach ache. Shyama Sastry's composition Duru Sugu uses music to pray for good health.

Raga chikitsa was an ancient manuscript, which dealt with the therapeutic effects of raga. The library at Thanjavur is reported to contain such a treasure on ragas, that spells out the application and use of various ragas in fighting common ailments.

Raga Chikitsa: Raga Therapy in India

Living systems show sensitivity to specific radiant energies – be it acoustical, magnetic or electro-magnetic. As the impact of music could be easily gauged on emotions and thereby on mind, it can be used as a tool to control the physiological, psychological and even social activities of the patients

Indian classical music can be classified into two forms: kalpita sangita or composition, which is previously conceived, memorized, practised and rendered and manodharma sangita or the music extemporised and performed. The latter can be equated to the honey-mooner's first night as it conceives both spontaneity and improvisation. It is fresh and natural as it is created almost on the spot and rendered instantly on the spur of the moment.

According to an ancient Indian text, Swara Sastra, the seventy-two melakarta ragas (parent ragas) control the 72 important nerves in the body. It is believed that if one sings with due devotion, adhering to the raga lakshana (norms) and sruti shuddhi, (pitch purity) the raga could affect the particular nerve in the body in a favourable manner.

While the descending notes in a raga (avarohana) do create inward-oriented feelings, the ascending notes (arohana) represent an upward mobility. Thus music played for the soldiers or for the dancers have to be more lively and up lifting with frequent use of arohana content. In the same way, melancholic songs should go for 'depressing' avarohanas. Although it is not a rule, most of the Western tunes based on major keys play joyful notes, while those composed in minor keys tend to be melancholic or serious.

Certain ragas do have a tendency to move the listeners, both emotionally as well as physically. An involuntary nod of the head, limbs or body could synchronize with lilting tunes when played.

Some Therapeutic Ragas

Some ragas like Darbari Kanhada, Kamaj and Pooriya are found to help in defusing mental tension, particularly in the case of hysterics. For those who suffer from hypertension, ragas such as Ahirbhairav, Pooriya and Todi are prescribed. To control anger and bring down the violence within, Carnatic ragas like Punnagavarali, Sahana *etc.* do come handy.

One of the unique characteristics of Indian music is the assignment of definite times of the day and night for performing or listening Raga melodies. It is believed that only in this period the Raga appears to be at the height of its melodic beauty and majestic splendor. There are some Ragas which are very attractive in the early hours of the mornings; others which appeal in the evenings, yet others which spread their fragrance only near the midnight hour.

This connection of time of the day or night, with the Raga or Raginis is based on daily cycle of changes that occur in our own body and mind which are constantly undergoing subtle changes. Different moments of the day arouse and stimulate different moods and emotions.

Each Raga or Ragini is associated with a definite mood or sentiment that nature arouses in human beings. The ancient musicologists were particularly interested in the effects of musical notes, how it effected and enhanced human behavior. Music had the power to cure, to make you feel happy, excited,keep you calm, balance your mind and so on. Extensive research was carried out to find out these effects. This formed the basis of time theory as we know it today.

Emotions, feelings and thoughts have been reported to be greatly influenced by music listening or participation. Emotional experience derived from music has a powerful effect on the formation of one's moral and intellectual outlook. Music activities enhance imagination and creative thinking.

Not only psychological impact, but also somatic or physiological impact of ragas has come to light in some recent works. For instance, stomach-related disorders are said to be cured with some Hindustani Ragas such as Deepak (acidity), Gunkali and Jaunpuri (constipation) and Malkauns or Hindolam (intestinal gas and for controlling fevers). Fevers like malaria are also said to be controlled by the ragas like Marva. For headaches, relaxing with the ragas like Durbari Kanada, Jayjaywanti and Sohni is said to be beneficial.

There is a growing awareness that ragas could be a safe alternative for many medical interventions.

Simple iterative musical rhythms with low pitched swaras, as in bhajans and kirtans are the time-tested sedatives, which can even substitute the synthetic analgesics, which show many a side-effect. They are capable of leading to relaxation, as observed with the alpha-levels of the brain waves. They may also lead to favourable hormonal changes in the system. (Crandall, 1986).

It is therefore felt that there is an urgent need for further detailed enquiry to be based on scientific parameters, which will go a long way in unearthing the goldmine on which the Indian musical system is resting now.

For this purpose, it is necessary that a group of exponents in Indian ragas join experts in medicine to help evolving a scientific system of raga therapy for the most common illness of the modern times: stress and stress-related disorders. Our leaders, professionals and managers all suffer from stress, thanks to the ever-increasing man-machine interface, resulting in the machine making the man to behave!

Music Therapy is helpful in both forms - sickness and wellness - of medical industry. In this both forms, music therapy helps to restore good health and helps to maintain the same. Active mode requires participation of the patients in the music therapy sessions, while the passive mode of music therapy requires mere involved listening. In the medical field, passive form of music therapy plays a dominant role in the betterment of the patients.

The active mode of music therapy is useful in Pediatric areas and in few of the Neurological problems.

In the Pediatric areas, the active mode of music therapy helps for the hyper active child to reduce the over activity in a given period of time and enhance the quality of concentration in child. With regard to speech difficulties in children this active mode of music enhances the quality of fluency in speech.

Music training also enhances the Verbal Memory improvement in children.

Many of the behavior problems in children and developmental delays and other problematic behaviors also may be attended through active mode of music therapy.

Music Therapy in expressive behavior, imagination development in children, and projecting the ideas while participating *etc.*, are all feasible in this active mode.

In the neurological areas the neurological aphasia; both the receptive and expressive aphasia may get the necessary stimulation, required to bring back the needed communication in patients. The lyrics in active music surely triggers the memory folders in the brains of patients and helps to revive the same.

The passive mode/form of music therapy may surely be implemented in almost in all areas of medical field as an alternative or as an adjunct or as a complement to medicine.

Does Music Therapy Belong to India?

Music Therapy prevails in the universe from the dawn of the civilization, since our music is universal and have authenticity in this refined field, we can claim this from of seeking solace belongs to us. The only deficiency we have is, our forefathers do not possess the tendency or attitude to document anything which they are doing for the betterment of the society.

What Happens when we are Listening to Music?

Scientifically, many things are happening in our body while we are listening the music, while we are participating in a live music session with body movements, the music which we are listening/participating many parts of the brain functions in a coordinated pattern and helps to enjoy the music, if we are well versed in music. The mind has the tendency to relate or to identify things in a known pattern, accordingly come to a result/conclusion through our expressed behavior. Rather if we let the music flow into us without getting into the nuances or intricacies of the particular piece which has been provided for the therapeutic purposes,does wonder with the patients,clients and participants. Since these informantion and other pertaining to brain are available only through brain injured patients, few research infromations are available with normal healthy subjects also.

Does Music have the Healing Effect?

Healing effect on individuals with music is a known fact. It has centuries of claim behind it. But the quality of exposure the music with its specifications and other procedures plays a very vital role in speaking about its healing aspect.

Why Instrumental Music?

When classical music with its notes being played through a string instrument the impact is complete. The specific notes being presented in the classical flair the impact will be good on the individuals. The music with its notes when essayed in classical format the frequency and the wavelength it emanates in the atmosphere and its impact enormous on the listeners. It helps in the quality of neuro transmitters secreted in brain and the behaviour of the individual. Music Therapy and its Essence Pleasant tunes transfers good vibrations in the atmosphere. Music acts on our mind before being transformed into thought and feeling. Music influences the lower and higher cerebral centers of the brain. Use of Music as a therapy helps search of an individuals personal harmony. Music therapy is an important tool in the treatment of both physiological and psychosomatic disorders. Music Therapy stimulates good vibrations in the nerves of the listeners. Music brings about a sense of mental well being in individuals. Music Therapy helps to clear the junked thought in mind. Which leads to have positive frame of mind. Music Therapy enhances the concentration level of children. Effect of music on the behavior of individuals is enormous. Music improves the capacity of planning. Musical training helps to express refined exhibition of emotions and clarity in cognition too. Music therapy stimulates beta cell activities. Music therapy enhances quality of neural Engram Music therapy enhances the quality of protein releases of brain chemicals. Music therapy enhances the quality of neurotransmitters. Music therapy conditions the heart. Music therapy reduces hypertension. Music therapy reduces unaccompanied explicit behavioral manifestations. Music therapy helps to restrain the emotional outbursts. Listening to music actually cause brain to perform better spatial reasoning. Music is part of human nature. Human brain processes music. Effects of music on human behavior and thought are powerful. Music enhances cognitive process. Music training and exposure increases the amount of brain that responds to musical sounds. Music during exercise produces physiological benefits. Musical interventions may reduce maternal depression and its effects on infants. Back ground music aids in developing memory. Memory recalls improves when the same music played during learning is played during recall. Music instruction is positively related to verbal memory. Music influences the perception of art. Lyrics and tunes are processed independently in the brain. The right hemisphere is used by musicians to solve the problem. Music promotes physiological and behavioural relaxation in neonates. Music education facilitates language development and reading readiness. Direct music participation enhances the development of creativity. Art activities foster positive attitude toward studies. Music education facilitates social development, personality adjustments and general intellectual development. Music alerts childrens brain waves. Music produces a kind of pleasure which human nature cannot do without. Music is the shorthand of emotion.

It is a listening therapy program using specially filtered classical music to improve ear and brain function. A link is established between the sounds we hear and our functioning in speech, learning, energy and stress. The classical music that one listens transforms the improvement of the ear function and also recharges the cortex of the brain. The target group listens to the specially composed music and sounds through headphones while doing home work, playing, sleeping and through private listening, The music should be listened to at least 30 minutes a day for six to eight weeks preferably without break in the listening sessions. Today we are confronted with pollution in every sphere of life situation – be at home, school, work place, factories, office and what not. The response of the ear, brain and nervous system is to shut down in varying degrees to this onslaught Stress, strain, anxiety, depressive state of mind, restlessness all pervade our system and pulls down the existing energies. Music Therapy is non invasive, non pharmaceutical and completely safe. There are no negative side effects. Musical healing treats the cause of a listening problem by stimulating and restoring natural ear and brain function.

How it Works

Hearing is physical and listening is psychological. Both are vital to our communication skills, establishing good relationship, socializing with everyone and learning intuitiveness. Out of the 12 cranial nerves, 10 are linked to the ear, indicating the importance of the Musical sounds to our nervous system.

Raga Chikitsa means "healing through the use of raga." Raga Chikitsa is defined as "the knowledge of how to use raga for the purposes of healing. Fundamental features ofRaga Chikitsa is the classification of the ragas based on their elemental composition (ether, air, fire, water, earth) and the proper use of the elements to balance the nature of the imbalance.

Ragas are closely related to different parts of the day acording to changes in nature and development of a particular emotion, mood or sentiment in the human mind. Music is considered the best tranqulliser in modern days of anxiety, tension and high blood pressure.

It is believed that the human body is dominated by the three Doshas – Kaph, Pitta and Vata. These elements work in a cyclic order of rise and fall during the 24 hour period. Also, the reaction of these three elements differ with the seasons. Hence it is said that performing or listening to a raga at the proper allotted time can affect the health of human beings.

Raga and its Effects

Kafi Raga – Evokes a humid, cool, soothing and deep mood.

Raga Pooriya Dhansari (Hamsanandi-Kamavardini) – evokes sweet, deep, heavy, cloudy and stable state of mind and prevents acidity.

Raga Mishra Mand – has a very pleasing refreshing light and sweet touch

Raga Bageshri – arouses a feeling of darkness, stability, depths and calmness. This raga is also used in treatment of diabetes and hypertension

Raga Darbari (Darbari Kanada) – is considered very effective in easing tension. It is a late night raga composed by Tansen for Akbar to relieve his tension after hectic schedule of the daily court life.

Raga Bhupali and Todi – give treamendous relief to patients of high blood pressure.

Raga Ahir-Bhairav (Chakravakam) – is supposed to sustain chords which automatically brings down blood pressure.

Raga Malkauns and Raga Asawari (natabhairavi) – helps to cure low blood pressure.

Raga Tilak-Kamod (Nalinakanti), Hansdhwani, Kalavati, Durga (Suddha Saveri) -evoke a very pleasing effect on the nerves.

Raga Bihag, Bahar (Kanada), Kafi and Khamaj – For patients suffering from insomnia and need a peaceful sonorous sleep. Useful in the treatment of sleep disorders.

Raga Bhairavi – Provides relief T.B, Cancer, Severe Cold, Phlegm, Sinus, toothache.

Raga Malhar – Useful in the treatment of asthma and sunstroke.

Raga Todi, Poorvi and Jayjaywanti – Provides relief from cold and headache.

Raga Hindol and Marava – These ragas are useful in blood purification.

Raga Shivaranjani – Useful for memory problems.

Raga Kharahara Priya – strengthens the mind and relieves tension. Curative for heart disease and nervous irritablility, neurosis, worry and distress.

Raga Hindolam and Vasantha – gives relief from Vatha Roga, B.P, Gastritis and purifies blood.

Raga Saranga – cures Pitha Roga.

Raga Natabhairavi – cures headache and psychological disorders.

Raga Punnagavarali, Sahana – Controls Anger and brings down violence,

Raga Dwijavanthi – Quells paralysis and sicorders of the mind,

Raga Ganamurte – Helpful in diabetes,

Raga Kapi – Sick patients get ove their depression, anxiety. Reduces absent mindedness.

Raga Ranjani – helps to cure kidney disorders,

Raga Rathipathipriya – Adds strengh and vigor to a happy wedded life. This 5-swara raga has the power to eliminate poverty. The prayoga of the swaras can wipe off the vibrations of bitter feelings emitted by ill will.

Raga Shanmukhapriya – Instills courage in one's mind and replenishes the energy in the body.

Raga Sindhubhairavi – For a Healthy Mind and Body, Love and Happiness, Gentleness, Peace and Tranquillity,suitable for singing and listening at late night (1 am – 4 am).

Raga Hameerkalyani – This particular Hindusthani coloured raaga, one with great therapeutic value relaxes tension with its calming effect and brings down BP to normal 120/80.

Raga Brindavana Saranga – For Wisdom, Success, Knowledge, Joyfulness and Greater Energy.

Raga Mohana – Useful for the treatment of migraine headache. It is suitable for singing and listening at evening (7 pm- 10 pm).

Ragas Charukesi, Kalyani (all time raga),Sankarabharanam(evening raga) and Chandrakauns is considered very helpful for heart aliments.

Raga Ananda Bhairavi- Supresses stomach pain in both men and women. Reduces kidney type problems. Controls blood pressure.

Raga Amrutavarshini – Ushana vyathi nasini (alleviates diseases related to heat) Raga Reethigowla– A raga that bestows direction when one seeks it.

Raga Madhyamavati – Clears paralysis, giddiness, pain in legs/hands, *etc.* and nervous complaints.

The power of musical vibrations connects in some manner all things and all beings and all beings in the universe on all plants of existence. The human body has 72,000 astral nerves (Nadis) which incessantly vibrate in a specific rhythmic pattern. Disturbance in their rhythmic vibration is the root cause of disease. The musical notes restore their normal rhythm, there by bringing about good health.

Listen with your heart and not your intellect.

Therapeutic Evaluation of the Classical Ragas

Singing or engrossed listening of Raga Bhairavi has been found to uproot the diseases of kapha dosha *e.g.* asthma, chronic cold, cough, tuberculosis, some of the sinus and chest related problems *etc.* Raga Asavari is effective in eliminating the impurities of blood and related diseases. Raga Malhar pacifies anger, excessive mental excitements and mental instability. Raga Saurat and raga Jaijaivanti have also been found effective in curing mental disorders and calming the mind. Raga Hindola helps sharpening the memory and focussing mental concentration. It has been proved effective in curing liver ailments.

Apart from the classical ragas played on musical instruments, the rhythmic sounds of temple bells and shankha (conch shell or bugle) produced during devotional practices have also been found to have therapeutic applications.

A research study in Berlin University showed that the vibrations of the bugle sound could destroy bacteria and germs in the surroundings. More specifically, it was found that if the shankha is played by infusing (through the mouth) twenty-seven cubic feet of air per second, within a few minutes it will kill the bacteria in the surrounding area of twenty-two hundred square feet and inactivate those in about four-hundred square feet area further beyond.

Dr. D. Brine of Chicago had treated hundreds of cases of hearing impairments/ deficiencies by making the patients play or listen to the sounds of shankha played rhythmically at appropriate (as per the case) pitch and intensity.

Music therapy is a scientific method of effective cures of disease through the power of music. It restores, maintains and improves emotional, physiological and psychological well being. The articulation, pitch, tone and specific arrangement of swaras (notes) in a particular raga stimulates, alleviates and cures various ailments inducing electro magnetic change in the body.

Music is basically a sound or nada generating particular vibrations which moves through the medium of ether present in the atmosphere and affects the human body. Music beats have a very close relationship with heart beats. Music having 70-75 beats per minute equivalent to the normal heart beat of 72 has a very soothing effect. Likewise rhythms which are slower than 72 beats per minute create a positive suspense on the mind and body. Rhythms which are faster than the heart rate excite and rejuvenate the body.

The history of Carnatic music says that the system of Mayamalava Gowlai was introduced by the blessed musician, Purandaradasar. This raga has the potency to neutralise the toxins in our body. Practising this raga in the early hours of the morning, in the midst of nature, will enhance the strength of the vocal chords.

In India, saint musician Thyagaraja is said to have brought a dead person back to life with Bilahari raaga, and Muthuswamy Dikshitar cured stomach aches with his Navagraha Kritis. In our own times Pt Omkarnath Thakur is said to have cured Mussolini of his insomnia with a song.

Kollegal.R.Subramaniam, a great musician wrote, "According to the Sangita Sastra, the 72 Melakarta Ragas control the 72 important nerves. If one sings with devotion adhering to lakshana and sruti sudhi, the Raga Devatas shower blessings, injecting the particular nerve in the human body which shall give needed relief in due course". By singing the 72 Melakarta ragas, stomach operation can be avoided and ailment cured.

Research has shown us that music does have healing effects. They stimulate the brain, ease tension and remove fatigue. The effect of Music Therapy may be immediate or slow, depending upon number of factors like the subject, his mental condition, environment and the type of Music, selected for having the desired effect. Music Therapy largely depends on individual needs and taste. Before using music as Therapy it must be ascertained which type of music is to be used. The concept of Music Therapy is dependent on correct intonation and right use of the basic elements of music. Such as notes [swara] rhytym, volume,beats, and piece of melody. There are countless 'Ragas' of course with countless characteristic peculiarities of their own. That is why we can not establish a particular Rag for a particular disease. Different types of Ragas are applied in each different case.

What is the Effect of Music on Brain Waves?

Music with a strong beat can stimulate brainwaves to resonate in sync with the beat, with faster beats bringing sharper concentration and more alert thinking, and a slower tempo promoting a calm and meditative state. The changes in the brain waves brings about changes in other bodily functions as well. Breathing and heart rate can be altered by music. That may be one of the reasons why listening to music promotes relaxation and prevents the effects of chronic stress.

Doctors at Kosice-Saca Hospital, the first private hospital in Slovakia, believe that music therapy helps newborns reduce stress and stay healthy. Shortly after birth, infants receive five 20-minute music sessions each day. Doctors found that while the tikes are tuned in, most of them fall asleep or lie quietly.

Effect of Music Therapy on Cancer Patients

It is my deep ambition to develop music for use in the pain and palliative wards to get relief from pain. Experts say " music reduces pain", and this is a proven statement as Music cause the body to release endorphins (hormones that counteract pain).

A Music Therapy program has the potential to improve a cancer patient's physical and emotional well-being by providing a distraction during treatment, thus decreasing stress, pain, and anxiety levels. Eighty percent (80 per cent) of cancer patients will sense their stress, pain, and anxiety levels decreased because they participated in Music Therapy.

Effect of Saptaswaras and Chakra Meditation

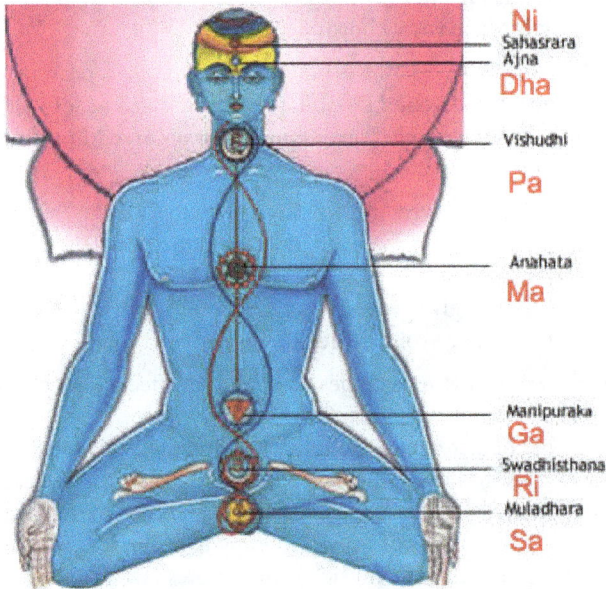

Ni
Sahasrara
Ajna
Dha

Vishudhi

Pa

Anahata

Ma

Manipuraka
Ga
Swadhisthana
Ri
Muladhara
Sa

The physical body is ensheathed by auric field in which seven major chakras (invisible to the naked eye) are present. The seven chakras are Sahasrara, Ajna, Vishudhi, Anahatha, Manipura, Swadhistana and Mooladhara. Each chakra is associated with an endocrine gland and controls specific organs. Each swara resonates with one major chakra. When each note is sung concentrating on the shruthi, vibration of the corresponding chakra can be experienced. According to an ancient Indian text, Swara Sastra, the seventy-two melakarta ragas (parent ragas) control the 72 important nerves in the body. It is believed that if one sings with due devotion, adhering to the raga lakshana (norms) and sruti shuddhi, (pitch purity) the raga could affect the particular nerve in the body in a favourable manner. The vibration of the notes activate a chakra and through the nadis emanating from the chakras, the organ at the side of the disease begins the healing process.

Music Learning should begin at a Young Age

Music is basically a sound or nada generating particular vibrations which moves through the medium of ether present in the atmosphere and affects the human system. So music is a power or universal energy in the form of ragas. Enhanced vibration of seven major chakras keeps the mind and body in good health. It increases concentration, memory, makes the mind disciplined and spiritual. Music can be a cure if the singer/listener choses the raga based on which chakras he wants to concentrate on. Hence while singing/listening, based on the dominant swaras in that raga, the corresponding chakras vibrate more. The singer/listener should concentrate on the chakra while singing/listening. Teaching music at the tender age has several advantages. The child is born with all the chakras. But it is the Mooladhara chakra, which starts functioning even at birth. As the child grows, the other 6 chakras start functioning one by one from Mooladhara to Sahasrara chakra and by 21 years, the development process is complete. So teaching music at a young age purifies the body, mind and soul and lays the foundation for a healthy future. Academic brilliance and morality are automatically inculcated in such children. To make our children lead a healthy life, teach them music at a tender age and make it part of curriculum. Music is also one form of meditation, which enhances the power of chakras by raising Kundalini shakthi.

Symphonies of raga have a definite soothing effect on the mind as well as on the body. Repeated listening to the particular raga being chosen for a particular disease produces a network of sound vibration. Energy from URF (universal energy field) to HEF(human energy field) transmitted by the strokes of the different tones of raga affects the CNS (Central nervous system) because the roots of the auditory nerves are more widely distributed and have more connections than any other nerves in the body.

Sapthaswaras and Saptha Chakras

The ascending and descending order in which the swaras is sung is called arohanam and avarohanam. "Sa" shadjam is the basic note of all ragas. It is not only the inter relationship between the notes that define a raga but also the relationship of these notes to the basic note (aadhara shadjam). A noteworthy feature is, even though the raga is built on the basis of a sequence of swaras, the drone of the tambura will be spelling out the aadhara shadja all the time, loud and clear so that the reliance is constantly maintained. So there cannot be a raga without the aadhara shadja. So this shadjam increases the vibration of the mooladhara chakra which is for survival. Kundalini shakthi – shakhti in coiled form exists at the tip of the backbone. Increased vibrations of Mooladhara chakra by repeating Shadjam raises Kundalini shakthi. Once Kundalini shakthi is awakened, the energy starts moving through shushumna nadi (spinal cord) towards the sahasrara chakra where Lord Shiva resides. For smooth passage of the shakthi, each note is sung in ascending order (as in arohanam). Then 'Sa' raises Kundalini shakthi at mooladara chakra, Ri – back swadhishtana chakra, Ga – back manipura chakra, Ma- back anahatha chakra, Pa – back vishudhi chakra, Dha – back ajna chakra, Ni – sahasrara chakra, Sa – cosmos. Now Avarohanam, notes sung in descending order connects cosmos and reinforces lord shiva's shakthi at the sahasrara chakra and the combined shiva shakti circulates through front chakras increases the divine power in corresponding chakras ie.Sa- Cosmos, Ni – sahasrara chakra, Dha- Ajna chakra, Pa – Vishudhi chakra, Ma-Anahatha chakra, Ga- manipura chakra, Ri –swadhishtana chakra, Sa – Mooladhara chakra and Ni- Mother earth. This swara garland by singing makes the body receive the energy from mother earth (ni lower octave) and Cosmos,(sa at upper octave). While sitting on the ground, energy from the mother earth reaches the mooladhara chakra increasing its vibration, which is solely for the survival.

Within the set of swaras, some play a major role and some less. There are some swaras which dominate and stay persistently -nyasa swaras. There are some ragas which have a special impact if they are initiated at a certain specific swaras. For instanse, a raga like Atana generally commences at the higher octave with the combination of the swaras – sa ri sa ni sa dha dha; or raga like anandhabhairaviin the middle octave with the swaras pa dha pa ma dha pa ma pa ma ga ri ga. or ritigowla with ni dha ma ga ri ga. Jiva swaras which lends life should be highlighted. They are the dhivata in Atana, antara gandhara in kalyani, madhyama and gandhara in hindola.

Musical Notes and Chakras

As mentioned elsewhere in this blog, the swaras of musical octave are related to Chakras in the human body with their respective Elements (Panchabhoothas) as follows:

Shadjam – The Mooladhara Chakra – Associated with the Energy of Survival- Element is Earth/Prithvi.

Rishabham–Swadhishtana Chakra – Associated with reproductive organs and life force – Element is Water/Jalam.

Gandharam– Manipooraka Chakra -Associated with power and mastery of self- Element is Fire/Agni.

Madhyamam–Anahatha Chakra– Associated with energy of compassion and love-Element is Air/Vayu.

Panchamam–Visuddhi Chakra– Directly related to Creativity and Communication – Element is Ether/Aakasam.

Dhaivatham–Ajna Chakra- The seat of Intellect and wisdom, Analysing and Reasoning -All Elements.

Nishadam–Sahasrara- Associated with spiritual enlightenment, Divine Wisdom and understanding and Union with God. It also integrates all the chakras with their respective qualities.

These seven chakras or the most important energy centers in the body become imbalanced due to several factors such as toxins and other impurities in our physical body and negative thoughts in our mental plane leading to several diseases. These 7 chakras can be brought back to balance by constantly developing the habit of positive thinking, getting out in the Sun, listening or getting involved into music by singing or playing instruments *etc.*

For instance, people having severe problem of Sore throat, Thyroid, Lost Voice, Mouth Ulcers, Tonsillitis, Stiff neck, Whooping cough, Laryngitis, Hoarseness or any infection or disease in throat area is said to be blocked in the Visuddhi Chakra. The physical parts associated with this Chakra are Thyroid gland, throat, upper lungs, digestive track, neck, shoulders, arms and hands.

These diseases can be caused due to toxins or impurities present in our food or caused from the environment. Just thinking out loudly, these can also be caused due to the mental impurities or negative emotions or poisonous thoughts that we "swallow" in our daily life speaking/thinking ill of others, the hatred and vengeance we carry with us through out our life time against certain people/incidents *etc.* All such thoughts deteriorates, undermines and disintegrates this energy center inviting physical diseases in the throat area.

The name of this chakra itself can be split as Visha + Shuddhi where "Visha" means poison in Sanskrit and Shuddhi refers to cleaning or purifying. This means purifying or getting cleared from all the difficulties and unpleasant experiences that we have suppressed or swallowed during the various phases of our life.

To get cleared of such problems with the throat we should either take a resolute determination to either spit out the poison (Visha) of negative or mean thoughts, harsh or rude words uttered to our own loved ones in the family/friends/neighbours etc, or to deeply swallow the poison of insult or bad treatment suffered by us by graciously forgiving the doer. Let us not forget the incredible power and intelligence within us that constantly responds to our thoughts and deeds. So we have to make sure that our thoughts and deeds arealways positive. This realisation or self- analysis will help us to clear the blocks in the Vissudhi Chakra which leads to enhance our abilities and skills to blossom. Such a person will have a talent for singing or speech together with a well balanced mind filled with calm thoughts. Musically Ragas Dwijavanthi and Desh are associated with this Chakra.

It has been scientifically researched that the probable cause for Cancer, the most dreaded disease could be deep hurt in one self, longstanding resentment or hatred towards a person (the resentment is so big that the moment the thought of other person comes to one's mind, he/she starts getting agitated/irritated), a deep secret or grief stored within ourself for a long time, *etc.*

The best solution is to lovingly forgive and release all pent up negative emotions firstlyand then choose to fill our world with joy and enjoy life's blessings called as family, children, parents, friends *etc.* Above all having implict faith and surrender to the SUPREME POWER definitely will help us to heal ourselves exponentially. When we make a sincere effort to do this, the whole Universe conspires for us and people whom we think would harm us (this is just psychological mostly) may even start thinking about us kindly and in good terms.

The Indian music which has its origin from the time of Vedas, is capable of removing all the blocks in the Sapta Chakras, thereby ensuring physical, mental, intellectual and spiritual well being to the human society.

Studies have shown that music has a profound effect on body and mind. Music with a strong beat can stimulate brainwaves to resonate in sync with the beat, with faster beats bringing sharper concentration and more alert thinking, and a slower tempo promoting a calm and meditative state.

The changes in the brain waves brings about changes in other bodily functions as well. Breathing and heart rate can be altered by music. That may be one of the reasons why listening to music promotes relaxation and prevents the effects of chronic stress.

Music is very much useful in warding off depression and anxiety. Thus, music promotes emotional well being and in maintaining positive mental health.

Particular raga for particular patient or disease:

Hypertension

Anandabhairavi raga is said to be proved to bring down the BP level.

Ananda Bhairavi has very soothing effect on hypertension.

Mental Illness

Sankarabaranam raga cures mental illness, soothes the turbulent mind and restores peace and harmony. Sankarabaranam, if rendered with total devotion for a stipulated period, can cure mental disorders said to be beyond the scope of medical treatment. Similarly Darbari Kanada and Mohanam also helps to soothe turbulent mind and migraine headaches.

Diabetes

Activation of The Manipooraka Chakra to Stimulate Secretion of Insulin Raga Bageshree, Chandrakauns.

There is a growing awareness that ragas could be a safe alternative for many medical interventions. Music improves the immunity of the body. It has been observed that medical treatment with Music Therapy has reduced the intake of antibiotics and pain killers over a period of time. In the West classical Instrumentals like compositions of Mozart are used in music therapy, however, simple iterative musical rhythms with low pitched tones, as in Instrumental Indian classical music are the time-tested sedatives.

Music will help improving mental and emotional health for collaborative learning for total life. It is therefore necessary that a group of exponents in Indian classical music join experts in medical field to help evolving a scientific system of music therapy for the most common illness of the modern times: Stress and Strain related disorders and even Cancer.

Music Therapy in Insomnia

Insomnia is a condition where people have trouble falling asleep, staying asleep or having a restful night sleep. Often people who suffer from insomnia fall prey to sleeping pills to get relief, which besides adverse effects create more problems.

It also causes problems when you interact with other medications or alcohol, which can be very serious or even deadly. Patients are prescribed with medications for a shorter period, but because these are addictive people tend to take them quite often and for longer. Pills can also cause high blood pressure, dizziness, nausea, confusion. It also can cause bizarre behaviour such as sleep walking, sleep binge eating and sleep driving, in which the person will not remember.

A good night sleep is necessary to maintain good physical and mental health. Sleep helps you to restore your body and mind, to help with memory and learning, also to keep you be in a better mood. When you are deprived from sleep, your body is more prone to infections, diabetes, high blood pressure, heart disease, also your daily work routine is hampered. It is very important to get a good 7-8 hours of sleep if you are an adult, and more than 8 hours if you are younger. Certain Ragas like Neelambari, Anandabhairavi, Yadukulakamboji, Kalyani etc has been found very effective for insomnia.

Music therapy uses binaural beats to stimulate your brain to a high level of relaxation, like a meditated state, to help you fall asleep and stay asleep. Binaural beats is music set to certain frequencies to stimulate a certain part of your brain that allows you to get some sleep and after the session wake up in a fresh energized mood.

28

Pranic Healing

Pranic Healing is based on the overall structure of the human body. Man's whole physical body is actually composed of two parts: the visible physical body, and the invisible energy body called the bioplasmic body. The visible physical body is that part of the human body that we see, touch, and are most acquainted with. The bioplasmic body is that invisible luminous energy body which interpenetrates the visible physical body and extends beyond it by four or five inches. Traditionally, clairvoyants call this energy body the etheric body or etheric double.

What Is Pranic Healing?

Pranic healing is an ancient science and art of healing that utilizes prana or ki or life energy to heal the whole physical body. It also involves the manipulation of ki and bioplasmic matter of the patient's body. It has also been called medical qigong (ki kung or ki healing), psychic healing, vitalic healing, therapeutic touch, laying of the hand, magnetic healing, faith healing, and charismatic healing.

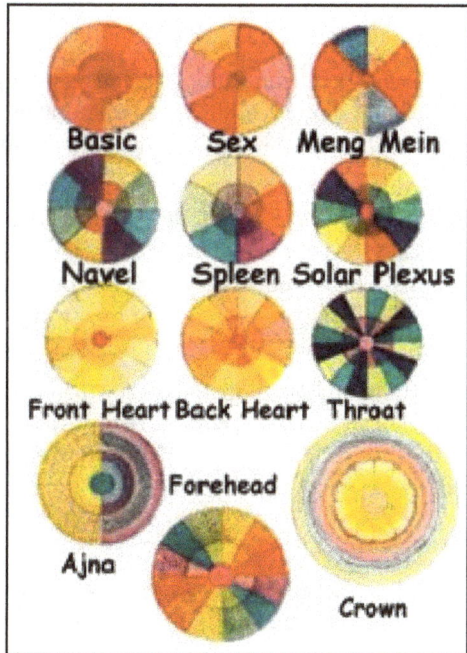

Basic   Sex   Meng Mein

Navel   Spleen   Solar Plexus

Front Heart   Back Heart   Throat

Forehead

Ajna

Crown

Two Basic Laws of Pranic Healing

Pranic healing is based on two laws: The law of self-recovery and the law of prana or life energy. These laws are quite obvious but strangely they are usually the least noticed or least remembered by most people. It is through these basic laws that rapid or miraculous healing occurs.

1. Law of Self-Recovery

In general, the body is capable of healing itself at a certain rate. If a person has a wound or burn, the body will heal itself and recover within a few days to a week. In other words, even if you do not apply antibiotic on the wound or burn, the body will repair or heal itself. At the present moment, there is no medicine available for the treatment of viral infection. But even if a person has cough or cold due to viral infection, the body will recover generally in one or two weeks without medication.

2. Law of Life Energy

For life to exist, the body must have prana, chi or life energy. The healing process can be accelerated by increasing life energy on the affected part(s) and on the entire body.

In chemistry, electrical energy is sometimes used as a catalyst to increase the rate of chemical reaction. Light can affect chemical reaction. This is the basis for photography. In electrolysis, electricity is used to catalyze or produce chemical reaction. In pranic healing, prana or life energy serves as the catalyst to accelerate the rate of biochemical reactions involved in the natural healing process of the body. When pranic energy is applied to the affected part of the body, the rate of recovery or healing increases tremendously.

What we call miraculous healing is nothing more than increasing the rate of self-recovery of the body. There is nothing supernatural or paranormal about pranic healing. It is simply based on natural laws that most people are not aware of.

Although science is not able to detect and measure life energy or prana, it does not mean that prana does not exist or does not affect the health and well being of the body. In ancient times, people were not aware of the existence of electricity, its properties and practical uses. But this does not mean tliat electricity does not exist. One's ignorance does not change reality; it simply alters the perception of reality, resulting in misperception and misconception of what is and what is not, what can be done and what cannot be done.

Children have more life energy than elderly people do. You notice that they move a lot from morning to night, hardly getting tired at all. When suffering from a fracture, who heals faster—the child or the elderly? The broken bone of a child heals very fast while that of an elderly heals very slowly; sometimes, it will not even heal at all.

Prana or Ki

Prana or ki is that life energy which keeps the body alive and healthy. In Greek it is called 'pneuma', in Polynesian 'mana', and in Hebrew 'ruah', which means 'breath of life'. The healer projects prana or life energy or 'the breath of life' to the patient, thereby, healing the patient. It is through this process that this so-called 'miraculous healing' is accomplished.

Basically, there are three major sources of prana: solar prana, air prana and ground prana. Solar prana is prana from sunlight. It invigorates the whole body and promotes good health. It can be obtained by sunbathing or exposure to sunlight for about five to ten minutes and by drinking water that has been exposed to sunlight. Prolonged exposure or too much solar prana would harm the whole physical body since it is quite potent.

Prana contained in the air is called air prana or air vitality globule. Air prana is absorbed by the lungs through breathing and is also absorbed directly by the energy centers of the bioplasmic body. These energy centers are called chakras. More air prana can be absorbed by deep slow rhythmic breathing than by short shallow breathing. It can also be absorbed through the pores of the skin by persons who have undergone certain training.

Prana contained in the ground is called ground prana or ground vitality globule. This is absorbed through the soles of the feet. This is done automatically and unconsciously. Walking barefoot increases the amount of ground prana absorbed by the body. One can learn to consciously draw in more ground prana to increase one's vitality, capacity to do more work, and ability to think more clearly.

Water absorbs prana from sunlight, air, and ground that it comes in contact with. Plants and trees absorb prana from sunlight, air, water, and ground. Men and animals obtain prana from sunlight, air, ground, water, and food. Fresh food contains more prana than preserved food.

Prana can also be projected to another person for healing. Persons with a lot of excess prana tend to make other people around them feel better and livelier. However, those who are depleted tend to unconsciously absorb prana from other people. You may have encountered persons who tend to make you feel tired or drained for no apparent reason at all.

Certain trees, such alpine trees or old and gigantic healthy trees, exude a lot of excess prana. Tired or sick people benefit much by lying down or resting underneath these trees. Better results can be obtained by verbally requesting the being of the tree to help the sick person get well. Anyone can also learn to consciously absorb prana from these trees through the palms, such that the body would tingle and become numb because of the tremendous amount of prana absorbed. This skill can be acquired after only a few sessions of practice.

Certain areas or places tend to have more prana than others. Some of these highly energized areas tend to become healing centers.

During bad weather conditions, many people get sick not only because of the changes in temperature but also because of the decrease in solar and air prana (life energy). Thus, a lot of people feel mentally and physically sluggish or become susceptible to infectious diseases. This can be counteracted by consciously absorbing prana or ki from the air and the ground. It has been clairvoyantly observed that there is more prana during daytime than at night. Prana reaches a very low level at about three or four in the morning.

Aura

Clairvoyants, with the use of their psychic faculties, have observed that every person is surrounded and interpenetrated by a luminous energy body called the bioplasmic body or aura. Just like the visible physical body, it has a head, two eyes, two arms, *etc.* In other words, the bioplasmic body looks like the visible physical body. This is why clairvoyants call it the etheric double or etheric body.

The word 'bioplasmic' comes from 'bio', which means life and plasma, which is the fourth state of matter, the first three being: solid, liquid, and gas. Plasma is ionized gas or gas with positive and negative charged particles. This is not the same as blood plasma. Bioplasmic body means a living energy body made up of invisible subtle matter or etheric matter. To simplify lhe terminology, the term 'energy body' will be used to replace the word 'bioplasmic body'. Science, with the use of Kirlian photography, has rediscovered the energy body. With the aid of Kirlian photography, cientists have been able to study, observe, and take pictures of small bioplasmic articles like bioplasmic fingers, leaves, *etc*. It is through the energy body that prana or life energy is absorbed and distributed throughout the whole physical body.

Benefits of Pranic Healing

☆ It can help parents bring down the temperature of their children suffering from high fever in just a few hours and heal it in a day or two in most cases.

☆ It can relieve headaches, gas pains, toothaches, and muscle pains almost immediately in most cases.

☆ Cough and cold can usually be cured in a day or two. Loose bowel movement can be healed in a few hours in most cases.

☆ Major illnesses such as eye, liver, kidney, and heart problems can be relieved in a few sessions and healed in a few months in many cases.

☆ It increases the rate of healing by three times or more than the normal rate of healing. These are some of the few things that pranic healing can do. All of these assume that the healer has attained a certain degree of proficiency.

Pranic Healing is Easy to Learn

Any healthy person with an average intelligence, an average ability to concentrate, an open but discriminating mind, and a certain degree of persistence can learn pranic healing in a relatively short period. Learning pranic healing is easier than learning to play the piano or painting. It is as easy as learning to drive. Its basic principles and techniques can be learned in a few sessions. Like driving, pranic healing requires much practice and time to achieve a certain degree of proficiency.

"In Ancient times, Pranic Healing could only be practiced by an elite few. My job was to develop a very effective healing system, which ordinary people could learn in just a short period. Anybody can practice pranic healing now. The knowledge of being able to deal with simple ailments is quite empowering."

— *Grand Master Choa Kok Sui*

29

Pyramid Healing

Pyramid is a geometrical shape that is formed by four equilateral triangles of same size on a square base in such a manner that they form an apex on that square.

"Pyramid" is a greek word consisted by two words '*Pyra*' + '*Mid*' where the term Pyra means "Fire" and Mid means "Center Core". Hence it is said that an Object which contains "Fire" (here the deeper meaning of Fire refers to the energy field) in its Center core or Nuclei. This shows that Egyptian people were fully aware of the deeper meaning of Fire and were using it accordingly thousands of years ago.

Though the Pyramids were discovered in many parts of world but Egypt is on the top name that comes to our mind while talking about Pyramids and their mysteries Besides Great Pyramid there are number of Pyramids found in Egypt but because of its Huge size and Mathematical and Geometrical Perfection it has attracted a big Popularity. The Base of this Pyramid is around 13 acres and it is levelled to a fraction of an inch. More than 2,600,000 blocks of heavy stones were used weighing from 2 to 80 tons each were put together so accurately that the joints are never more then 1/4th of an inch wide.

The Great Pyramid is so perfectly oriented towards true North, that it is only 2 minutes off. The Best modern effort with all our modern infrastructure and engineering techniques and equipments, to creat structure in True North is Paris Observatory which is 6 Minutes off and far behind from ancient Egyptians which is a miraculous example of Egyptian Perfection.

The Pyramids are like puzzle for mankind and a very little is known and more about them are still on the way. They have generated a curiosity in our minds about their power of preserving and their energy. Pyramid has a special property to deflect any type of cosmic radiations that falls on its apex downwards through its base line at the bottom where this deflected cosmic radiation, with the help of magnetic field of Earth's Gravitational Force, creates a new and very powerful bio energy field. Secondly as pyramid deflects all radiations that falls on its apex through its bottom from all of its four sides the inner centre of pyramid remains unaffected and safe and surrounded by a powerful bio-energy field on all sides, which helps to preserve the things and objects kept in the pyramid for a long-long time.

The History of Pyramid Therapy

The shape of the pyramid had intrigued historians, flummoxed math students and awed many. Few know that the pyramid shape was vividly described by Indian Sages much earlier in India than Egypt. Even thousands of years ago the temples of India had a pyramidal top called gopurams. The base of the 'Yagna kund' was given this shape. Thus, naturopaths say that pyramid therapy had been practiced from times immemorial.

Pyramid Vastu

Pyramid Vastu is a practical art to harmonize-mind, body and spirit with the environment, by just placing pre-programmed "Pyramid Yantra" at appropriate locations: to achieve health, happiness and prosperity. It is a powerful science of creating balance and harmony by core level corrections, it is based on the essential principles of subtle anatomy and law of the universe. Here you are utilizing your own hidden capabilities to achieve a better tomorrow. Pyramid vastu is ideal for correcting vastu or feng shui defects without physical alteration, shifting or breaking home or work place. It deals with the roots or vital force, behind all action within it. It is totally innovative concept of subtle level correction with pyramids based on mind over matter. It deals with our inbuilt power and ability of mind-body-emotion. This powerful intention at the core level induces energy within us.

Pyramid Set

This heavy copper pyramid set contains pyramid top, pyramid plate and 9 pyramid chips made from copper. Effect of this copper pyramid is very fast and more accurate. With few simple calculations with your date of birth you may find key to your energy and vitality and feel the difference in your life.

Pyramid Water

Working on principle of bioresonance on a subtle energy level (similar to homeopathy), when taken either internally or externally, pyramid water harmonizes all the processes in our organism. And the effect begins exactly in the location where it is needed the most. Such water is extremely rich with oxygen and negatively charged ions, which makes it easy to assimilate. This in turn increases water concentration levels in cells and the overall water level in the organism.

Various advantages of pyramid water:

1. Increases water concentration in cells
2. Assists in removal of toxins and "cleaning" of organism
3. Rejuvenates skin
4. Increases vitality
5. Speeds up healing processes and rehabilitation of organism
6. Strengthens immune system
7. Anti-stress effects

8. Healing without interference with organism (self-healing)
9. Helps heal skin, digestive disorders

Pyramid Meditation

Meditation done inside a pyramid, or underneath a pyramid, is called as pyramid meditation. The pyramid shape itself is being seen as a supernatural source of power or energy of the universe. The use of a pyramid in meditation can give rise to feelings of calmness, wellbeing and a more open and a positive attitude. The best results are achieved by sitting in a north facing direction. For better health, one must sit facing the East. If the Meditation is being done for having wealth in future, he/she must sit North-East facing direction. It is better to do healing by Reiki, Crystal Healing, Color Therapy, *etc.* Who have done pyramid meditation, describe as experienced a total relaxation of their body, followed by the shutting out of unnecessary external stimuli and irrelevant thoughts and finally achieving an altered state of consciousness which allows them to concentrate on deeper inner or emotional levels.

Pyramids provide most effective high-energy environments for meditation. Pyramids help to reduce the level of stress and tension in the physical body.

Meditation done inside a Pyramid is thrice more powerful.

Several experiments conducted in man-made pyramids have revealed generalized pyramid powers. These can be broadly classified as:

Preservation

Pyramid energy preserves fruit, milk and other perishables, taste of coffee, wine, fruit juices, *etc.*, is improved; used razors, knives get sharpened; acts as a room freshener, foul smells disappear.

Healing

Wounds, boils and bruises heal quicker; reduces over-weight and increases resistance to diseases; gives relief to and cures asthma, toothaches, migraine, common cold, high B.P., arthritis, palpitation of heart, epilepsy, insomnia *etc.* Drinking pyramid energized water cures conjunctivitis, other eye problems; helps digestion; gives the skin a healthy and youthful glow. Spiritual empowerments are felt when done inside a pyramid. Dreams appear clearer and meaningful.

There are two types of pyramids : Pyramids metallic and non-metallic.

A metallic pyramid harnesses the electromagnetic field of the earth, and tends to concentrate its energies onto the surface of the pyramid, where the metal is. These pyramids should be aligned to magnetic North to have the greatest effect. Non-metallic pyramids, being composed of non-conductive materials like wood and stone, consist of the majority of the pyramids around the earth: from the Great One in Giza, to the ones in China and Turkey, to those that exist in North and South America. These pyramids are said to harness the rotational inertia of the Earth, and because of this, they should be aligned to True North. Non-metallic pyramids tend to concentrate their energies into the centre of the pyramid, and thus the reason for its name "pyramid" or "centre core."

In non-metallic pyramids, like the Great Pyramid, the greatest concentration of energy is located at what is called the King's Chamber. This is also the geometric centre of the pyramid, located at one-third up its height from its base.

Pyramids act as an amplifier of one's senses especially thoughts and emotions, thus if you are focusing on a thought or an idea that you want to manifest into your life, it will amplify that thought accordingly. If one is trying to clear one's thoughts, as is done in many forms of meditation, the intent of that thought is amplified, and one can go into a deeper state of focus and peace much more quickly and easily. Pyramids are one of the gradually rising forms of naturopathic therapies for all kinds of ailments and diseases. There's said to be a close relationship between the external forces and the overwhelming energy found inside a pyramid. When kept in a room, a small pyramid can purify the polluted air, kill germs and keep the dwellers sound and healthy. In some of the most path breaking researches, the energy field created by the unique shape of pyramids has been touted to be more powerful than solar energy.

The Science behind Pyramids

The geometry of the pyramid liberates a very strong pranic field or energy force field inside it. Due to this property pyramids are beneficial in increasing the bio-energy of human being. It is claimed to increase the bio-energy level of vegetables, foods, water, milk *etc.*, when kept inside a pyramid for a certain period, and that by consuming these products that energy is absorbed by the human body.

How to Build a Pyramid

Cardboard, plywood, plastic, fibre glass, stone, PVC or any other materials can be used to make a pyramid, except magnetic materials like iron. The pyramid must be aligned as each of the four sides facing one of the cardinal points, that is, north, east, south and west. A straight line is drawn through the centre of the pyramid. A compass can be used to align the pyramid to magnetic north but true north would be more accurate. A well constructed and properly aligned pyramid, with proper measurements and placed at the proper place satisfying the conditions of alignment emits a mysterious power at 1/3 of the height from the base of the pyramid.

Pyramid Principles

A pyramid has often been symbolic of duality. The principle of pyramid therapy is that the apex of the pyramid releases energy which is hot in nature and the base gives a cooling effect. Therapists align and place small pyramids over the affected organs for 10-20 minutes everyday. It is also recommended that while using a pyramid, one should lie on a wooden bed and there should be no contact with the floor or any metal, otherwise the treatment's efficacy will be lost. Alternately, a pyramid can be aligned and placed in a room.

Pyramids have been used to evoke a sense of calmness and tranquillity next to a meditative state. The curative properties are said to benefit people with skin disorders and women with menstrual problems it is also said beneficial for persons with auto immune diseases.

Studies in USA have shown that growth rate of plants increase by 15 per cent when grown under pyramids. The Soviets in their Pyramid Research Department of Leningrad University even have instruments to measure pyramid power. The Pyramids reportedly help increase crop yield by 3 times, correct faulty vision, deafness, bestow longevity, increase the IQ of children and increase vigour and vitality also improve sex life.

30

Qigong

The word Qigong (pronounced Chee-gung) is a combination of two words: "Qi" means air, the breath of life, or vital energy of the body, and "gong" means the skill of working with, or cultivating, self-discipline and achievement.

The art of Qigong consists primarily of meditation, relaxation, physical movement, mind-body integration, and breathing exercises. Practitioners of Qigong develop an awareness of Qi sensations (energy) in their bodies and use their minds to guide the Qi. When the practitioners achieve a sufficient skill level (master), they can direct or emit external Qi for the purpose of healing others.

Qigong has evolved from many sources throughout the East. Although China is seen today as being the origin of both ancient and modern Qigong, all of the Asian countries have histories filled with examples of these traditional forms and styles of Qigong. From India, the monks traveling to China thousands of years ago introduced many methods into the Chinese culture.

From Buddhist traditions, Qigong methods promoted a sense of acceptance and ways of harmonizing life as a reflection of the greater unfolding of one's purpose in the world. Taoist (Dow-ist) Buddhist monks often prefer forms of Qigong that help achieve balance and promote longevity as a way of prolonging life and achieving optimum health. From the martial arts world, Qigong is used to develop both internal and external strength for fighting and self-development.

The emerging field of Chinese medical Qigong is rapidly spreading throughout the world and utilizes the energy stored in and transmitted through the healer to aid in the treatment of many acute and chronic diseases. This form is referred to as Qi-emission.

The ancient Chinese saw disease and natural disasters as signs that an individual or a tribe of people had fallen out of harmony with Nature. The cure for the Navajo was to re-establish a correct relationship with Nature, with society, and within the individual through ceremony, including sand paintings, chants, prayers and dances. To achieve a similar healing goal, the legendary Daoist emperor Yü the Great, of the early Xia dynasty (2,000 - 1,600 B.C.), ecstatically danced the movements of a bear to harmonize heaven and earth and to stop the floods and pestilence in his kingdom. His shamanic dance, known as "The Pace of Yü," is still practiced by Daoists today.

Diagram of Steps Used in the Pace of Yu.

The Chinese character for healer, yi, depicts a feathered shaman doing an ecstatic dance and holding a quiver full of arrows. The arrows, presumably, represented spiritual power, or righteous Qi, to drive off evil influences; later this concept was extended to the use of acupuncture needles. The shamans were women as well as men. They would go into ecstatic trance, and would often journey to the spirit world or channel divinities to diagnose the cause of the problem; they would then pray and dance to treat the diseases.

The early Daoist healers postulated that by connecting to the natural powers through dance and movement they could restore outer harmony and balance with the forces of nature. It was not long before they transferred this same reasoning to the microcosm of their own bodies. Not to wonder, many of the earliest Qigong healing forms, were derived from the movements of animals. The Qi Gong Classic (Dao Yin Tu), discovered in the tomb of King Ma in 1973 and dating back to the second century B.C., illustrates in manuscripts written on silk over 45 Qigong postures with descriptions of the movements as well as the names of the diseases which they treat; over half of these postures are animal movements. One of the ministers of Huang Di (2697-2598BC), the Yellow Emperor and patriarch of Chinese medicine, was the shaman Zhu You, who advocated exorcist prayer over the use of needles and herbs to treat illness. According to Kenneth Cohen, an American Qigong master and author, "Some scholars believe that Zhu You practiced External Qi Healing at the same time that he prayed for patients. This is remarkably similar to the synergism of non-contact healing and prayer in Native American and other indigenous healing traditions. The Yellow Emperor's

Classic states that in ancient times most illnesses were treated according to the methods of Zhu You....Professional 'prayer healers' (zhu) were once widespread in China. They may have formed a specialized branch of shamanism." Another of the Yellow Emperor's ministers was Qi Bo, one of the doctors to whom Huang Di posed his various medical questions, and whose famous dialogues were recorded as the Huang Di Nei Jing (The Yellow Emperor's Classic of Internal Medicine). Qi Bo embodied, for his time, a radical and modernistic theory of health and illness: that disease was not so much a matter of spirits, ancestors, karma, disharmony with the gods, with nature or with society, but was more a matter of External Factors (heat, cold, wind, damp, fire, and dryness), Internal Factors (the seven pathological emotions of excess anger, joy, worry, pensiveness, sadness, fear and shock), and Neither External nor Internal Factors (overwork, excess sex, improper diet, environmental toxins). Over time, this medical philosophy, appealing to the intellectual Confucian scholars and literati within the imperial court, and represented in the views of Qi Bo, won out over the intuitive shamanistic and spiritual views of Zhu You. Nonetheless, Zhu You's approach never disappeared; instead

1. Tiger
2. Deer
3. Bear
4. Monkey
5. Bird (Crane)

Qigong Postures from Daoist Physician Hua Tuo's famous Wu Qin Xi (Five Animal Frolics).

it co-existed as a parallel medical approach to the "scholar's medicine", and has been widely practiced since, primarily by Daoist and Buddhist priests, village healers and martial arts masters.

Most practitioners incorporated both philosophies of medicine into their practice. For example, Hua Tuo (110-207A.D.), the renowned Daoist physician, told his disciple Wu Pu, "The body should be exercised, but not to excess. Exercise improves digestion and keeps the meridians clear of obstructions. In this way, the body will remain free of illness. A door hinge does not rust if it is frequently used. Therefore the ancient sages practiced Daoyin.

Hua Tuo devised Daoyin method called the Wu Qin Xi (Five Animal Frolics). It can eliminate sickness and strengthen the root." this series of now famous Qigong exercises for his patients based upon the movements of the crane, bear, monkey, deer and tiger. These animal "dances" or Qigong exercises served to harmonize the flow of Qi within the inner universe of his patients. Each set of animal exercises relates to the principle of the Five Elements and works to strengthen, balance and harmonize the associated internal organs.

Over the years, many other doctors, martial artists, monks and priests have contributed to the body of exercises now known as Qigong therapy. There are over 3,600 different known Qigong systems are prevailing now.

Categories of Qigong Therapy

Qigong can be easily subdivided into three well recognized and widely accepted modalities used in current medical practice:

1. Physical therapy for fitness, health maintenance, and the treatment of specific disorders;

2. Stress management exercises and relaxation techniques; and,

3. External Qi Healing (Chinese Touch Therapy).

Qigong for Physical Therapy

The Qigong systems used for physical therapy are the most widely known and recognized Qigong exercises. The importance of prescribing these exercises cannot be underestimated in practice. Diseases like heart attack, cancer and stroke are the primary causes of death in present era. Their prevention and treatment are brought about by movement therapies and exercise. Qigong exercise, in contrast, is slow, gentle and encourages deep breathing, stretching, and movements that are beneficial and healing to the joints, organs and bones. The Qigong is aimed not only at strengthening the muscles and the cardio-vascular system, but also specifically focuses on balancing the systems of the body, building, circulating and conserving Qi, strengthening the bones, tendons, joints, nervous system, internal organs, glands, and the reproductive system as well.

Qigong Physical Therapy for Specific Problems

There are also specific Qigong exercises for almost every type of ailment: muscular-skeletal problems, internal organ problems, and for many other specific diseases and conditions. For example, two 20th century Shanghai physical therapists, together with Qigong and martial arts masters from the Shanghai Physical Culture Institute, developed a form of

Qigong called *Liangong Shr Ba Fa* (18 Refinement Methods) which combines Western physical therapy knowledge and traditional Qigong forms into six exercise sets to treat, respectively: neck and shoulder problems, lower back problems, knee and hip problems, joint problems of the upper and lower limbs, tennis elbow, and internal organ disorders.

Grasping Knee to Chest from 3rd Set of Liangong to Benefit Hips and Knees.

Another system, called Shu Xin Ping Xue Qigong, is specifically used for the treatment of heart diseases. It has been proven beneficial for angina, hypertension, and congestive heart failure.

There exist many other Qigong forms for treatment of specific diseases such as obesity, ulcers, eye problems, epilepsy, Raynaud's syndrome, asthma, emphysema, arthritis, diabetes, herpes, and many other chronic health problems.

Stress Management

Modern science has acknowledged that stress and its debilitating effects are either primary or secondary causes in the onset or prognosis of over one hundred diseases.

Stress management include methods like breathing exercises, visualization, meditation, progressive relaxation, and physical exercise.

Breathing Exercises

The word "Qigong" itself means "breathing exercise." The word "Qi" is used to refer to the life energy circulating through the acupuncture meridians, the "inner breath of life"; in other contexts Qi refers to the breath or air breathed in normal respiration. Respiratory Qigong therapy is called tu gu na xin, or simply tu-na, meaning, "Expelling the old, drawing in the new."

Our breathing is a bridge between our conscious and sub-conscious mind. Breathing is regulated by our autonomic nervous system; it goes on whether we are conscious or unconscious, awake or asleep. Yet unlike most autonomic functions, breathing can also be easily regulated by our conscious intent. By controlling the pace and quality of our breathing, we can affect deep changes in our physiological functions.

Breath is also the link between our body and mind. Oriental medicine speaks of The San Bao (Three Treasures): Jing, Qi, and Shen. Qi is in the middle. From a stress management perspective, the ability of breath control to influence both physical and mental states is vitally important. Since the mind and body being inter linked, consciously control the breathing process which can have a pronounced impact on both physical and mental tension. By gently guiding and allowing the breath to adopt the qualities of breathing exhibited during states of deep relaxation (breathing should become quiet, deep, smooth, even, soft, and fine), one can thus induce the accompanying physical and mental states of relaxation.

Breathing exercises have proven to be effective in reducing anxiety, depression, irritability, muscle tension and fatigue, and are also used in treatment and prevention of agoraphobia, hypertension, breath holding, hyperventilation, shallow breathing, and cold hands and feet.

Passive Progressive Relaxation

One passive progressive method is Fang Song Gong relaxation. It involves deep breathing combined with auto-suggestion messages such as, "I am quiet....I am relaxed." It can be practiced standing, sitting or lying down, and is easily taught in clinic.

Healing Imagery and Visualization

Other qigong methods, such as Liu Qi Fa or Liu Zi Jue (The Six Healing Sounds) involve healing imagery and visualization of the organs, their Five Element colors and positive qualities. These have proved effective in treating many physical and emotional maladies, including cancer. Moreover, patients experience an increased sense of well-being and peacefulness and patients feel energized.

Qigong Meditation

Qigong meditation is not limited to sitting practice. The Daoists speak of the *Four Human Dignities:* standing, sitting, walking, and lying down. Each of

these postures has its own forms of meditation. Most of us are familiar with sitting meditation, walking or moving meditations such as Tai Chi, or lying down relaxation techniques such as the "corpse pose" in yoga. However, in the qigong tradition, the most important of all of these is standing meditation. Usually referred to as zhan zhuang (standing firm like a post), standing meditation is particularly acts as rejuvenative, and is an absolutely essential foundation not only for promoting one's own health, but also for accumulating Qi to heal others and for martial arts practice.

Walking meditation is very relaxing as well as strengthening, and helps teach one to integrate meditation into the activities of one's daily life. It is also very soothing and relieves the stiffness after long periods of sitting, especially after seated meditation practice or following long periods in front of a computer. It has some features of active progressive relaxation methods, in that the natural weight shifting from one leg to another causes an alternating tension and relaxation in the muscles. This alternating action also acts like a second heart; it helps to pump the blood out of the legs and assists the venous blood return to the heart.

Lying down meditation is valuable for entering the deepest states of relaxation, since no muscular effort is required. It is especially recommended for people too ill to sit or stand. Daoist and Buddhist yogis practice lying down meditation for dream yoga, and as preparation for being able to continue meditation while dying, when they might be too weak to sit up.

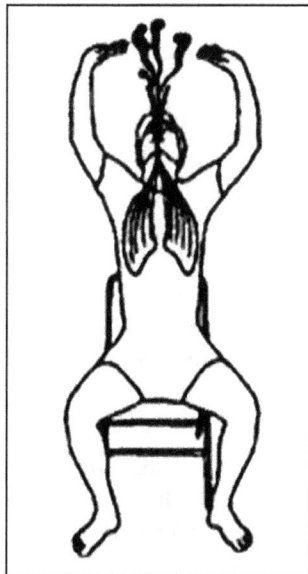

The Lung Sound, SSSSSS, (First of the Six Healing Sounds Qigong).

There are lots of varieties of qigong meditation, differing in both outer posture and inner focus. Since different types of meditation work more effectively for different personality types.

In Classical Chinese medicine, Astrology and Feng Shui, personality is differentiated according to the Five Elements (wu xing): metal, water, wood, fire and earth. In Tantric Buddhism, individuals are classified according to the Five Buddha Families with which they are most strongly associated.

The deep abdominal breathing combined with slow graceful motion helps to oxygenate the body, improve blood circulation, move lymphatic fluids, and release skeletal muscle tension. Qigong exercise deepens breathing and opens the chest, rejuvenates Zong Qi (Ancestral Qi, Pectoral Qi), moves Zhen Qi and removes stagnation, moves the Blood, strengthens the muscles and tendons, builds Wei Qi (Defensive Qi), rejuvenates Heart Qi and calms the Shen (spirit).

External Qi Healing

During the era of China's Cultural Revolution (1965-1975), concept of Qi was officially suppressed and were considered archaic, throwbacks to the superstitions of feudal times. Even acupuncture, although officially sanctioned by the State as a vital part of China's overall health care delivery system, was stripped of all its traditional theories; attempts were made to empiricize point functions and to describe maladies in modern Western medical terminology. Qigong exercises were also officially discouraged, with the exception of Tai Chi.

Towards the end of the Cultural Revolution, practice of Qigong encouraged, presently, hundreds of qigong clinics and hospitals operate across China with official sanction.

Qigong Anesthesia

On June 21, 1980 at Shanghai No. 8 People's Hospital, a unique surgical operation took place which made world news. A qigong master, Lin Hou Sheng from the Chinese Medicine Research Institute stretched out his right hand and pointed his index and middle fingers at Yin Tang (an acupuncture point) between the eyebrows of the patient. Through his fingertips he emitted wai qi (externally projected qi) from a distance of about 3 centimeters on a 29 year old female patient. After three minutes, he nodded to the surgeon who then picked up his sharp scalpel and commenced a surgical operation on a thyroid tumor.

The patient received no additional anesthesia, remained conscious throughout, and did not show even the least sign of pain during the 140 minute operation. When a walnut sized tumor was removed and shown to the patient, a smile lit up her face.

This was the tenth thyroidectomy in little over a month performed with qigong anesthesia at the hospital. Lin, age 41 at the time, has used emitted qi to treat successfully such varied conditions as stomach ulcers, hypertension, urinary incontinence, and protrusions of lumbar vertebra. Since the mid 1980's, due in part to the attention generated by Lin, renewed interest in Qigong and Qigong healing developed into a national fad in China.

More recently, Dr. Wan Sujian, a Chinese army doctor and Director of the Institute of Chinese Daoist Medical Qigong in Beijing, has gained worldwide recognition for his success in treating thousands of paraplegic and quadriplegic patients with External Qi Healing. Dr. Wan's army hospital has also searched throughout China for children who exhibit special Qigong healing abilities and has brought them to the hospital for further training as Qigong therapists.

External Qi Healing (Wai Qi Zhi Liao) is not usually a primary health care choice for most people. It is mainly resorted to when other conventional treatment methods have failed. The fact that External Qi Healing is successful when nothing else works points to its special value and importance as a limb of Oriental medical practice. A Scientific Look at Qi Emission

The Shanghai Atomic Nucleus Research Institute of the Chinese Academy of Sciences has identified Lin Hou Sheng's "Wai Qi" as a low-frequency modulated infrared radiation. Other Qigong research in China on emitted Wai Qi from other Qigong healers has measured not only infrared radiation being emitted from their hands, but also low levels of electric energy, electro-magnetic, magnetic, and low frequency modulated infra-sonic sound (8-14 hertz).

Robert O. Becker of New York has verified that the body responds best not to high currents of electrical stimulus, but to extremely low levels; so low level current (measured in millivolts and nanoamps) is capable of enerating low level electrical stimulation to treat complicated fractures which are otherwise difficult to heal.

Dr. Becker also received one of the first NIH, Bethusa, USA's research grants to study acupuncture in the early 1970's. His research determined that the acupuncture meridians are electrical conductors, with the surrounding skin displaying greater conductivity and lesser resistance than the skin in non-meridian locations.

Qigong and Acupuncture

External Qi Healing can be effectively combined with acupuncture therapy in the clinical practice. Instead of stimulating the needles through manual manipulation or electro-stimulation, one can project qi directly into the inserted needles.

Some Qigong acupuncturists point their palm or fingers directly at the needles, without physical contact. Others transmit a healing field which stimulates the inserted needles from a distance, causing them to vibrate in some cases.

31
Reiki

Reiki (pronounced Ray-Key) is a method of natural healing based on the application of Universal Life Force Energy. How might giving something like Reiki can make any changes in a persons body or attitudes? This isn't a very well understood question and, well, the answers tend to reduce to hand waving. And so it goes that the public at large finds it difficult to accept this technique when the medicine they are familiar with is able to explain itself so well, or so it seems. While western medicine is able to explain a lot of things and put up a dazzling array of words with precise meaning, in actuality the functioning of the body remains largely a mystery. Reiki has a somewhat unique history. The many books on Reiki give similar histories of Reiki, and these are summarized belowAll of the histories of Reiki come from the verbal stories passed on from Mrs. Takata, with little or no hard evidence of Reiki from before World War II remains in Japan, to the knowledge of the mainstream of Reiki practitioners. The lack of hard evidence gives opportunity for skepticism. For instance admission records at Loyolla University should be available for Dr. Usui if he indeed attended there, yet William Rand claims that such a request proved fruitless. There is also the recent discoveries by Dave King of Edmonton Alberta who, while traveling in Japan, came upon a lineage of Reiki practitioners who learned from Usui and do not involve Dr. Hayashi or Mrs. Takata in their lineage. More interesting things are being learned through that avenue. With roots in Japan prior to World War II, it is not surprising that some documentation was lost. Apparently the survivors of Dr. Hayashi lost to the war the resources allowing them to continue the clinic he founded and perhaps stopped practicing Reiki. If it were not for Mrs. Takata learning Reiki before the war and bringing it to America, this healing technique could well have been lost to the world. Such a skin of the teeth saving of Reiki perhaps lost some valuable memories, knowledge and continuity had the lineage not been squeezed through one person. We can only hope that practice, study and intuition will bring back any lost knowledge and practices. Lost knowledge, particularly the evidence to support the following history, does give rise to possible skepticism. Still, Reiki speaks for itself on every use. The energy is real and easily experienced. Once one has experienced the energy, particularly if one is an attuned Reiki practitioner, it is always there and easily demonstrates its truth. Whatever the truth and reality of the claims in the history given below, the ability to perform Reiki so easily came from somewhere and is, for me, the ultimate proof that this path of developing my healership is wise.

Reiki is a Japanese word meaning "Universal Life-Force-Energy". The "Ki" part is the same word as Chi or Qi, the Chinese word for the energy which underlies everything. Reiki is a system for channeling that energy to someone for the purpose of healing. It was discovered by Dr. Usui in the late 1800's, a teacher or perhaps dean of a Christian school in Japan. Reiki is one of the more widely known forms of energy healing. Energy Healing involves direct application of Chi for the purpose of strengthening the clients energy system (aura). Chi is the term used by the Chinese mystics and martial artists for the underlying force the Universe is made of. Mystics in all cultures have talked about the physical universe being made of an underlying form of something, much as modern physics research is now coming to understand the Universe is made of energy which is subject to (or affected by) thought. Just as modern physics says this energy is affected by thought the mystics also say this underlying form is affected by thought, going so far as to claim we create our own reality from our thinking and the thoughts we share between each of us every day.

Stop and reflect for a moment. Imagine the implications of the universe around us made from energy which can be shaped and manipulated by thoughts. This is the implication of both Quantum Physics and ancient Metaphysics. Might this make some diseases easier to explain, especially those which have no apparent physical cause? Might this make miraculous cures easier to explain? For example there was research study done a couple years ago showing that prayer improves the health of those who are prayed for, the focus of intent in prayer sends thoughts out in the form of "I wish such-and-so to happen", so if the world is energy subject to the power of thought then prayers must work (depending on how clearly and carefully you hold and elaborate on the prayers or thoughts). Remember this while reading here about Reiki.

Reiki is very easily learned, very simple to use, and beneficial for all. It is one of many forms of healing through the use of the natural forces which were given the name Chi by ancient chinese mystics. Some forms of healing using Chi energy forces are Chi Gong, Pranic Healing, Chelation (as taught by Brabara Brennan and Rosalyn Bruyere), and Polarity Balancing. All (apparently) use the same energies, with the difference being techniques of application and an energy quality commonly known as vibration. (This term vibration is widely used in the "New Age" community to refer to the range in experience between the dense or low vibration of the physical world to the high vibration of the highest spiritual expression; while the term is inadequate to truly describe the differences, and while Reality is probably not a simple bipolar scale of experience, the term is descriptive of how the energy feels). Reiki uses Chi or Prana, Orgone, Odic, as they are all names for the same force.

A Reiki healing is very simply performed. The practitioner places his or her hands upon the person to be healed with the intent for healing to occur, and then the energy begins flowing. The Reiki energy is smart since the Universe is a very smart place indeed. The energy knows where to go, and what to do once it gets there, or else is being directed by a higher intelligence. The energy manages its own flow to and within the recipient. It draws through the healer exactly that amount of energy which the recipient needs. All this happens without direct conscious intervention by the healer. The healers job is to get out of the way, to keep the healing space open, and to watch/listen for signs of what to do next.

Reiki is capable of healing anything because it works at very fundamental levels of reality. Even though the capability is there, this is not what always happens. The limits to Reiki seem to be in the recipients willingness to cast off old habits and patters, to accept change and to accept healing. The level of reality where Reiki operates is the underlying energy structure of matter, as the physical matter we see around us is a solidified form of energy (Remember Einsteins famous equation which says that each unit of mass is equivalent to a certain amount of energy). At the level where Reiki functions, anything can be changed because all is fluidlike and is very malleable. Or so is the theory and so is the experience of Reiki practitioners. Emotional difficulties are just as healable as physical ones since emotional issues are even more directly present in the energy structures.

In this context, the word Healing has a somewhat different meaning from the widely accepted meaning. The widely accepted meaning for healing seems to be curing of symptoms, for that seems to be what medical doctors and the like look to do in their practice. The other meaning for healing, used in the practice of Reiki as well as other related areas, is the return to greater wholeness. There is an ideal form each of us has, this ideal form being the highest and clearest expression of who we are. Pain or disease comes from any deviation between the persons current form in the 3D physical world and this ideal form. Healing, then, is to bring this physical form into closer alignment with the ideal form.

There should be some attempt at understanding how Reiki, and other methods of using Life Force Energy, works. Anybody who has spent much time with Reiki will have seen very interesting occurrences happening and experienced changes in their life. Perhaps it will have challenged their belief system, for it did mine, or it may have fit nicely with their already existing beliefs. While some say it is merely enough to accept what has happened without looking a gift horse in the mouth, many people like to peek under the hood and see what's happening.

The first thing to discuss is the model for our experience of the world.

There is an overall energy field that is vast, wide, and encompasses all that is. This energy field is both vastly more complicated and vastly simpler than the physical world we see with our eyes, taste with our mouths, smell with our noses, hear with our ears, and touch with our skin. The energy field permeates all the physical things we see, and the physical things we see are sort of a condensation of the energy in this field. The term condensation is imprecise and incomplete, but is the best term I know.

That which we can sense with physical senses is limited. The organs have a specific range within which they operate, for instance the eyes operate in the light frequencies between red and violet, the ears between 20 and 20,000 Hz, and so forth. There is a wide range of possible vibrations of physical matter which our physical senses cannot perceive because it is outside the capability of our physical bodies. Science and Engineering have done wonders with extending the range of our sensing, but not matter how widely sensing of physical phenomenon is extended no proof of such an energy field is found. The energy of this energy field that permeates all that is is not physical (in the sense of being 3D, whatever that means) and cannot be directly sensed by any physical 3D instrument, biological or otherwise, given current understandings. The recent developments in Quantum and SuperString theories may lead to instruments capable of directly sensing the less condensed states of energy and thus directly measure these energy fields.

By condensed energy the picture in mind is something like a stalactite, the thing which hangs from the ceiling in a cave. The ceiling of the cave, in this analogy, is the highest vibrational level of this energy field, and the tip of this stalactite-shape is an object in the 3D physical reality we see with our eyes. For every object there is a continuum of energy states between the vibrational level of 3D physical objects and the vibrational level of the whole universe. The stalactite is this continuum, and at various places along this stalactite one is at different vibrational levels.

We believe this continuum is the same model as discussed by the theosophists (Helen Blavatsky, Alice Bailey, Barbara Brennan, etc, discussing the different bodies (physical, etheric, auric, mental, soul, etc). These authors claim there are discrete levels to this spread of vibrational states between 3D physical reality and oneness, as the theory of discrete energy bodies is explained to them, watching their own energy field form into discrete energy bodies.

The level closest to the physical body is termed the etheric body. The purpose of the etheric body is to be a template from which the physical body is grown, and the etheric body holds both our personal most perfect form as well as the template for our current form. If one, for instance, breaks a bone then the break can be felt in the etheric body even after the physical bone has knitted back together, because the etheric body is remembering the break. It has been observed that the pains often times associated with old breaks in bones are resolved by working with the etheric body and retraining it to mesh more closely with the bones most perfect shape, to help the etheric body forget about the break. It is the memory of the break which seems to cause these old pains, and by releasing the memory of the break the pains can be no more.

This shows us something about the energy fields and the continuum of energy vibrational levels between the 3D physical world and oneness. Namely that what shows up in the physical world manifests in some way in the higher vibrational levels as well. It has been observed that some symptoms begin with processes in the higher vibrational levels eventually manifesting in the physical body as a disease or some other symptom. Other things, such as broken bones, begin with physical disturbance (the broken bone) which causes disturbance in the higher vibrational levels as well.

So then this way of creating healing, by applying Universal Life Force Energy, is to direct the energy that makes up the universe to the people, places or things where there is difference between the current form and the ideal form. For instance the ideal form of a bone is to be strong and whole, not broken and weak. The application of Life Force Energy then would tend to erase the broken and weak pattern replacing it with a strong and whole pattern. Depending on the depth to which this pattern changing is accepted determines the speed with which the pattern is changed. If one is only able to accept a little change at a time then healing will require much time, or if one is able to accept lots of change the healing can be instantaneous.

Reiki is a method of applying Universal Life Force Energy, hence the name. Reiki is only one such method, there being many other methods of doing this. The different methods of applying this energy take different forms, some methods taking a lot of input and interaction from the practitioner, while others (such as Reiki) take care of directing themselves and work pretty much automatically. In essence using Reiki is to simply turn on a fire hose and let some higher wisdom or power direct the flow of the energy to where it is needed. Otherwise what has been said above applies fully to Reiki.

Do you see yourself entering healing from a desire to be healed? If so do not let this stop you from becoming a Reiki Practitioner for having Reiki can be a great help in self healing. At the same time, in being a healer you are placing yourself at the service of your clients, and during their sessions it is the clients issues which are primary.

In everything we choose to do there are multiple purposes for taking that action. As you progress through Reiki training it is good to be aware and explore your reasons for learning Reiki.

There are several levels to Reiki, and not all of us are called to teach or to offer Reiki treatment on a professional or semi-pro basis. Maybe your purpose turns out to be for your own self healing, or to assist family members, and that's as far as it goes. Maybe your purpose turns out to meld Reiki with other treatments such as Massage. Maybe your purpose is to explore the world of energy healing, and Reiki is one of the stepping stones to greater understanding. Or maybe your purpose is truly to offer Reiki, and that the Reiki practices are the perfect path of spiritual awakening for you.

The only way to find out is to start down the path and see where it leads.

Here's a thought to consider: Many Reiki teachers offer weekend trainings in which the students become certified as a Reiki Master. In one weekend. Is this the right speed for you, or is it too fast? I wish to suggest that it's too fast of a training.

What does the term Master mean, anyway? There are two meanings, one is the term Reiki Master, and the other is the dictionary definition for Master. In Reiki, a Reiki Master is one who is attuned with the "Master" attunements and has been shown the attunement procedure. Clearly one can learn the facts and mechanics of Reiki, of giving Reiki attunements, and be given all the required attunements, and if the class is fast paced enough it can be finished in a weekend. They can be called a Reiki Master but have they achieved mastery in that period of time?

Lord knows, the world is crying out for healing and for more people to take up healing work. If you feel honestly the call to take a fast track to becoming a Reiki Master, by all means go for it. You will not make a mistake any path you take.

There are many Reiki teachers who claim to teach Reiki at a distance. I suppose they are operating under the theory that if Reiki can heal at a distance, then one can send a Reiki attunement at a distance. There is a certain logic to this.

On the other hand, consider what you receive from a teacher who is in the same room with you. The main difference is the quality of communication, hence the level of information the teacher can convey to you.

There is a well known concept in Linguistics about levels of quality of communication. The worst sort of communication is something like smoke signals or waving flags or Morse Code, where the words are highly encoded and require great training to both send and receive. The next level might be typewritten text, with handwritten text slightly better. The next level might be a telephone call. The next level might be talking from one room to another. The next level might be talking while in the same room. While the last level might be while making love. I think you get the point, that each level of communication quality gives more and more information, and its easier for clear and complete communication to happen.

The way we aquire deviations from our ideal form is to accept limitations into our life. Most of this comes from early childhood because that is the phase of life where we are the most open and inquisitive about life. A limitation may be a parent yelling "BE QUIET!!" enough times that the child learns to not speak. Another limitation may be a limp that continues longer after the physical injury has healed, maybe with phantom pains. These limitations are behavior patterns, eating patterns, physical limitations, imagined physical limitations, psychological, mental, or emotional ways of being, living, expressing or loving that is not in alignment with our personal highest expression.

In any healing the goal is to find the limitation, recognize the pattern, recognize where it came from, and let it go. Reiki accomplishes this by providing the recipient enough energy to step above (metaphorically) to see all that and have the courage to let go. This usually does not happen consciously as a result of Reiki, but sometimes it does happen that Reiki gives the recipient the conscious awareness of the pattern and recognizing where it came from. Our lives are a constant flow of patterns of activity (such as the pattern used to accomplish eating breakfast) and it is our choice to have these patterns remain stuck in limited expression, or to release the old patterns and try on new and shinier ones.

Reiki is also a gateway shining pure love into the universe. It is this love which allows us to transcend our wounds and help us remember our true nature.

Principles of Reiki

Just for today

    do not worry.

Just for today

    do not anger.

Honor your parents, teachers

    and elders.

Earn your living honestly.

Show gratitude

    to everything.

— Dr. Mikao Usui

Healing is not the popular conception of removal of symptoms. Healing is to recover a greater experience of unity with the divine harmony, and our true selves. Healing is returning to a state of alignment with your Higher Self or true way of being.

It is western medical science that has made Healing be thought of as the removal of symptoms.

In its simplest form using Reiki is simply the practitioner placing their hands on the recipient with the intent of bringing healing, and willing for Reiki energy to flow.

There is a set of hand positions traditionally taught which give good coverage over the recipients entire body. It is not necessary to follow those positions, they are merely taught as a starting position from which the practitioner can learn. If there is a specific area of concern the practitioner can keep his/her hands right there for as long as necessary.

Since there is no time and space to limit spirit, Reiki can operate without regard to limitations of space (at least). A level II Reiki practitioner can bring healing to a recipient regardless of distance.

In doing remote healings the recipient is to be objectified somehow, and the symbol drawn in the vicinity of the object representing the recipient. By objectified I mean that it is easier for the conscious mind to deal with objects than abstract things. Therefore when doing a remote healing it is easier on the healer to visualize sending the energy to something than otherwise.

There are many ways of visualizing or objectifying the recipient. For instance the healer might intend that their knee is the same as the intended recipient, which will create the necessary connection. Or they might draw in the air a shape to represent the recipient. Or use a teddy bear, pillow, or some other physical object as a stand in. The only limit is the healers imagination. For the healers intent is primary in bring forth the Reiki energy.

What is Reiki and How you can Become a Reiki Master

Reiki was developed as a way to promote healing and reduce stress. "Laying on the hands" is the motto behind Reiki techniques.

The stress appears when the life energy that flows through our body is very low. If this unseen life energy is high then we are capable of doing almost anything. If this energy is high so will our capabilities and we'll be more capable in dealing with day to day stress and problems. Reiki comes from two Japanese words - spiritually guided life force energy and wisdom.

Most patients that went through Reiki treatment said that it was a nice and wonderful experience. Since this treatment applies to the whole body it creates a feeling of security, peace and confidence. It's the best natural way to treat our emotions and bodies. Reiki meditations work and are effective no matter what your illness is or what problems you may have. There are no side effects and is can be used in combination with medical treatment.

The Reiki techniques are quite easy to comprehend and learn and you can find Reiki classes almost everywhere. You don't have to be young or healthy in order to do this. You can learn how to heal others as well as yourself regardless of your training, intellect or background.

Reiki is not a religious thing so if you want to experience the Reiki treatment you don't need to quit your current religion. Reiki is more a spiritual treatment and you'll be able to heal even if you don't believe in God. All the Reiki practitioners promote harmony with every thing, person and life form that surrounds us. Only this way the healing comes naturally, since it comes from inner harmony and peace. It will make you more balanced and you'll see its effects right away. You just need to believe in it for this.

The Reiki healing process works like this: The healer lays his hand just a few inches above the patient's body and move their hands until they "sense" where is located the problem.

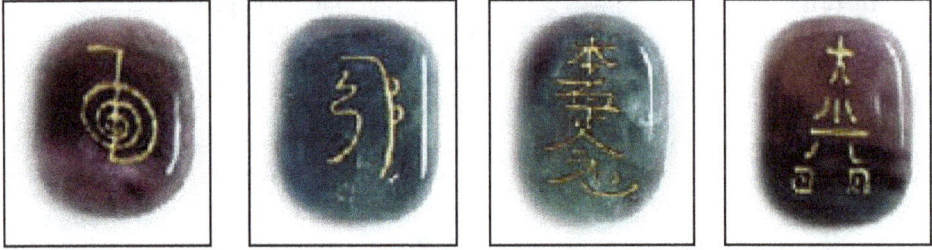

After that they'll ask the patient to concentrate its energy into that area while they'll do the same. This process will help in healing the mind and the body of the patient.

They say that your emotions are to strong or your mind is too stressed that your body will suffer as well. This technique can be used along with medical treatment in hospitals and can help the person recover faster in both ways: body and mind.

Reiki is a form of spiritual practice proposed for the treatment of physical, emotional, mental and spiritual diseases. Reiki was developed in early 20th century Japan to receive the ability of 'healing without energy depletion'. The name Reiki derives from the Japanese pronunciation of two Chinese characters that are said to describe the energy itself: 'rei' (meaning 'soul', 'spirit' or 'ghost') and ki (Chinese qi, meaning breath or 'life force energy' in this context). Reiki is a therapy that the practitioner delivers through the hands, with intent to raise the amount of ki in and around the client, heal pathways for ki, and reduce negative energies. Reiki can be practiced in several ways: on its own, along with conventional medical treatments. In the latter case, Reiki is a type of "distant healing." Practitioners perform Reiki most often in offices, hospitals, clinics, and private homes. The practitioner and client determine the number of sessions together. Typically, the practitioner delivers at least four sessions of 30 to 90 minutes each. Practitioners use a technique similar to the laying on of hands, in which they claim to be channels for energy ("Ki") guided by a universal spirit or spiritual nature ("Rei") flowing through their palms to heal a person wherever they may need healing. They also believe that if ki's flow is disrupted, the body's functioning becomes disrupted, and health problems can occur. The concept that sickness and disease arise from imbalances in a vital energy field is the foundation not only of Reiki but of some other CAM therapies, such as traditional Chinese medicine (in which the energy is called qi or chi) and homeopathy (vital force). Reiki appears to be generally safe, and serious side effects have not been reported. The person may have symptoms such as a feeling of weakness or tiredness, a headache, or a stomach ache. Reiki practitioners believe that these are effects of the body releasing toxins. They advise the client on how to deal with such symptoms if they occur, such as by getting more rest, drinking plenty of water, or eating a lighter diet.

Theories and Practices of Reiki

Reiki (pronounced 'ray-kee') is commonly perceived to be (only) a form of spiritual healing. However Reiki is more than this: Reiki is now also increasingly regarded as a belief and behaviour approach towards personal fulfillment, whose principles and philosophy can explain and achieve better life-balance and well-being; enable the furthering of self-development and discovery, as well as being a remarkable healing and recovery treatment. Based on very ancient Eastern teachings, Reiki methodology and history has given rise to confusion, speculation and mystery, although in recent years, happily, the consistency and detail of information about Reiki has progressively improved, to the point today where Reiki is an accepted form of 'alternative therapy. The Reiki energy is an"intelligent" energy, which "knows what to do," or "where it is needed the most." Thus, Reiki adherents say, if the recipient needs it and is ready to heal, the Reiki energy will go where it needs to for healing. If the intended recipient does not accept the energy on some level, the energy will not be absorbed. Some schools teach that Reiki "spirit guides" keep watch over Reiki energy and assist the practitioner. It is said by them that any intention to do harm will block the flow of Reiki energy. It is also taught that Reiki energy enters through the 1st (root or Muladhara) chakra at the base of the spine, fills the aura, becomes centered in the 4th (heart or Anahata) chakra, and flows out trough the practitioner's hands. In a Reiki session, the practitioner asks the recipient to lie down and relax. The practitioner then is said to act as a channel for Reiki energy, purportedly allowing "Reiki energy" to be channelled through the practitioner to wherever the patient is thought to require it.Reiki healers say that their energy can be used for healing either in physical proximity or from a distance. There is no belief system attached to Reiki so anyone can receive a Reiki treatment or a Reiki attunement.

What does Reiki Do?

Reiki helps to harmonize body, mind and spirit for yourself or anyone you want to help.

Reiki: A Powerful and Gender Healer

☆ Relieves pain

☆ Clears toxins

☆ Promotes creativity

☆ Relaxes and reduces stress

☆ Heals holistically

☆ Promotes natural self-healing

☆ Enhances personal awareness

☆ Adapts to the natural needs of the receiver

☆ Treats symptoms and causes of illness

☆ Releases blocked and suppressed feelings

☆ Aids meditation and positive thinking

☆ Balances the energies in the body

☆ Balances the organs and glands

☆ Strengthens the immune system

Is Reiki a Religion or any other Belief System?

No, Reiki is not a religion and you do not have to believe in anything special. Reiki is suitable whatever faith or beliefs you have. The only thing you need to have is a sincere wish to help yourself (or others). Most healing systems are based on intent and so is Reiki.

Can Reiki be Used on any Illness?

Yes, Reiki can be used on any illness but this does not mean everyone will be healed using Reiki. Reiki will help restore the balance in the body and can relieve or even remove symptoms. Reiki has successfully been used on conditions such as reducing stress, relieving pain, headaches, stomach upsets, back problems, asthma - respiratory problems, PMT, menstrual problems, sinus, anxiety and many many more. Remember that you as an individual have to have the firm intention of getting well before any real healing can take place. (Please observe that Reiki should never be used instead of medical treatment, it should be used as a compliment and a way to become and remain healthy.)

What will I Feel or Experience during a Treatment?

Once again I will have to say that it differs from person to person and possibly also between different sessions. The energy flows wherever it is required (spiritually guided) and can normally be felt as a warm sensation or tingling in the body. Receiving Reiki is a very relaxing and soothing experience! Usually when receiving Reiki you will feel very relaxed and you may have a "floating" feeling. You may feel heat, a tingling or cold under the practitioners hands. These and other feelings come from what the Reiki energy is actually doing in and around your body *i.e.* adding energy, removing excess energy, balancing, removing blockages *etc.* It is also possible that you experience different emotions or thoughts you had forgotten might pop up. Remember that all healing starts a process of release, this could be emotions or even toxins in your body that you need to get rid of. You might initially even feel worse than before but this is soon passing. You can help your body by drinking lots of water.

What is the difference between Healing and Curing?

If you look up the two words in a dictionary you will probably find that they mean more or less the same. In real life we seem to have given the two words slightly different interpretations. Healing is when you go to the root of the problem and restore the body's balance. Curing is what is normally considered done when you actually remove the problem (but maybe not the cause). When you have surgery you can remove the "sick" organ or tissue but this might not solve the problem as the procedure often does not solve the underlying problem. One can also be healed but not cured, *i.e.* your body is in balance but you still have a medical problem.

What does a Reiki Treatment Consist of ?

You relax, on a couch or seated while the healer holds his hands on or above you. A treatment can last an hour or longer depending on the treatment required. In the western world many practitioners use the standard hand positions and commonly a full treatment is given covering all the important organs of the body. There is no pressure on the body making it ideal for treating all ages and conditions, sometimes hands are even held away from the body.!

How many Treatments should I have ?

There is not really a set number. Every person and his/her problem is different. Some people only require one or two sessions while others need several. If you are just interested in seeing what Reiki can do for you then just take one or two sessions. If you have a chronic problem you want to work with then make a plan. Suggestion: Start with one session per day for 4 days and then follow up with one session per week. Remember that a Reiki initiation in itself is a powerful healer.

Can Children Receive Reiki ?

Yes, anyone can receive Reiki. A session for a child will normally be shorter than for an adult. It is also possible to attune children to Reiki. The first level, Reiki I, can be given even to small children. Older children can be given Reiki II and learn the use of the Reiki symbols.

Can I Stop Using Medications ?

NO! Reiki should never be seen as a substitute for medical care. Reiki is a complementary therapy that can help your healing process and in many cases improve both your physical and mental health. Remember, Reiki cannot hurt you!

How does One Learn Reiki ?

Anyone can "learn" Reiki. The word learn is not quite right as most of us cannot just read about Reiki and then do it. The ability to use Reiki is normally given via an attunement or initiation. When attending a Reiki course the participant gets attuned by a Reiki master through a simple process, this opens him/her to receive and utilize more of the Universal Life Energy.

32

Rudraksh Therapy

Rudraksha is considered to be the most potent manifestation of the Cosmic Force. Hence Rudraksha is the object of veneration and also the source to reach the higher self. These beads are the seeds of the Rudraksha fruit obtained from Rudraksha trees. The Rudrakasha tree is botanically known as ELAEOCARPUS GANITRUS ROXB. Its English name is UTRASUM BEAD TREE. Rudraksha trees are mostly found in South Eastern Asian Islands of Java, Sumatra, Borneo, Bali, Iran, Java, Timor (Indonesia) and parts of South Asian Kingdom of Nepal. Around 70 per cent of the Rudraksha trees are found in Indonesia, 25 per cent in Nepal and 5 per cent in India. Considered a major stress reliever, reducing circulatory problems and of course as the best beads, the berry (Elaeocarpus Ganitrus).

There are clefts called MUKHIS on the surface of the beads. The number of mukhis on the surface of a Rudrakasha bead helps in determining its quality. According to the number of mukhis the rudrakasha bead ranges from single face to a several faced bead. Asians have used Rudraksha beads traditionally. Asian Yogis and Monks found that merely wearing the Rudraksha beads gave them astonishingly tremendous amount of tranquility, concentration that helped them meditate for a long period of time with spectacular control over their mind.

Importance of Rudraksha

Rudraksha is often believed to symbolize the link between earth and heaven, and though Hindu scriptures like the updated Jabal Rudrakasha Upanishad have made the berries their own, modern scholars and gurus have described them as being of no particular religion or rather non religious and have pointed out that they can be used as a mala, rosary or Tasbih (sibah).

Though mired in obscurity so far, the Rudraksha is slowly but surely making its way into the lives of the health conscious human beings. The Rudraksha helps in relieving from blood pressure and lessening stress. Soak it in a glass of water overnight and drink it the next day to relieve stomach disorders. Dip it in any vegetable oil for 21 days, and apply it on your aching joints and feel the relief you get. Of course, most people who have tried this have done so along with their regular medicine, but one must approach the Rudraksha with a mixture of open mindedness and faith, rationalism as well as emotion, to get positive feedback that is why many who put it under their pillow have sworn that it cured them of their insomnia. Ancient medical texts claim that the Rudraksha can prevent aging and can prolong life. The beads are supposed to be anti-pyretic, anti-paralysis and can help balance the vital chakras of the human body that control bile, wind and phlegm. It also has the power to heal those suffering from hypertension and heart ailments. Besides, as it is completely without any side effects it cannot do you any harm. That explains why the old medical literature allows its use for menstruating women while banning other beads and malas.

Among the 38 varieties mentioned out of which 21 are most conspicuous the rarest is the one faceted or one - mouthed (the one with a single cleft). Wearing it is considered to be auspicious during lunar and solar eclipses, or on full or new moon days. A person attains unlimited pleasure only by seeing Rudraksha. The persons wishing devotion and salvation should wear it after purifying themselves.It removes their numerous pains, sorrows, and calamities. Rudraksha fulfills all the wishes. It is peerless in the universe. If a person wears Rudraksha with love and faith, he gets rid of all sins and attains the supreme goal. As long as a man keeps wearing Rudraksha, he does not fear untimely death. He cannot die without completing his span of life. At the time of death he gets the true knowledge of Lord and His abode too.

One should wear Rudraksha by all bits of efforts. Thousand of persons who have worn Authentic Rudraksha beads have repeatedly confirmed that it has given them considerable relief from Blood pressure, Stress, Hyper tension, Depression and other mind related problems including neurotic conditions. Rudraksha has been found to boost the confidence and inner strength of the wearer significantly.

The Rudraksha Beads have also been boosting the luck and prosperity of the wearer significantly. It is quite understandable as with enhanced self-esteem and confidence one achieves greater success in all ventures. Surely Rudraksha are the wonder beads helping us lead a more successful and complete life. Buddha wore them, Gandhi wore them, and Dalai Lama wore them. So did Yehudi Menuhin, Osho Rajneesh or far that matter many Hollywood and Bollywood artists. Probably the best of the berries ever produced in the world, the Rudraksha, courtesy a growing interest in alternative therapies, is catching on like never before.

Medical and Scientific Value of Rudraksha

Rudrakasha beads have several amazing powers due to their electromagnetic character. This Electromagnetic character empowers the bead to cure the human body medically as well as spiritually. Electrical Properties: The human body can be considered as a complex Bioelectronic Circuit consisting of the whole of Nervous System and all other organs residing inside the human body. Number of Electrical impulses are generated by the human body due to different reflex actions taking place, continuous pumping of heart for blood circulation, neurons and nervous system *etc.* The ability to send out subtle electrical impulses and Inductive vibrations and act as a ielectric as in a capacitor to store electrical energy. This is termed as Bioelectricity. All the work of our sense organs depends on the subtle flow of Bioelectricity current. Thus all the actions that our body performs are effectively controlled.

The Bioelectric current is produced due to the difference in the energy levels of different body parts. This smooth flow of Bioelectric current causes the properly controlled functioning of the body parts. Bioelectric Circuit-There is a third element in the Body and Brain called the Bio Electronic circuit Interface, that of the mind.

Any activity that can produce stress or maladjustment can throw the streamlined activity, the Electronic circuit of the Body and Mind out of gear. Human beings and all living beings are prone to stress continuously in the continuous fight for survival and prosperity.

In modern age with intense competition the stress levels have increased tremendously. Almost every individual has problems of stress and stress related ailments like insomnia, alcoholism, depression, maladjustments, heart diseases, skin diseases *etc*. Any Doctor will confirm that almost 95 per cent of the ailments are Psychosomatic or stress related *i.e.* originating from mind. When there is stress or maladjustment corresponding stress signals are sent to the Central Nervous systems, there is an increased activity or abnormal of Neurons and Neuro transmitters. The magnitude of change will depend on the cause and specific case. When such a thing occurs and it occurs continuously, streamlined flow of electrical signals throughout the Mind-Body interface is disrupted and it makes us feel uncomfortable and we are not able to act with our full efficiency. Our Blood circulation becomes Non-ideal and we feel various illnesses. Unfortunately this happens all the time. Rudraksha beads act as a Stabilizing Anchor. Rudraksha beads' electrical property can be broadly categorized into: Resistance: There is continuous and subtle flow of bioelectrical signal throughout the body due to potential difference between parts of the Body. Rudraksha beads of particular Mukhis or Facets have a definitive Factor of Resistance. It is measured in Ohms. When these beads resist the flow of bioelectrical impulses a specific ampere of current flow is generated depending on the factor of resistance. This acts in tandem with heartbeat, streamlining it and sending out specific impulses to brain. These impulses stimulate certain positive brain chemicals. Making us feel better, more confident, poised and more energetic. It has been observed that specific Mukhis or Facets of Rudraksha beads send out specific signals acting on a particular brain chemical and thereby by effecting specific positive changes in personality.

What are the Precautions that Need to be Taken when Wearing Rudraksha ?

Rudraksha should be taken off during sex, menstrual cycles, during visit to funeral grounds and also while visiting a newly born baby. It is advisable to take off the beads while taking bath as excess soap may dehydrate the natural oil content of the beads.

What is the Average Life Span of a Rudraksha Bead ?

A good quality Rudraksha can last for several years. If taken good care of, it can be passed on from generation to generation.

Can the Family Members Keep Interchanging Neck Mala, Japa Mala or Rudraksha Beads ?

A sort of binding develops between the wearer of the Rudraksha bead and the bead itself therefore it should not be interchanged within the family members. But one is free to inherit the bead from the later generation.

Should Rudraksha be Worn in a Thread or Silver or Gold Metal ?

Rudraksha can be worn in anything according to the person's faith and is best if it is worn in a metal be it Silver, Gold, or Copper. When worn in metal the Electromagnetic powers of the bead increases. Thus increasing its effect.

What are the Tests for the Genuineness of a Rudraksha Bead ?

There are many tests suggested to test the genuineness of a Rudraksha bead. But these tests cannot be relied upon always.Therefore the best way to check the genuineness of a Rudraksha bead is by consulting a reliable person.

Are Rudraksha Beads Used only for Meditation? Is it Meant Only for People of Certain Background ?

Besides giving the user peace of mind, focus and concentration Rudrakasha beads prove to be a great help in curing physical illnesses. They are also said to have an anti ageing property. The rudraksha beads are found in number of varieties according to the number of facets on their surface. All these varieties have different Electromagnetic properties and thus show different effects. By wearing these beads one gets tranquility, peace of mind, focus and concentration. When worn around the neck they are said to control the blood pressure, hypertension, insomnia *etc.* Sometimes they are very effective in curing ailments of fever, pain etc, when taken as an oral drug. People of all religions, castes or creed can use these beads. There is no age or sex bar for their use until they are used under certain precautions.

Can One Wear the Combination of Different Mukhis at the Same Time?

A person can wear Rudraksha beads having a single type of Mukhi or a combination of several mukhis. Wearing a combination of several mukhis will bring all the different effects of them simultaneously to a person. Remember that wearing Rudraksha brings no harm to an individual.

How Long it takes to Feel some Effect ?

A person can wear Rudraksha beads having a single type of Mukhi or a combination of several mukhis. Wearing a combination of several mukhis will bring all the different effects of them simultaneously to a person.Remember that wearing Rudraksha brings no harm to an individual.

What are the Astrological Purposes of Rudraksha ?

In a horoscope of a native, there are 9 main planets. Each planet has its own nature and character. There are 12 houses in the horoscope. The snapshot of the situation of these 9 planets in 12 houses (at the time of birth of the native) is called a Horoscope. 12 houses of the horoscope represent different aspects of the life of the native. These houses contain different numbers from 1 to 12. These numbers represents different Signs (*e.g.* 1 represent Aries, 5 represents Leo and 12 represent Pisces). Each sign has the lordship of particular planet (*e.g.* Aries, Leo and Pisces are under the Lordship of planet Mars, Sun and Jupiter respectively). Now, in astrology some planets are friends to each other, some planets are enemies to each other and some planets are average to each other. When a planet is situated in the sign of his friend planet, he use to give good results related to that house and if a planet is situated in the sign of his enemy planet, he use to cause tribulations related to that house. In general, astrologers use to suggest gems to the native to overcome the tribulations caused by the malefic planets or to enhance the positive results given by the benefic planet. But it is often seen, that a particular gem doesn't suit the native and starts giving tribulations. But as the Rudrakshas are believed to be always beneficial, it can be used for both kinds of planets, may it be benefic or malefic.

☆ 1 and 12 faced (Mukhi) Rudraksha is related to planet Sun.
☆ 2 faced (Mukhi) Rudraksha is related to planet Moon.
☆ 3 faced (Mukhi) Rudraksha is related to planet Mars.
☆ 4 faced (Mukhi) Rudraksha is related to planet Mercury.
☆ 5 faced (Mukhi) Rudraksha is related to planet Jupiter.
☆ 6 faced (Mukhi) Rudraksha is related to planet Venus.
☆ 7 faced (Mukhi) Rudraksha is related to planet Saturn.
☆ 8 faced (Mukhi) Rudraksha is related to planet Rahu.
☆ 9 faced (Mukhi) Rudraksha is related to planet Ketu.

Ganesh Rudraksha

It is a moorti roop of Ganeshji wherein a trunk like mounting is natural in the bead of Rudraksha. Having resemblance to Lord Ganesha, this bead is used to seek the blessings of Lord Ganesha to remove the obstacles in life.

1 Mukhi Rudraksha

Enlightens the Super Consciousness, provides improved concentration and mental structure changes specific to renunciation from worldly affairs. The wearer enjoys all comforts at his comand but still remains unattached. Ruling planet Sun. Affects 7th Chakra on top of head named Sahasrara Padma Chakra. Affects the neurophysiology and consciousness through the Pineal Gland, Pituitary Gland, the Optic Chaiasma and the Hypothalmus. Recommended for Headache, Heart Disease and Right Eye Defect.

Ruling God : Shiva Ruling Planet : Sun Beeja Mantra : Om Hreem Nama.

2 Mukhi Rudraksha

This represents Ardhanareshwar a joint image of the Lord Shiva and Goddess Parvati (Shakti) and blesses the wearer with UNITY it could be related to Guru-Shishya, Parents-children, husband-wife or friends. Maintaining ONENESS is its peculiarity. Ruling planet Moon. Recommended for Left Eye Defects diseases of Heart, Lung, Brain, Kidney and intestine.

Ruling God : Ardhnareeshwar Ruling Planet : Moon Beeja Mantra : Om Namah.

3 Mukhi Rudraksha

This Rudraksha represents "Fire" God. The way fire consumes every thing and remains pure, the wearer too gets free from sins or wrongs of life and return to purity. Ideal for those who suffer from inferior complexes, subjective fear, guild and depression. Its ruling planet is

Mars. Recommended for Blood defect, plague, small pox, blood pressure, weakness, disturbed menstrual cycle, spontaneous abortion and ulcer.

Ruling God : Agni Ruling Planet : Mars Beeja Mantra : Om Kleem Namha.

4 Mukhi Rudraksha

This represents God Brahma. The wearer gains power of creativity when blessed. Very effective for students, scientists, researcher, scholars, artists, writers and journalists. Increases memory power, wit and intelligence. Its ruling planet is Mercury. Recommended for mental disease, paralysis, yellow fever and nasal disease.

Ruling God : Brahma Ruling Planet : Mercury Beeja Mantra : Om Kleem Namah.

5 Mukhi Rudraksha

This repressents Lord Shiva, the symbol of auspiciousness. The wearer of Five Mukhi mala gains health and peace. Five Mukhi Rudraksha monitors blood pressure and cardiac ailments. Five Mukhi Rudraksha mala is also used for japa. By wearing the mala the wearer's mind remains peaceful. There is no suspcion about the fact that the wearer of Five Faced Rudraksha mala never gets untimely death. Its ruling Planet is Jupiter-Affects all major Chakra Points. Recommended for Bone Marrow,Liver, Kidney, Feet, Thigh, Ear disease of fact and Diabetes.

Ruling God : Kalaagni Rudra Ruling Planet : Jupiter Beeja Mantra : Om Hreem Namah.

6 Mukhi Rudraksha

Represents Lord Kartikeya. Saves from the emotional trauma of worldly sorrows and gives learning, wisdom and knowlege. Affects understanding and appreciation of Love, Sexual Pleasure, Music and Personal Relationships, Rulling planet Venus. Six Mukhi affects the neurophysiology through the sacral plexus, sexual and reproductive organs, prostrate gland and gives protected emotional response as it relates to physical and mental comfort levels. Recommended for problems with Eyes. Reproductive Organs, Urinary Tract, Prostate, Mouth and Throat, could be worn on right hand.

Ruling God : Kartikeya Ruling Planet : Venus Beeja Mantra : Om Hreem Hoom Namah.

7 Mukhi Rudraksha

This Rudraksh represents Goddess Mahalaxmi. Good health is blessed to him who wears seven faced Rudraksha. It should be worn by those who are suffering from miseries pertaining to body, finance and mental set-up. By wearing Seven Faced Rudraksha man can progress in business and service and spends his life happily. Its ruling planet is Saturn. Affects entire psychological and physical neurophysiology regarding malefic effects of Saturn which are weakness, colic pain, handicapidness, pain in bone and muscles, paralysis. long term disease, impotency, worries and hopelessness.

Ruling God : Mahalaxmi Ruling Planet : Saturn Beeja Mantra : Om Hoom Namah.

8 Mukhi Rudraksha

This Rudraksha represents Lord Ganesha. Remove all obstacles and brings success in all undertakings. It gives the wearer all kinds of attainments. Riddhies and Siddhies. His opponents are finished *i.e.* the minds or intentions of his opponents are changed. Its ruling planet is Rahu (Dragon's head) Affects Medulla oblongata, Hypothalamus, Limbic System Recommended for Lung, Feet, Skin and Eye.

Ruling God : Ganesh Ruling Planet : Rahu Beeja Mantra : Om Hoom Namah.

9 Mukhi Rudraksha

This Rurdraksha represents Goddess Durga (Shakti). When invoked, the mother Goddess blesses the wearer with lot of energy powers Dynamism and fearlessness, which are useful to live a life of success. Its ruling planet is Ketu (Dragon's tail) Affects Pineal Gland, Pituitary Gland. Recommended for diseases of Lung, fever, eye, pain, bowel pain, skin disease body pain. It could be worn on left hand.

Ruling God : Durga Ruling Planet : Ketu Beeja Mantra : Om Hreem Hoom Namah.

10 Mukhi Rudraksha

It has no ruling planet, and pacifies all negative planetary energy. This contains the influence of ten incarnations and the ten directions. It works like a shield on one's body and drives evils away. When invoked, the wearer's family will be nourished endlessly generation after generation. It has no ruling planet and is held to sublimate all the malefic effects of all the planets and is recommended for entire neurophysiology.

Ruling God : Vishnu Ruling Planet : None Beeja Mantra : Om Hreem Namah Namah.

11 Mukhi Rudraksha

Represents Lord Hanuman. When invoked, blesses with wisdom, right judgement. Powerful vocabulary, adventurous life and success. Above all, it also protects from accidental death. The wearer becomes fearless. It also helps in Meditation and removes the problems of yogic practices. It has no ruling planet. Recommended for nerve Energy and Maintenance of entire neurophysiology.

Ruling God : Hanuman Ruling Planet : None Beeja Mantra : Om Hreem Hoom Namah.

12 Mukhi Rudraksha

This represents Lord Sun. The wearer gets limitless administrative capacity. He gets the quality of the sun - to rule and to move continuously with brilliant radiance and strength. Good for ministers, politicians, administrators, businessmen and executives. It is miraculously effective. Ruling planet is Sun. Affects 4th chakra at the Heart Center named Anahata Chakra. Removes worry, suspicion and fear. Increase self imageand motivation. Affects the neurophysiology through the Cerebral Hemisphere, Heart, Lungs and Skin. Recommended for problems with Heart Disease, Lung Disease, Skin Disease, Hiatus of Stomach and Bowel Problems.

Ruling God : Sun Ruling Planet : Sun Beeja Mantra : Om Krom Sarom Rom Namah.

13 Mukhi Rudraksha

Represents Lord Indra. When invoked, showers all possible comforts of life one can ever desire. It gives riches and honour and fulfills all the earthly desires and gives eight accomplishments (Siddhies) and the god cupid (Kamadeva) pleased with the man who wears it. Being pleased cupid fulfills all the worldly desires. Ruling planet Venus. Affects Celiac Plexus and Prostate. It is helpful for meditation and spiritual attainments and also causes material upliftment.

Ruling God : Indra Ruling Planet : Venus Beeja Mantra : Om Hreem Namah.

14 Mukhi Rudraksha

14 Mukhi Rudrakshs is the most precious divine gem- Deva Mani. It awakens the sixth sense organ by which the wearer foresees the future happenings. Its wearer never falls in his decisions. It protects from ghosts, evil spirits and black magic. It provides the wearer safety, security and riches and self power. It is a very powerful antidote for Saturn miseries and provides miraculous cures to several ailments. Recommended to be worn on chest, forehead or right arm.

Ruling God : Hanuman Ruling Planet : None Beeja Mantra : Om Namah.

Gauri Shankar

Two naturally joined Rudrakshas, called Gauri Shankar is regarded as the Unified form of Shiva and Parvati. It makes the husband and wife identify each other. Therefore it is regarded the best thing for peace and comfort in the family. If a man worships Gauri Shankar at his worshipping place, the pain and suffering and other earthly obstacles are destroyed and the peace and pleasure of family are increased.

Ruling God : Shiva and Parvati Ruling Planet : Moon Beeja Mantra : Om Shree Gauri Shankaraya Namah.

33

Shiastsu

Shiatsu a TCM therapy is originated in Japan and traditional Chinese medicine, and has been widely practiced around the world since the 1970s. Shiatsu means "finger pressure," which describes the technique. In shiatsu, pressure with thumbs, hands, elbows, knees or feet is applied to pressure points on the body. This form of massage also focuses on rotating and stretching limbs, joints, and pressure points, or meridians, as they're called in traditional Chinese medicine.

Shiatsu, is a unique form of holistic healing touch, it offers us a way of supporting our body at all times of our life. It works with the body's own resources. This is both its strength and weakness. If the body is really sick and out of balance, then sometimes more interventionist measures, such as those offered by conventional medicine are needed. However whatever is happening, our body can always be supported to process it in some way. The only time it can not really be included is during emergency medical procedures. It can support both before and after as there are no reactions with drugs as it is only ever supporting the body in its processes.

Shiatsu could be considered a form of massage, which is often done through the clothes and incorporates simple points and holds. Its essence is simple to learn and effective. Shiatsu includes awareness of body posture, breathing and exercise. Like acupuncture, Shiatsu stimulates the body's vital energy (known as Qi or Ki). Shiatsu is calm and relaxing in nature, yet dynamic in effect; the body begins to re-adjust itself and healing takes place. The receiver is supported to become more aware of their body/mind as an integrated whole, on either a conscious or subconscious level. They become aware of areas of tension or weakness on either a physical or emotional level and through this process healing occurs.

As well as the points and meridians of acupuncture, work with the physical body, muscles, joints, blood and so on, is included. Massage type strokes like kneading or effleurage are part of shiatsu. It is characterised by extensive use of pressure techniques over acupoints often done using thumb or palms. The pressure varies according to the person, the area of the body, and what the work is being done for. It can be very deep, and help ease out physical tensions. It can be very light and feel soothing. Breathing and visualisation may be included. Usually there are some stretches and mobilisations, so it can feel a bit like Thai massage or having yoga done to you. It is often done on a futon on the floor rather than a massage table. The practitioner will suggest suitable self care stretching or postural awareness exercises as appropriate.

Shiatsu has its origins thousands of years ago in Japan and was more recently formalised into its modern form over a hundred years ago. It draws on much of traditional Chinese knowledge for its theoretical base, using the same meridians and points as in acupuncture and tuina. It is now quite widely practised in the UK and throughout the world.

Shiatsu is constantly evolving as our understanding of the body evolves and different styles draw upon other bodywork traditions, including massage, cranial-sacral and soft tissue work. Some practitioners support the integration of change by including within the session other modalities such as exercise and breath awareness, dietary therapy, psycho-therapeutic and meditative practices.

Suzanne Yates did her original training in Healing-Shiatsu with Sonia Moriceau (link to the Orchard). This approach was developed out of Sonia's extensive years of training and practise in Zen Buddhism. By understanding the whole being, through the breathing pattern, posture and mental attitude, practitioner and client can reach to the origin of the dis-ease, be it mental or physical.

The theory behind shiatsu is that our bodies are made up of energy, called qi, and that energy gets blocked and causes suffering. Shiatsu massage helps remove the blockages by realigning meridian points, which balances the qi and eases the body and mind. When one balances qi, or vital energy, healing occurs in the body. The nervous and immune systems are both stimulated by applying pressure to the meridians, providing relief for both body and mind. Shiatsu also restores the circulatory system, improving blood flow throughout the body.

The body has twelve meridians, named according to its corresponding organ: lung, large intestine, stomach, spleen, heart, small intestine, bladder, kidney, heart governor, triple heater, gall bladder and liver. The functions of these organs have a broader definition in Eastern medicine.

Shiatsu has many benefits to the body and mind. Here is a list of some of those benefits:

Restore and maintain the body's energy, especially helpful to those suffering from fatigue and overall weakness

Improves circulation

Reduces stress and tension as well as anxiety and depression

Relief from headaches

Promotes healing from sprains and similar injuries

Helps bring relief to arthritis sufferers

Reduces problems with stiff neck and shoulders as well as backaches (including sciatica)

Coughs, colds, and other sinus and respiratory problems

Helps those dealing with insomnia

Aids in treatment of such various things as digestive disorders, bowel trouble, morning sickness, and menstrual problems.

All forms of work with body, *viz.*, exercises, stretches, breathing, different types of massage and body work support our health and well being. In most traditional cultures their importance was considered a fundamental part of health care and we have devalued their importance to our detriment. It is becoming increasingly clear that it is not possible to separate the body from the mind. How we are physically in our body affects how we feel. If we have poor posture or hold tension that will affect our emotions. Touch is our first sense which develops. Already at 8 weeks after conception we have reflex responses to touch before we can hear or see: senses which only develop much later. We experience the world through touch and we store our memory of the world in our body. Any form of supportive touch can be immensely healing, as any form of abusive or violent touch can be immensely damaging. This can help explain why body work can have such a profound effect not just on the physical body, but on the emotional and even spiritual level.

All forms of bodywork: massage, osteopathy, cranial sacral work, rolfing (structural integration), pilates and yoga (to name but a few) support the body in this way. Shiatsu, and any modality which includes awareness of meridians and energy (Qi/Jing) additionally offer a very specific way of understanding how our whole experience is stored in the body.

What exactly are the meridians of Chinese medicine and how do they work? There are various theories ranging from tissue connections, nerve and hormonal relationships. However, it is believed that the key to understanding the meridians lies in understanding how our bodies developed. Many of the connections of the meridians make sense when we understand how we developed embryologically. Three key meridians (Conception, Governing and Penetrating Vessel) lie along the physical mid-line of the body. We develop from the mid-line and our core strength, both physical and emotionally, comes from the mid-line.

Sadly the power of touch and its simplicity has been largely forgotten in the development of modern medicine. Pregnancy and birth are times when women are offered an opportunity to experience their body in an intense and powerful and often healing way. This aspect is often neglected in the medical approach to looking for what might be wrong with the body: which of course is also important to be able to recognise. Shiatsu, massage and touch can offer powerful tools in supporting women to contact the wisdom of their bodies in pregnancy and birth and for parents to bond with their developing child.

Since Shiatsu is working with the whole person, rather than simply focusing on conditions, most people, ill or healthy, and of all ages from babies to the elderly can potentially benefit from it. Shiatsu is extremely useful in enhancing health and vitality and many people use it as part of a stress management or preventative health care programme. Shiatsu is also excellent if you are feeling unwell but are suffering from no known medical condition. However, if you do have a medical condition for which you may or may not require orthodox medical treatment, shiatsu can still offer a support to your body. It may simply be stress reduction or relaxation, or just being more accepting of where you are at.

People come to shiatsu for all kinds of reasons and they may come with specific ailments ranging from the acute to the more chronic from the more physical to the more emotional. They may come presenting with structural problems such as bad necks, backs or poor posture, as well as conditions like menstrual difficulties, skin disorders, digestive problems and migraines or with more psychological issues such as depression or stress. Often people seek out shiatsu during major times of change like adolescence, infertility, pregnancy, the menopause and adjusting to later life.

As shiatsu is simply working to support your body, then it can work alongside other approaches, whether they are more physically based or emotionally based.

Shiatsu session begins with some time for sharing any relevant issues, whether physical or emotional, so that the treatment can be tailored to your needs. This is followed by 40 -50 minutes of hands on work, a short rest and then feedback. Suggestions may be worked out together for exercises or activities which support the work of the session.

The work is usually done on a futon, a light cotton mattress on the floor. If people don't want to, or are not able to, lie down, sitting or other positions can be used. It is recommended to wear loose fitting clothes; tracksuit bottoms or light cotton trousers are ideal. Avoid having a heavy meal before the session. It is advisable to rest for at least one hour afterwards, as the process continues after the actual session is over. The effects may be experienced immediately, or after several days. The whole process is very individual and is tailored to meet each person's desires. In acute conditions, such as frozen shoulder or back, or severe depression, people will probably come weekly until they begin to feel better. Some massage therapists are also shiatsu practitioners and some shiatsu practitioners use massage.

Shiatsu is usually done clothed, although some massage therapists include shiatsu in work directly on the skin with oil. Sometimes people decide that they want to remain clothed. Other times they might prefer the more physical and oil based approach of massage. Shiatsu includes more static holding techniques: sometimes people want a more physical dynamic approach, although some shiatsu can also be quite physical. Shiatsu tends to include more stretches and mobilisations in the way that Thai massage does but some massage therapists include a lot of stretches in their work. The basic difference is that shiatsu includes use of the meridians and so can address emotional issues in a more specific way than massage or other techniques.

34

Siddha Medicine

The world 'Siddha' comes from the word 'Siddhi' which means 'an object to be attained' or 'perfection' or 'heavenly'. Siddhi generally refers to Ashtama siddhi *i.e.*, The eight great supernatural powers which are enumerated as Anima *etc.*, Those who attained or achieved the above said powers are known as Siddhars.

Siddha (citta) is one of the codified traditional medicines recognized and supported by the Ministry of the Heath and Family Welfare. It shares the main concepts of ayurveda but while the texts of ayurveda are written in Sanskrit and this medicine is widespread throughout all India, the textual corpus of siddha is in Tamil language and this medicine is practised in Sothern parts of India, and in some Asiatic and Arabic countries which accommodate a large Tamil community. Siddha is taught in traditional ways: by hereditary transmission within the family from parents to children (paramparai) or from a master to a disciple (kurukulam).

Siddhis are also construed as powers which are attained by birth, (according to previous Karma), by chemical means or power of words or by mortification or through concentration. As for instance-Kapila, the father of the great Sankhya philosophy is a born Siddha. Concentration on the elements beginning with the Gross and ending with the Superfine enables one to get mastery over the elements; and this was practiced by a sect of Buddhists who concentrated on a lump of clay with a view to see its fine ethereal particles.

In Ayurveda, such classes of persons were called Rasayanas on account of their proficiency in the knowledge of Alchemy and Rejuvenation.

Nagarjuna an Indian metallurgist and alchemist, born at Fort Daihak near Somnath in Gujarat in 931. He wrote the very first treatise Rasaratnakara that deals with mettalic Siddha preparations. Besides, there are innumerable works on this subject written by different authors in Tamil among the following may be mentioned *viz.*, Agastiya, Tirumular, Siva Vakkiyar, Yugimuni, Terayar, Punnakisr, Sudamuni, Mechamuni, Pulippani, Sattamuni, Bogar, Vara Rishi, Ramadevar, Idaikkadar, Konganava *etc.* The medical treatises by these sages contain so many honestly recorded factors simple, short and perspicuous that they have been justly estimated as the most practically useful and valuable acquisitions to medical science.

The Siddhars were a class of popular writers in Tamil in all its branches of knowledge; and many of their works were written in what is called high Tamil. The Kavi or poetry in which the medical and other scientific tracts have been composed is much admired by those who have made it their special study. The Siddhars were further the greatest scientists in ancient times. They were man of highly cultured intellec tual and spiritual faculties combined with supernatural powers. They contain a large number of valuable formulae and exhibit further minute enumarations of morbid symptoms.

This Siddhars are universally supposed to have lived at a very early period; and we cannot ascertain their exact period of existence as their school also ceased to function long long ago. Agastya Siddha who is the chief of the Siddhars' school is said to have been a celebrated rated philosopher and physician who laboured amongst the Tamils in Southern India. Some of his works are still standard books of Medicine are Surgery in daily use among the Siddha medical practitioners.

Macrocosm vs. Microcosm

Man is said to be the Microcosm, and the world the Macrocosm; because what exists in the world exists in Man; or in other words there is nothing in Macricism of Nature that is not contained in Man. So man must be looked upon as an integral part of universal Nature and not as anything separate or different from the latter. Further, the forces in the Microcosm or man are identical with the forces of the Macrocosm or the world; or to put it more plainly-the natural forces acting in and through the various organs of the human body are intimately related to the similar or corresponding forces acting in and through the organisms of the world.

In the organisms of man, these forces may act in an abnormal manner and cause diseases thereby. Similarly, in the great organisms of the Cosmos, they may act abnormally likewise and bring about diseases on earth and its atmospheric conditions such as earthquake, storms, lightings, rain-fails resulting in the elements constituting the blood of a man corresponds to the quality of the invisible influences rediating from Mars. If the scull-essentials that characterise the influences of Venus do not exist, the natural instincts that cause men and animals to propagate their species would cause to operate; because all beings in the Universe are sympathetically connected with the only one universal principle of life from Venus resulting in love between two persons of the opposite sex.

The following are the instances in which every sign of the Zodiac has an aspect towards some particular part of the human body:

1. Aries relates to the neck Libra relates to the kidneys
2. Taurus the neck and shoulders Scorpio genitals
3. Gemini arms and hands Sagittarius lips
4. Cancer chest and adjacent parts Capricorn Knees
5. Leo the heart and stomach Aquarius legs
6. Virgo the intestine, Pisces the base of feet, the stomach umbilicus

Like the signs of the Zodiac each of the planets has jurisdiction over some parts of the body. A few instances shown below will be enough to exemplify the manner and the way in which the Seven Planets exercise special power over some part of the body to cause disease or diseases according to their influences on the three humours in the system:

1. Saturn: Presides over bones, teeth, cartilage's, ear, spleen, bladder and brain and gives rise to Quatrain fever, leprosy, tabbies, paralysis, dropsy, cancer, cough, asthma, phthisis of the right ear, hernia, *etc.*

2. Jupiter: has jurisdiction over blood, liver, pulmonary veins, diaphragm, muscles of the trunk and sense of touch and smell.

3. Mars: has power over bile, gall-bladder, left ear, pudenda and the kidneys; and brings about fever, jaundice, convulsions, haemorrhage, carbuncle, erysipelas, ulcers *etc.*

4. Venus: presides over the pituitous blood an semen, throat, breasts, abdomen, uterus genetalia, taste, smell, and pleasurable sensations; and causes gonorrhoea, barrenness abscesses or even death from sexual or poison.

5. Mercury: has jurisdiction over the animal, spirit, over legs, feet, hands, fingers, tongue, nerves and ligaments and produces relapsing fevers mania, phrenitis, epilepsy, convulsion, profuse expectoration etc, or even death by poison, witchcraft and so on.

Eastern Physiology

Nature is the material cause not merely of the outer Universe but also of out body with all its grosser and subtler divisions and components, its instruments of knowledge and action and the proclivities and tendencies, in which the soul dwells even, as in a cottage.

The question that naturally arises is, what constitutes the human body according to the theory of Siddhantists and the following is the answer to it:

There are 96 Tatwas postulated by the Siddhantists' school; and a simpler for, of the account is herein given; and this requires to be carefully studied. A careful and precise definition of these Tatwas has to follow: but this we do not attempt here for want of space. The human body is composed ninety-six Tatwas or constituent principles in Nature including elements, bodily and mental organs, faculties, matter etc, and they are as shown below:

1. The five elements
2. The five object of senses
3. The five organs of action
4. The five organs of perception
5. The four intellectual faculties
6. The ten nerves
7. The five state of the soul
8. The three principle of moral evil
9. The three cosmic qualities
10. The three humours(wind, bile, phlegm)

11. The three regions (sun, moon, and fire)
12. The eight predominent passions
13. The six station of the soul
14. The seven constituent elements of body
15. The ten vital airs
16. The five cause of sheaths of the soul
17. The nine doors or vents of the body

In Tamil language, small tracts called Kattalai exists: and they define and describe these Tatwas which are variously enumerated as 19 or 25 or 36 or 96. Rev. Hoisington has translated one of these tracts, as also Rev. Foulkes of Salem, but both these books are unfortunately out of print. Both Siddhantins and Vedantins(idealists) accept the number 36 or 96; but they differ in several particulars. Thirty-six when still more analysed given rise to Ninenty-six. The enumeration of these Tatwas beings from the lowest and the grossest, which is the earth.

Tract nos. 1 to 5 making up a total of twenty-four are called the powers of the soul; whereas Nos. 6 to 17 making up a total of 72 which together with the above said 24(in 1 to 5) constitute 96 Tatwas.

The Siddhars school fully recognises these ninety-six Tatwas and further add that the human body is composed of 72,000 blood-vessels, 13,000 nerves, 10 main arteries, 10 vital airs(Prana), all together in the form of a net-work; and it is, owing to the derangement of the three humours becomes liable to 4448 diseases. This is well explained in the following verses from Iswara's Meignana Nadi:

Of this ten vital airs, five play an important role in the physiological unetions necessary for the preservation of the physical body and they are:

1. Chief Prana - regulates the respiratory system.
2. Apana - helps excretions from the lower organs, evacuation and generation.
3. Vyana - principle of circulation of energy throughout the entire nervous system.
4. Udana - regulates the function of higher organs of the brain.
5. Samana - the principle of digestion on assimilation.

Tatwa is the primordial and eternally existing basic essence.

Each elements playing its own part goes to bring about the harmonious working of the human and other animals bodies. There are, nine gates(ten in women) described for the play of forces of the five senses in the human body in which lives the soul commonly known as Jivatma a miniature representation of Paramatma, the universal soul in contra- distinction to the former. The human body is therefore considered to be a temple of God.

There are in our body several supports to the soul for the existence and continuation of life; and these supports are closely connected by Prana. Siddhars attach much more importance to this Prana which is the Life principle of the Universe absorbed and specialised by every human being. This Prana stimulates the two very active centres *viz.* the brain and the heart. The positive matter flows along the vertebral column and is gathered up in the Medulla oblongata and this flow we call Sushumna, and it stimulates the spinal column with all its ramifications; Pingala is the channel for the current which work in the right half of the body through the right sympathic system and Idakala is the channel for the currents working in the left half of the body or the left sympathetic. These channels of life-forces are called in Tamil Nadis. The forces of Prana which diverge from either way from these Nadis are only the ramifications and the nervous system is but the plexuses or webs(physical) for the play of the force of Prana through the physical body. As the Prana courses itself through, the lungs inspire, and as it recedes the process of exhalation sets in.

The three Nadies Ida, Pingala and Sushumna meet in six different places known as Shadadharam the six nerve plexses. Each of these plexuses is round like a wheel and hence they are called Chakras or Padmams. Every plexus pulsates with the vibrations of the great stream current of Prana which Sushumna absorbs the Great Life principle. These major Chakras in their turn cause the smaller ones or the minor nerve-centres to function. The main seat of the pranic force is the heart; and this is made to function by the force of the Great Energy. The Great Energy playing on the (sushumuna) gives the motive power which enables the respective parts of the human organism to function.

Sushumna is one of the passages of the nerve-center at the top of the vertebral column running down through the spinal cord. Within this Sushumna is a hollow called Chitra and in this hollow the Divine spirit dwells. This spirit is in all probability the seat of vitality and of life. The other two Nadies are called Idakala and Pingala of which Idakala coiling round Sushumna enters the right nostril, and the other Pingala in like manner enters the left nostril.

The Five Elements

According to Hindu science, there are five elements in Nature. They are the original bases of all the corporeal things which when die out or destroyed resolve themselves again into elements. This is in fact just what takes place in the case of dead bodies of animals and thing. All earthly beings live, move, grow and die to be resolved into the five elements again after death. It has already been pointed out that there is a very close and intimate connection between the external world and the internal man. The human body is composed of five elements *viz.*, earth, water, heat, air and ether and is a small world in itself; and so the five elements lie at the root of the external world and the internal man. They are also found in all bodies by the processes of transmutation and union and the following are instances of such transformed conditions:

☆ Earth into bone, flesh, nerves, skin and hair.

☆ Water bile, blood, semen, secretion and sweat.

☆ Fire hunger, thirst, sleep; beauty and indolence.

☆ Vayu contraction, expansion and motion.

☆ Akasa interspaces of the stomach, heart, neck and the land.

In their natural existence, they are crude more or less mixed up and apt to be changed the one into the other. They are the fundamental principles of creation, preservation and destruction in the Universe. They are so very closely connected with one another that they borrow their qualities one from the other and thus each of them has two specific properties of which one is retained as original belonging to itself and the other is that which comes to it from the others. Therefore, as fire is to air, so is air to water and water to earth; and again, as the earth is to the water, so is water to air, and air is to fire; and this is the root and foundation of all bodies and their wonderful functioning.

Every element will be found mixed up thus with the other five elements. One element cannot be viewed dissociated from the other elements. Where there is one element there are other elements also present. This is briefly explained as follows:- Elements are in themselves divided into two halves or parts *viz.*, Physical and Subtle and this subtle is again divided into two equal parts of which one is retained as such and the other part is sub-divided into four equal parts. The process of combination of each of these parts with the retained half in the others is known as five-fold combination, as for instance:

1/2 of Akasa is integrated with the other sub-division of the four elements *viz.*, air, fire, water and earth *i.e.*, 1/2 of Akasa + 1/8 of the other elements constitute - physical ether and likewise 1/2 of air - 1/8 of the other four elements constitute physical fire and so on. In this way we get the five elements in mutual combination but with the designation of that which is predominant in each. This is what is called the theory of Panchikaranam. Siddhars hold also that if one is aware of the secret doctrine of the five elements, one metal may be changed into another on the ground that all substances spring or emanate from some Primordial matter; and so there can be no classification as elements and compounds.

The elements above referred to are the subtle elements and not the gross ones. It is clearly laid down in ancient treatises that the earth is derived from water; water from fire, fire from air, and air from ether (sky)-vide Upanished; and so these elements supposed to be originals are not in themselves really elements; but they are twice compounded and each changeable into others. So the Siddhars assert that none of the so called elements which enter into the composition of all living bodies is by itself pure; and that the only purest and original one in this world is the Soul and the rest are all only compounds.

The three physical elements of the external world, *viz.*, air (wind), heat (fire) and water are selected in Medical Science as they form the three fundamental principles on which the constitution of human beings has been based. A detailed account of these three elements known as humours as they enter into the body is given separately under 'Humoural Pathology'.

Humoural Pathology

Humoural Pathology explains that all diseases are caused by the mixture of the three cardinal humours *viz.*, Wind, Bile and Phlegm, and that the relative proportion of these humours are responsible for a person's physical and mental qualities and dispositions. The three humours under references are called in Tamil 'Muppini' and in Ayurveda 'Tridosha'. They are the three fundamental principles and essential factors in the composition and constitution of the human body. These three humours *viz.*, Wind, Bile and Phlegm represent respectively the air, the fire and the water of the five elements which form the connecting link between Microcosm or man and Macrocosm or world.

The external air corresponds to the internal Vayu; the external Heat corresponds to the internal Pitta; and the external water corresponds to the internal Phlegm (Kapha). Man is thus linked with the external world; and any change in the elementary condition of the external world has its corresponding change in the human organism; and it is upon this interchange of influences that the Tridosha theory and the doctrine of Humoural Pathology are based.

According to the Siddhars' Science, the three humours in their normal order occupy respectively the lower, middle and upper parts of the body and maintain their integrity-the Vayu in the regions of the pelvis and the rectum; the Pittam in the region of the stomach and the internal viscera and the Phlegm in the region of the breath, throat and head. It is also said that the characteristics of the three humours in the constitution of man is either hereditary or atavic. In scientific parlance, Vayu comprehends all the phenomena which come under the functions of the central and the sympathetic nervous system; Pitta, the functions of thermogenesis or heat-production, metabolism within its limits, the process of digestion, colouration of blood, excretion and secretion *etc.*, and Kapha, the regulation of the heat and the formation of the various preservative glands. Thus we see that the Indian medical science is based on morbific diathesis; and that human dispositions are inseparable from the three humours. In fact, there is no substance in the universe which does not own its formation to humours in large or small degree.

The Siddhars' Materia Medica also is based on Humoural Pathology. It asserts that all substances of the animal, the vegetable and the mineral kingdoms contain one or more of these three humours in their composition; and that therefore diet should play an important role in the maintenance of these humours in men and women in preventing diseases or aliments; and that the patient should seek the advice of a physician in the matter of diet in the course of treatment.

The three humours maintain the upkeep of the human body through their combined functioning. When deranged, they bring about diseases peculiar to their influence; when in equilibrium freedom from disease; and when one or the other of the humours combine in such a way as to get deranged by aggravation dimunution *etc.*, disease or death may be the result. The humours by themselves are not the producers of diseases in their functioning; but they give rise to diseases if they are vitiated by other factors; and hence we see that humours and diseases are altogether different and have no connection in their normal condition. Humours may be said to be the component parts of the human organism, and diseases the outcome of external factors that put those organisms out of order.

According to the fundamental principles of Humoural Pathology, no disease can be local and absolutely unconnected with the other parts of the organism. If the physician tries to cure a disease, he should necessarily concentrate his attention upon the why and wherefore of the vitiated humours resulting in that disease. The principal rules to be followed in cases of irregularity of the three humours is either to augment the loss or deficiency, to pacify the aggravation or to reduce the increment of Doshas, Vata, Pitta and Khapa are in the proportions of $1 : 1/2: 1/4 : 4 : 2 : 1$ respectively. The normal degree or force of pulse also is to be in the same order. Any change in these proportions is sure to bring about disease or death; but the maintenance of their normal proportion gives vitality to the organisms and assures the preservation of health and longevity of life.

It has already been stated that these three humours form these three fundamental principles in the composition and constitution of Man; and so the physiological doctrine on which they are based is also exactly the same as that of Pathology.

Let us now examine part played by each of these humours in the system:

Vayu (wind) forms the vital force of the human body and is present everywhere in the system. It is believed to be self-begotten in its origin and identical with Divine Energy *i.e.*, the Almighty. It is unconditioned, absolute and all prevading in its nature; and forms the lifeforce of all animated beings. Although it is invisible, its presence is manifest everywhere. It always takes a transverse course and is known by its two attributes namely, sound and touch. It is the root cause of all disease and the king of all sorts of aliments. It is very prompt in its action and it pass through the whole system in a rapid current.

Pitta (Heat) is the human organism is nothing but heat as it possesses all the characteristics of the external fire, such as burning, boiling, heating, *etc.* It produces the internal heat necessary to maintain the integrity of the human body; and any increase or decrease in this, produces a simultaneous action in the organism. The chief function of bile lies in metamorphosing the chyle to a proto-plasmic substance like the sperm in men and ovum in women. It corresponds to metabolism or cell sub-division. Heat may be said to include both bile and metabolism of tissues as well as the bodily heat which is product of the latter. It is also viewed by some that Pitta is the name for the heat incarcerated in the liquid bile - the principal agent in digestion and in purging out of the waste matter in the form of urine and feces. The origin of bile is in the liver. In the heat, bile brings about the realisation of one's desire; in the eyes the catching of the images of external objects; and in the skin, the absorption of the lubricating substances that are applied to the skin. It is blue in its normal colour and yellowish in its deranged condition; and it turns into an acid when deranged or vitiated. Pitta (heat) in its normal state remains in the lymph, chyle, blood and saliva but chiefly in the stomach. It gives sight to the eyes, beauty to the skin and cheerfulness to the mind. Its derangement causes sleeplessness, indigestion, red boils, jaundice, chlorosis, ulcers, catarrh, dropsy, haemorrhage, acidity, eructation, delirium, perspiration, thirst, bitter taste in the mouth, burning sensation in the body especially palms and soles, *etc.*

Khapa (phlegm) supplies the body with moisture even as Pitta furnishes it with heat and imparts stability and weight to the body. It adds to the strength of the body, increases the firmness of the limbs and keeps them united, preventing their disunion. It helps digestion by moistening and disintegrating food with its humid essence. It imparts to the tongue the power of taste and helps the sense-organs like the eyes, ears and the nose in the performance of their respective functions. Its derangement causes excess of thirst, dull appetite, throwing out of phlegm in cough, goitre, Urticaria *etc.* Meals taken before digestion, day-sleep, taking sweats, molasses *etc.*, generally aggravate Phlegm.

The existence of these three humours in the human system in due proportion is well indicated by plus without which no correct diagnosis of disease if possible to get oneself well acquainted with the inner working of a disease, inner vision, intuition and the spiritual sense are absolutely necessary as it cannot otherwise be easily judged from a material point of view. It is only spiritual knowledge endowed with inner vision that will enable a physician to diagnose a disease at sight and suggest forthwith remedies therefore. A Physician who is incapable of entering into spirit of his patient by the light of knowledge, intellectual faculties and imaginative penetration would be of no use for the diagnosing or treating of human diseases.

Humoural Pathology was in vogue at one time over a great part of the Globe. Even in Europe, it was believed in by all sects and theorists evidently about 400 B.C. Hippocrates, the father of the Greek medicine was the first who had a leaning towards it; but it was Plato who developed it and Galen who defended it very zealously. It was only at the commencement of the 18th century that a change had come in the then prevailing doctrine explaining the essential humoural nature of disease. Humoural Pathology which ascribed all disease primarily to a morbid condition of the fluids, had prevailed in all schools of Medicine up to the time of Hoffmann who argued that solids were more often the primary seat of diseases than even of the fluids. He thus revived the doctrine of solidism which gradually gained credit. The fall of the Galenic School paved the way for the growth of the Western Medicine of the present day. Humoural Pathology is still generally believed in and acted upon throughout Asia.

Some alchemical authors of the West held that the three invisible fluids which by their coagulation formed the physical body of man, were but symbolised sulphur, mercury and salt, sulphur representing the aura and the ether; mercury, the fluids, and salt, the material and corporeal parts of the body. They believed that in each organ the three substances remained combined in certain proportions. They were also of opinion that the said substances which contained in all things, if held together in harmonious proportions constituted health; their disharmony, disease and disruption, death.

The Alchemy of the middle ages interpreted these three substances in a different way although they followed closely the above principle. The three substances were according to them the three forms in which the Universal Primordial Will was manifesting throughout Nature-sulphur representing the expansive power, the centrifugal force *i.e.*, the soul or light in all things; 'mercury', the life or the principle which manifests itself as vitality; and 'salt' the principle of corporification or contractive and solidifying quality. The school also asserts that the physician should be thoroughly familiar with the substances by studying them in the light of nature and not through depending on his own imagination. It will be found that all the alchemical masters who have written on the magnum opus have so employed the said three symbolical expressions as to make them understood only by the adepts and not the profane.

Siddha also tells us that a man generally takes 15 breaths a minute; and this makes 21,600 (15 X 60 X 24) breaths a day; and at this rate, he can live for a period of at least 120 years, taking into consideration the fundamental principle on which respiration is based *viz.*, The force or energy lost in every exhalation operating to a length of space extending to 12 inches is regained or recouped by inhalation only to a partial extent; because the operation in this case extends only to 8 inches, thus losing every time the force or energy required for supplying the difference of 12-8 or 4 inches; and consequently as much force or energy that ought to enter the body *i.e.*, lost in every process of inhalation taking place.

In eating or vomiting, the breath forced out extends to 18 inches; in walking, to 24 inches; in running, 42 inches; cohabiting, 50 inches; in sleeping, 60 inches and so on. It is for the purpose of saving such loss or losses caused by indulging in the action mentioned, that Yogis take up a silent posture and suppress their breath in such a way and to such an extent that they are able to live for any number of years as desired by them without disease or death, devoting their life all the time for the good of mankind. Such a kind of spiritual breathing is sure to develop spiritual powers of consciousness much sooner that the ordinary process of evolution. It is a fact well-known to the Hindus that Siddhars and Yogis remained in a state or trance known as suspended animation and continued to be so far a longer period of time without breathing. At first sight this may appear a physical impossibility; but it is now generally accepted by the Western physiologists who explain it saying that the skin may to some extent perform the function of the lungs just in the same way as a frog breathes without lungs, while its respiration is carried on sufficiently through the skin. Though naturally the capacity to perform the above said function is very small, yet it is capable of considerable increase as in cases of diseases of the lungs such as Asthma, wherein the needs of the body excite them to perform these functions or by special training as Yoga by practice.

Prana

The esoteric breath is spoken as Prana; because in ordinary breathing, we absorb a normal supply of Prana just enough to maintain our life keeping up the body in a healthy condition, and so it is found necessary for our readers to know something about this Prana. Prana is the Universal Principle of Energy which is absorbed and stored or assimilated and specialised in one's system in the ordinary course. It is the essence of all force of energy useful for the proper functioning of the human body. It is taken in by the system along with the oxygen which is found in its purest state in the atmospheric air. It is also in the water that we drink, in the food that we take and in the sun-light, we bather in; and in fact it pervades all nature and as such is everywhere found in all things. It is in all forms of matter, it is in the air, but it is not the air. It is more subtle than ether, and so can penetrate where the air cannot reach. It is of the nature of the dynamic force of electricity and all the life depends on it for its sustenance. It is the vital magnetic force absorbed by every human being in various ways; but it can be made flow forcibly under proper and favourable physical conditions. Susceptible persons will feel it as an electric current; because it is the life-giving energy in all things.

Now, let us see what role is played by this prana in respiration; it is but the esoteric breath. Human breath is nothing but a physical manifestation of Prana. As oxygen is taken up by the circulatory system, so is Prana taken up by the nervous system and is spent as nerve-force in the act of thinking, desiring *etc.* Regulation of breath enables one in fact to absorb a greater supply of Prana to be stored up in the brain and the nerve-centres to be used whenever necessary.

The extraordinary powers attributed to advanced Siddhars is due largely to knowledge and intelligent use of this energy stored-up in Kundalini (Serpent power). It should be borne in mind that every function of the bodily organs is dependent on nerve-force which is supplied by Prana emanating from the Sun and circulating in space. Without this nerve-force, the heart cannot beat, the lungs cannot breath the blood cannot circulate and the various organs cannot perform their respective normal functions. This Prana not only supplies electric force to the nerves, but it also magnetises the iron in the system and produces the aura as a natural emanation.

The Science of Pulse

The Science of pulse forms a very important branch of the Indian system of Medicine; and hence an elaborate description of it is given here.

In ancient times, there were no appliances like the modern stethoscope, sphygmograph, endoscope *etc.*, for extending sensual perceptions into the interior of the body; nor were there apparatuses and contrivances for the test of urine, faeces, sputum (phlegm) *etc.*, to enable the physicians to observe, record and compare the phenomena of health and disease to such a degree as to bring medicine neared and neared to perfection, to deserve the name of what is now called 'science', the physicians in those days had to depend almost entirely on pulse in the matter of diagnosing diseases; and in this, they were pre-eminently successful.

The word pulse means the beating of an artery felt with the tip of the finger or fingers at the wrist; its rate and character go to indicate a person's condition of health. It is also understood as the beating, throbbing or the rythmical dilation of the arteries as the blood is propelled along them by the contraction of the heart in the living body. The term pulse in medical practice is usually applied to the beat or throb felt in the radial artery at the wrist, though it may be felt over the temporal, carotid, lunar, bracial, femoral and other arteries. The Science of Pulse is no doubt based on the theory of Tridosha; and so it cannot be easily understood unless one thoroughly acquainted with the working of the three humours in the human system. There are innumerable arteries spread out from head to foot in the human body, and amongst them pulse is felt in twenty-four; and out of these, the one in the right hard and other in the right foot are considered consequential, reliable and important.

According to Tirumular's work on pulse, the following constituent parts forming the fundamental principles in the human body seem to play an important role in variations of pulse on account of their interpenetrating nature.

The ten vital airs.

The three nerve-channels Idakala *etc.*

The six nerve-plexuses.

The three regions of the body named the Sun, Moon, and the Fire.

Pulse is sub-divided as follows according to the nature action and other characteristics, *viz.*:

Frequent Pulse - one which is faster in rate than normal

Goat - Leap Pulse - irregular and bounding pulse

Strong Pulse - one that is hard or wiry

Weak Pulse - a pulse with no strength

Cordy Pulse - a tense and firm pulse

Full Pulse - one with a copious volume of blood

Hard Pulse - one which is characterised by very high tension

Slow Pulse - one which is abnormally slow in rate

Dropped beat Pulse or Intermittent Pulse - one in which various beats are dropped

Undulating Pulse - a pulse giving the sensation of successive waves

Normal Pulse - pulse beating at a normal rate

Sharp Pulse - a pulse in which the artery is suddenly and markedly distended

Feeble Pulse - a pulse in which the force of the beat is very feeble

Formicant Pulse - a small, nearly imperceptible pulse

Wiry Pulse - a small tense pulse

Collapsing Pulse - a jerky pulse with a full expansion, followed by a sudden collapse

Unequal Pulse - a pulse in which some of the beats are strong and others weak

Decurtate or Mouse-tail Pulse - a pulse which gradually tapers away in strength

Abrupt or quick Pulse - a pulse which strike the finger rapidly

It is done by pressing with the physician's three fingers (index, middle, and ring) of his right hand, at a place two fingers in length just below the root of the thumb *i.e.*, a little above the wrist. The physician should feel the pulse three times by holding and letting loose the hand of the patient and then diagnose the disease with great care and caution.

According to the most commonly accepted view, the natural order in which is the forces of the three humours are indicated and are to be observed is 1) the pulses showing wind (Vayu) in the first place above the wrist is felt underneath the fore-finger 2) that of bile (Pitta), below the middle finger and 3) that Phlegm (Kapha), the third, under the ring-finger, c.f.

The three kinds of Doshas (humours) are ascertained from the three kinds of movements of the pulse-swift, middle and slow, felt by the pressure of touch of the three fingers on the radial artery. Examination of the pulse furnishes the best criterion of the phenomena and progress of a disease. It also helps a physician to force-tell the attack of a disease with its prognosis long before it has taken possession of the patient's system just in the same way as a chiromancer would do with regard to events before they actually come to pass by examining the streaks of the palm.

The radial artery at the wrist which is usually chosen, shows the precise character of the pulse. In feeling the pulse, the physician has to note its impressibility, frequency, regularity, size and the different impressions it produces through the fingers. The pulse no doubt signifies whether a particular disease is due to Vayu (air), or Pitta (heat) or Kapha (water) or whether it is due to the influence of any two combined or whether it is due to the concerted action of all the three; and whether the disease is curable or incurable.

The physician as pointed out already, must be endowed with a spiritual perception without which it is impossible for him to arrive at a correct diagnosis of the exact nature of the inner working of complicated and obstinate diseases by simply feeling the pulse in accordance with the rules and principles of the science. It is his observation, thoughtfulness and imaginative penetration into the deep recesses of the patient's organism that will enable him to form a true estimate of his patient's condition. If he cannot penetrate into the spirit of the patient., no success can be achieved; and for real success, meditation and concentration are necessary. The nature and condition of the pulse in different diseases are fully described in the Siddhars' science, the extraordinary pulse-rates that go to indicate incurable or chronic diseases; approaching death, the prognosis diseases etc, are all well explained therein. The general explanation regarding the cause of pulse is that it is due to the dilatation of the arterial walls which travel in the form of a wave from the larger to smaller arteries and that the differences in the beating of pulse are due to those humours in the blood in circulation; but according to Siddhars' science it is also due to something else which is explained already as arising from the motive powers of the three vital currents.

The Nature of Pulse: The Physician should carefully study the nature of the beating which he feels under his fingers; as the beats are described in an intelligible manner in various ways according to the force, rate and movement by comparing them with the movements of those of animals like horses, snakes, frogs, etc, and birds like peacock, fowl *etc*. The idea of this comparison should be well-borne in mind at the time of feeling the patient's pulse for purposes of corroboration.

Diseases and their Cure

Disease, according to modern science is only a departure from a state of health and more frequently a kind of disturbance of the healthiness of the body to which any particular case of sickness is assigned. According to Siddhars' Philosophy, diseases in man do not originate in himself, but from the influences which act upon him. As already stated, man is compared to the world because the elements that exist in the world exist in man as well; and therefore any change in the elementary condition of the external world has its corresponding change in the human organism. There is the feeling of oneness between the external and the internal world of man; and it is upon this oneness that the doctrine of Humoural Pathology *i.e.*, the theory of Tridosha is based. This may occur through different causes, *viz:*

Derangement of the Three Humours

Air, bile and phlegm are considered the three supports of the human system because they are the three fundamental principles in the composition of the human body. When the harmony of the said humours get deranged owing to a relative increase or decrease of one or more of the principal humours, disease or death will be the result which would be well-indicated by the pulse.

Astral Influences

All the influences that come from the Sun, the planets and the starts act on human beings. If evil elements exist in the sphere of one's soul, they attract, such astral influences as will develop diseases. Astral influences do not act directly upon the physical bodies of men and animals, but upon their vital essences in which all the elements are contained; and this is what is called in Tamil as 'Graha Dosha' in Alchemy it is refereed as planetary influences, children are more liable to such diseases than adults. Moon exercises a vary bad influence over diseases in general, especially during the period of New Moon; and it is for this reason that patients in our country ailing from serious illness are afraid of its virility during the approach of the New Moon or the full Moon; it may bring about at times, even lunacy, paralysis or other brain affections, stimulation of sexual passions, injurious dreams, dropsy, hysteria, *etc.* Mars causes women's suffering from want of blood and nervous strength. A conjunction of the moon with other planets such as Venus, Mars, *etc.*, may make Her influence still more injurious. These influences from the planes are fully dealt with under Astrology. Diseases may often occur without any assignable cause; and sometimes people get cured without the administration of any medicine. Such happenings are attributed to the planetary changes of whose action upon the human system.

Poisonous Substances

Impure and injurious elements enter the human system in various ways such as through food, drink, inhalation or absorption by the skin of poisonous air or vapour and so on. There are impurities of various kinds of all about us; and what may be healthy for one person may be injurious to another, according to the astral influences on the person concerned and to the hidden virtues and vices contained in things in general. Rheumatism, gout, dropsy, and many other diseases are often caused by the accumulation of poisonous elements.

Psychological Causes

A disease state of the body is often the result of the diseased state of the mind. ("Men's sana in corpore sano"). This class of diseases includes all evils are caused by passions, evil desires, disordered thoughts and morbid imaginations. Such physiological changes in the physical body, as for instance-shame causes a blush in the face and fear produces paleness; fear causes diarrhoea; anger or envy, jaundice, Violent emotions which produce hysteria, miscarriage, apoplexy, spasms, malformations of the fetus *etc.* also come under this class. All diseases in so far as they are not directly due to external mechanical causes, are to mental conditions. The majority of diseases are due to moral causes; and so the treatment in such cases ought to be of a moral kind, and should be instructional. In the application of such remedies, care should be taken to see that they correspond to those states of mind which we wish to induce in the patient.

Many are the diseases caused by the abuse of psychological powers resulting in boils or blisters all over the body, atrophy of organs, derangement of the mind, loss of vitality, inflammation or enlargement of the kidneys, and so on.

Spiritual Causes

Morbid imaginations may create hunger and thirst, produce abnormal secretions and give rise to diseases. The power of the true Spiritual Will is known only to a few advanced Occult students. It is a power which may affect the whole body and product or cure all kinds of diseases. He who has a strong Will power will have strong Spiritual power. Evil influence exercised by one person may affect another not only when his body, because the forces created in the sphere of one's mind may be projected by powerful suggestions into the mental sphere of another through the medium of an image of wax set up or used for the purpose of attracting the evil spell; and this is a common thing known in the practice of Sorcery, Witchcraft, Black-art, *etc.*

Diseases Originating from the Soul

This includes all diseases originating from the above mentioned five causes. All diseases are no doubt the effects of previously existing causes. Some originate from natural and others from spiritual causes. Spiritual causes are those that have not been created by Man during his present life, but what he had created during the formed existence; and this is what we call in the popular language chronic diseases or off-shoots of Karma. For such causes there is no remedy, but that of waiting patiently until the evil force is exhausted and the law of Karma has had its effect by due adjustment. Even if the just retribution for our sins could be evaded at one time, it could only be postponed; and the evil would return again with accumulated interest and increased force. If the patient's time for redemption comes, then will he find an efficient physician through whom his soul will get the needed gentle relief.

Cure of diseases: Modern Science has only two kinds of cures *viz.* Medicinal and Surgical; but Indian Science, while accepting both the kinds, contemplate even other kinds of healing such as Pranic healing, Mental healing, Spiritual healing, Thought force healing, Suggestive healing, Metaphysical healing, Magnetic cure, Water cure, Colour cure, Insufflation *etc.* all physical.

The philosophy of thus curing diseases is quite different from what is really understood by the Westerners. All influences, terrestrial and astral converge upon man; and they are invisible just like heat, light and electricity; but how can a physician recognise the manner in which they act and still less prevent or cure diseases that are caused by such actions, if he is not acquainted with the influences exercised by forces in the astral plane.

Every metal and every plant possessed certain qualities that attract corresponding planetary influences; and unless one knows the influence of the stars and the conjunction to planets and the qualities of drugs selected for the purpose, then, one will not be in a position to know what remedy to give for attracting such influences as may act beneficially upon the patient. Our physicians pay no attention to the positions of the planets and so they seldom cure several aliments.

To cure diseases is an art which cannot be acquired by the mere reading of books but which must be learnt through experience. Neither Academies and Institutions, nor Colleges and Schools can turn out physicians for curing diseases but they can only grant diplomas or titles and turn out doctors which they are actually not. A theory which is thus not confirmed by practical work should be abandoned. Modern methods of treating and curing diseases are to a great extent looked upon and employed as if they were means by which man by his cleverness tries to cheat Nature out of Her dues and acts against laws of Nature with impunity. To many persons calling themselves doctors or physicians, these are merely systems for making money to fill their pockets with and gratifying their greed. Many hundreds of years ago, the Siddhars who were the greatest Philosophers and Scientists of their age, have spoken elaborately all about the qualifications of a true physician; explained how physician should conduct themselves towards their fellowmen and have also criticised their untoward behaviour towards their patients. It is therefore for the readers to judge, whether or not the same logic and principles find just the same application to-day because the physicians have entirely deserted the path indicated by Nature and built up an artificial system convenient to themselves. It is clearly stated in the Siddhars' Science that a physician who has no faith and consequently has no spiritual power in him cannot be anything but failure, even though he might have graduated from all the Medical Colleges or Academies of the world and knew by heart the contents of all the medical books that have been printed and published up-to-date. The character of the physician may act more powerfully upon patients, that a strong belief, undoubted faith and deep love for the physician, conduce much toward their health even more perhaps than the medicine itself so much so that they would be able even to change the qualities of the body of the sick especially when the patient responses implicit confidence in the physician.

A powerful faith and Will are certainly bound to cure where doubt has failed. There should be entire harmony between the physician and his patient. Wonderful cures may be effected by changing internal causes from which the outward effects grow; and this can only be done by spiritual knowledge and power. Therefore a physician should have true knowledge, and not merely plenty of information gathered from books or other sources; while a patient on the other hand should have a certain amount of faith and vitality without which no cure can be effected.

What Constitutes a Good Physician

Writers of antiquity have handed down to us the qualities which they considered requisite for constituting the good physician and the following is the extract of a translation from Agastiyar, "He must be a person of strict veracity and of the highest sobreity and decorum, having sexual intercourse with no women except his own wife. He ought to be thoroughly skilled in all the commentaries on Medicine and be otherwise a man of good sense and benevolence; his heart must be charitable, his temper calm, and his constant study should be how to be useful and to do good to the public" "When a sick person expresses himself peevishly or hastily, good physician is not thereby to be provoked to impatience; he should be mild and courageous and should cherish a cheerful hope of being able to save his patient's life; he should be frank, communicative, impartial and liberal; and yet ever rigid extracting an adherence to whatever regimen or rules he may think it necessary to enjoin upon the patient.

The following is a summary of those characteristics that constitute a true physician:

☆ The Physician ought to be an Alchemist or the son of an Alchemist. He should understand the Chemistry of Life; and must have every natural qualification for his Occupation. The pseudo-physician bases his art only on books; but the genuine physician depends for success on his knowledge and skill. He should exercise his art not for his own sake or benefit, but for the sake of his patients; and his power should rest not merely upon medicine but also on Spiritual Truth.

☆ A physician should be a philosopher acquainted with the laws of external Nature. A knowledge of Nature is the foundation of the Science of Medicine; and it is taught by the four great departments of science *viz.*, Alchemy, Physical Science, Philosophy and Astronomy.

☆ He should be an Astronomer; and this means he should know the mental sphere wherein man lives, will all its stars and constellations; the influences of the seasons of heat and cold, of dryness and moisture, of light and darkness and so on, as also the organism of Man. A physician who knows nothing about Cosmology will know little about diseases. He should know what exists in Nature and upon earth, what lives in the five elements and how they act upon men.

☆ He should be well versed in physical science, should know the action of medicines and learn by his own experience how to regulate the diet of the patient, the ordinary course of a disease and its premonitory symptoms.

☆ A True Physician should be able to do his won thinking, and should not mechanically employ the thoughts of others. He should be the product of Nature and not of mere speculation or imagination.

☆ A Physician who knows nothing else about his patient but what the latter has told him, knows very little indeed.

☆ He should not depend too much on the accomplishments of the animal intellect in his brain; he should listen to the Divine Voice which emanates from his soul and learn to understand the same.

☆ He should have a knowledge which cannot be acquired by reading books, being a gift of Divine Wisdom.

✩ He should be wedded to his art as a man is to his wife and should love it with all his heart and soul for its own sake and not for the purpose of making money or realising his own ambition.

✩ He must have the faculty of intuition, *i.e.,* a knowledge of his own and not a knowledge borrowed or purchased from others.

✩ He should try to relieve his patient from suffering, but on no account delay his treatment for extracting money.

✩ He should not venture to treat a patient without arriving at a correct diagnosis, also he will be committing a great sin which will not only affect him but also visit his future generations.

Ancient Chemistry

Although Chemistry has come to be known as an exact science within a comparatively recent period, yet its origin dates back to the earliest times of philosophical study. The word 'Chemistry' is closely associated with that of Alchemy. In a book on "Chemistry is Modern Life", from the pen of the renowned Swedish Chemist named Svante Auguste, the author has done ample justice to the claims of India as the land in which the beginnings of Chemistry as a Science can first be traced in the history of human civilization.

In India, Chemistry had been known as a science auxiliary to Medicine which was practised openly after the beginning of the Chemistry era in the Buddhistic monasteries where the priests were found engaged in curing all sorts of diseases; and they believed not only in a specific compound but also in the utterance of a specific religious formula considered necessary for the physician's healing power; and it is thus that chemistry took a religious impress.

Chemistry as defined now, means the science which relates to the peculiar properties of matter an of the elementary substances, the proportions in which the elements combine, the ways and means of their separation, the laws which govern and affect them and all the connected and allied phenomena; but the simpler definition is-it is the Anatomy of natural bodies by fire.

In the Siddha System, Chemistry had been found developed into a Science auxiliary to medicine and Alchemy. It was found useful in the preparation of medicines for curing all sorts of suffering, spiritual as well as corporeal and also in transmutation of baser metals into gold. The knowledge of plants and minerals was of a very high order.

We also find application in Medicine of so many chemical products of their times proving that they were of course the first to prepare valuable medicines by chemical or other extraordinary methods unknown to the present day. The process of preparing Seynir and distilling several kinds of acids were not unknown to them since the distilled products had been to them of much help in using as solvents. Unfortunately, the Siddhars Science Terminology is difficult and highly technical in character and what is even deplorable, they are not fully expressive.

There knowledge of poisons also was not in any way inferior; it had been very exhaustive and surpassing as seen from the description of the properties of poisons furnished below. According to their science, poisons are divided into two main classes, viz: Natural or Native and Artificial or Synthetic, each of which is further sub-divided into 32 kinds making up total of 64 kinds.

The Siddhars had done several classifications and preparations of some metals and metallic compounds like, Calx or Calcined Oxide, red-oxide, carbonate, *etc.*, prepared from chemical salts, mercury, sulphur, arsenic zinc, vermilion, corrosive sublimate, sal-ammoniac, bichlorid of mercury, *etc.*

Cleansing process of metals like lead, copper iron, mercury, *etc.*, and compounds like orpiment, sal-ammoniac, borax, vermilion, corrosive sublimate, sub-chloride of mercury, *etc.*

Countless Chinese priests came to India and studied Tamil Medicine of whom Bogar and Pullippani are to be remembered specially. Ramadevar also learnt the Siddha system of medicine and propagated it in several countries such as Arabia, Egypt and so on.

These chief articles of import in those days from China into India will go to show not only the commercial relations that existed between the two countries, but also the close similarly of and relationship between the two systems of medicines as is apparent from the use of their drugs in India in the preparations of Tamil Medicines, as a substitute for indigenous drugs. That chemical knowledge ought to have been transferred from India to the accidental countries is very plain from the fact that Siddhars like Bogar, Rama Devar alias Jacob and others, had visited Arabia, Turkey, China and other places spreading the knowledge as they went along; one this has already been touched upon before.

Alchemy

That Alchemy has applied to the imagination of Man from Centuries is evident from the prominent part it has played in the ancient science and form legends of the past. In India, unlike other countries, its origin, growth and development are interwoven with a phase of religious activity. It is regarded as a Divine and Sacred Science and Art enveloped in mystery and could only be approached with reference, faith and due piety.

Although no one seems to have ever witnessed any person effecting conversion of inferior metals into gold, still the idea which had taken a firm hold of the imagination of many would never leave them even now, on account of the man's innate avarice and desire to become rich. Many are the families that have been ruined on account of this mad thirst for making gold; and therefore, it has been thought necessary to deal with the subject somewhat elaborately so as to put all on the alert as to the false notions enshrouding the subject. It is a science by which things may be decomposed and recomposed, and their essential nature changed, raised higher or transmuted into one another.

According to Siddha, it is the grand touchstone of natural wisdom; and it is purely of a Spiritual Origin; and hence for one To be an Alchemist, he should necessarily be a Spiritualist. Therefore, the employment of strong will, benevolence, charity, and above all purity of mind are the essential qualification of an Alchemist. Like a born physician, he should be a born alchemist, taking birth at a particular constellation of stars, according t his former Karma. Which again only means what was left undone or left done imperfectly in the former birth is bound to be accomplished or completed perhaps in the present birth.

As it is an art not to be practised for material advantages, it is not intended for materialists who can never expect any fruitful result in their attempts. Siddhars have written most profoundly and with utmost critical accuracy, Yet obscurely; but they all describe the thing sought for indirectly. Some say they are forbidden to reveal the process;

while others have declared it plainly and intelligibly leaving out some little points which they have kept for themselves. The different parts of the magunam opus have to be found our by a comparison of the works of several authors — one of them describes the materials, another, their preparations, a third, their calcinations, a fourth, the rules *etc.*, for regulating heat and so on. This arrangement is one of the most serious obstacles in the way of understanding alchemical processes in addition to the many other difficulties that have been thrown in the way with the set purpose of concealing the art, aim and processes and thwarting the attempts of the uninitiated.

In the first place, the alchemists have deliberately made use of an elaborated system of signs for materials, astrological signs. Sun and Moon, and so on, to indicate the seven metals. In addition to these, there are signs for every important substance known to them and also for various pieces of apparatus employed by them in common use. A vocabulary of words was also made use of, each representing an alchemist's ideas quite different from its meaning in our ordinary speech as for instance the three salts fetus three months old; an unctuous substance covering the skin of the tetus; liquor amni; alchemy *etc.* Unless a student is inspired with a genuine and dauntless enthusiasm, this technology is bound to deter him from advancing further in the study of this art. Siddhars have thus, enshrouded their operations with symbolism; and have given their materials fantastic names so as to conceal their identity from those outside the mystic cult. Even the symbols are unfortunately not employed with a uniform signification. It is to the credit of the Siddhars that they never sold their secrets; but were always ready to communicate them gratuitously to a chosen few whom they deemed to be worthy recipient. But here also, it has to be observed that they never communicated the whole of an operation to anybody at any one time and at any one place; and the same caution was invariably observed throughout in their works even as it was the case with alchemical writers. Even adepts in that science held its secrets inviolate and did not associate with any but their trusted collegues. It is not even definitely known what methods they employed in the science and what raw materials they chiefly used — whether of mineral, animal or vegetable origin.

The Siddhars were also aware of the several alchemical operations divided onto several processes such as — clacination, sublimation, distillation, dissolution, fusion, separation, conjunction or combination. coagalation, cibation, fermentation, exaltation, *i.e.*, the action or process of refining gold, fixation, *i.e.*, the action or process of refining gold, fixation, *i.e.*, bringing to the condition of being non-volatile, *i.e.*, to the state of resisting the action of fire purification, incineration of metals, animation or vivification, fabrication, liquification, extraction and so on. Some modern scientists and pharmaceutical chemists boast of having discovered some of the above processes under different circumstance.

That they were very much interested in the mineral side of Alchemy is evidenced from the fact that medicines prepared from minerals and salts were often freely used in this art and that Mercury occupied the central place in Alchemy. 'Muppu' was chief believed in as a Universal Salt for calcining metals and other metallic compounds and as such compounded in all medicines. Even the caustic alkali preparation from Fuller's earth played an important role; and whenever this alkali acted too strongly, it was generally moderated by the addition of sour-gruel (acetic acid).

The Siddhars were also aware of the mystic process of killing a metal which means depriving a metal of its characteristic physical properties such as its colour, lustre etc; and a list of such agents are already given under 'chemistry' (supra). Mercury is said to be Siva's generative principle and its efficacy extolled when it has been subjected six times to the process of killing. Many are the volumes written by Siddhars; some of them original and genuine while others of later day production, probably written by their followers. Some people are inclined to think or are led t believe that the later generations took up the visionary and fantastical side of the older alchemy, compiled mystical trash into books and fathered them on Agastya, Konganava, Sattaimuni and other great alchemical writers. Even the style employed in those books, is a farrago of mystical metaphors, full of technical terms and code words without any clue whatever to their interpretation.

Having dealt so elaborately with this subject, it is now our purpose to know what it is that turns or transmutes baser metals into gold. it is well known that al substances spring from some primordial matter called by so many terms—the Muppu of Siddhars, the Protyl(of Sir William Crooke), Travisa (of Bernad) and the Primum ens (of Paracelus); and all these indicate the same.

All philosophers agree that, is the first matter is found, we may proceed without much difficulty Where is it to be found from is then the question? The answer is that it is found in ourselves. Than how to draw or attract the secret matter of the stone out of us? Not by any common and easy means, surely! The secret is — our soul has the power to do it, when the body is free from any pollution and the heart void of malice and offence. The soul is them a free agent and has the power to act spiritually and magically upon any matter whatsoever, and, therefore, the first matter, *i.e.*, the prima materia of the lapis philosophorum is in the soul and the extracting of it is to bring the dormant power of the pure living, breathing spirit and the Eternal soul into active functioning. Thus, it can be seen that Alchemy in one of its phases is really psychochemistry.

The common preparations of Siddha medicines are, Bhasma (caclined metals and minerals), Churna (powders), Kashaya(decoctions), Lehya (confections) Ghrita (ghee preparations) and Taila (oil preparations). Siddha have specialized in Chunna (metallic preparations which become alkaline), Mezhugu (waxy preparations) and Kattu (preparation that are impervious to water and flames).

The chief cause of any ailment is due to the three humors, Tridoshas and mainly due to Azhal Kurtrum (Pittam or bile, acidic nature) exhibited in the blood stream.

The following herbs are recommended for the effective treatment of "Vettai Noi".

1. Aragumpul (*Cynodon dactylon* Pers.)
2. Karisalinkanni (*Eclipta alba* Hassk.)
3. Musu Musukkai (*Mukai scavrillia*)
4. Thoodhovali (*Solanum trilobatum* Linn.)
5. Jeeragam (*Luminum cyminum*)

Arugampul is used to cure: heampotese, phlegmatic-respiratory problems and disease related to eye and head, epilepsy, sleeplessness, liver cirrhosis, ulcerated wounds, diarrhea. The herbs act as emollient, astringent, diuretic and styptic (causing contraction of organs or tissues, astringent qualities).

Karisalinkannee (Karisalai or Potralai Kayanthakari) has regenerative effects and rejuvenative properties, on the human metabolic activities. It activates the liver, spleen and bone marrow in boosting immunity against diseases. The herbs act as hepatic tonic, chologogue (stimulated flow of bile into duodenum), purgative and deobstruent. The Karisalai has as its contents, reducing sugar, sterols, sulphur, gloveocides- proteins, phenols, tannic and sapomine. It further more acst as a Kaya kalpha herb improving intelligence, endowing wisdom and provides a healthy lustre to complexion.

The overall benefits of the above herbal medicines cure most of the diseases that arise out of the disorder of the liver and spleen. It can be summarised as liver, spleen tonic, boost stamina/libido, increases alkaloids (increase alkalinity in body- most diseases arise out of an acidic environment in the body) and glucosides and cures uraemia (raised levels of nitrogenous waste compounds in the blood, excreted in by the kidney -which results in nausea, drowsiness, etc).

Musa Musukkal is used to cure many respiratory system illness of cough, chest pains, *etc.* It is used as a expectorant and has an astringent qualities.

Thoodhovali-another Kaya kalpha herb increases strength by toning the respiratory system and bone marrow function, thereby removing the defects arising out of Silethamam the 3rd humor. It acts as a stimulant tonic and expectant, allowing the body to be purified to increase the metabolic activities. This leads to an increase in body weight, builds muscle mass, and increases libido.

Jeeragam -has carminative and astringent properties. It is a remedy for illness like stomach ache, dysentery, asthma, bronchitis, disintegrates stones formed in liver, spleen and renal tract.

It improves digestive capacity aiding increase in body weight. Jeeragam acts on all parts of the abdomen, especially on the liver, spleen, urino-genital system, bone marrow, and blood -respiratory organs.

Thus, these five herbal medicines, Aragumpul, Karisalinkanni, Musu Musukkai, Thoodhovali, Jeeragam were used to eradicate most of the disease of the liver, spleen, bone marrow, blood, respiratory organs, urino-genital organs, resulting in good improvements in the general condition of the body.

As an adjunctunct to the above herbal medicines, sandal wood products are used. The most common is as, massage oil or incense. Sandalwood has properties of disinfectant, astringent, cooling, diuretic and diaphoretic.

Other Siddha medicines often prescribed and administered are as follows:

1. For purification of blood: Kanthaga Rasayanam, Paranki Pattai churam, Palakaria Parpam
2. For reducing fever: Linga chenduram, Gowri Chinthamani, Thirikadugu Churnam, Rama Banam, Vadha, Piththa, Kaba Sura Kudineer
3. For persistent diarrhea: Thair Sundi churnam, Kavika churnam, Amaiodu Parpam
4. Revitalizers and rejuvenators to the disabled immune system of the body: Orilai Thamarai karpam, Serankottai Eagam, Thertran Kottai leyham, Amukkara,

5. Antiviral drugs: Rasagandhi Mezhugu, Murukkanvithtu, Masikai, Edi Vallathathy mezhugu

6. Restoration to the disturbed mind: Vallarai

1. Herbal Preparations

Serankottai Nei (herbal ghee), Mahavallathy leyham, Parangi rasayanam

2. Herbo Mineral Preparations

Gandhak Parpam, Gandhaka rasayanam

3. Herbo Mercuric Preparations

Idivallathy mezhugu, Poorna Chandrodayam

(Mercurial compounds used in Ayurveda/Siddha the mercury is undergoes a 18 step process of oxidation. The end results is an ore or derivative of mercury.)

4. Herbo-Mercuric-Arsenial Preparations

Rasagandhi mezhugu, Nandhi Mezhugu, Sivandar Amirtham, Kshayakulanthan Chenduram.

Apart from these medicines other Siddha medicines mentioned by other practitioners can be also be used with the above.

Kalpa Drugs

Most every source on Kaya Kalpa will define the term for you as "transformation (kalpa) of the body (kaya)."

Kaya Kalpa and Rasayana (Rejuvenation Therapy)

Rasayana, or Rejuvenative medicine, is one of the main eight branches of Ayurvedic Medicine.

These herbs or their preparations are used in rejuvenation therapy; haritaki, amalaki, saindhava, nagara, vacha, haridra. Pippali, vella and guda; followed by preparation of various rasayana formulas, including Brahma rasayana (an herbal confection recipe like chyawanprash), and various amalaki and pippali recipes – amalaki is indeed prominent in the recommended rasayanas.

Kayakalpa Medicines

The word 'kaya' means body and 'kalpa' means stone. The word 'kayakalpa' means sturdy as a rock and ageless. [Here we have a slightly different definition.] Some good kalpa herbs are Ginger (*Zingiber officinalis*), Kattukai (*Terminalia chebula*), Amukkura (*Withania somnifera*), Keelanelli (*Phyllathus niruri*), Date Palm (*Phoenix dactilyfera*), Seran Kottai (*Semicarpus anacardium*), Vembu (*Azadiracta indica*), Tulsi (*Ocimum sanctum*), Lemon (*Citrus media* Varacida) and Vilvam (*Aegle marmelos*).

Kalpa drugs are manufactured from inorganic compounds also. For instance Ayasambeerakarpam is made from certain metals soaked in lime. Ayabringaraja Karpam is made from iron processed in lime juice. Poorna Chandrodayam contains gold, mercury, and sulphur. Kalpa drugs have rejuvenating powers and are believed to retard the aging process.

Kalpa Yoga

The eight steps of Kalpa Yoga are identical to the eight stages of yoga prescribed by Patanjali.

These are:

1. Iyama (observances) 2. Niyama (restraints) 3. Asana (physical postures) 4. Pranayama (breathing exercises/control of the vital breath) 5. Pratiyakara (withdrawal/detachment of the senses) 6. Dharana (concentration of the mind) 7. Dhayana (sustained meditation) 8. Samadhi (merging of the conscious mind with the Superconsciousness).

The practice of yoga holds the promise of slow aging and a long and healthy life." Deha sidhi (perfection of the body and mind of the individual) is the primary aim of using processed mercury. The first testing phase of examining the processed mercury by the transmutation of base metals into noble metals is called lauha sidhi (perfection in achieving transmutation of metals).

The kayakarpam (or kaya kalpa) treatment consists in the following steps:

1. Preservation of vital energy by influencing internal secretions and blood circulation through controlled breathing and Yoga practice.

2. Conservation of male semen and female secretion, to use it in regenerating processes.

3. Use of a 'universal' salt known as muppu/muppuu, prepared by special processes to induce rejuvenation.

4. Use of (calcinated) powders prepared from metals and minerals, such as mercury, sulphur, gold, mica, copper, iron, *etc.*

Tanvantiri (cf. *Tanvantiri vailtiyam*) mentions Indian gooseberry (*Emblic myrobalan*), Aloes, root of Vilvam (Bael, Crataeva) *etc.*; Pokar talks of Asparagus, Aloes, root of Bael, *etc.*; Tirumular mentions fresh and dry ginger; Teraiyar prefers lime fruit, holy Basil, Margosa, *etc.*; and a text known as Nanacastirattirattu speaks of five plants which purify the body and induce rejuvenation (among them Cloves, Pepper, and Cumin are identified). As mentioned above, these remedies, while considered powerful, are relatively harmless and without side-effects, and may be tried by anyone."

'Soma'

'Soma' is a mysterious plant elixir, or perhaps group of plants, that is referred to a number of times in the classical texts, In the Rig Veda Soma is the juice of a milky climbing plant (Asclepias acida) extracted and fermented to form a beverage, liked by the gods and priests. ... the soma or moon plant is said to be produced on the Mujavat mountain. The juice is described as sweet, pungent flavored and exhilarating. It is said to confer immortality. The plant has been identified as Acido Asclepsias.

Some consider it to be Semitia Genia, while others think it to be Sarcostema Viminalis and some opined it is Pon seenthil -Ye.

The list of kalpa drugs given by different Siddhars slightly varies from one another. To study the kalpa drugs in detail the works of Bohar, Machamuni, Konkanavar, Karuvurar (these are 4 of the 18 Siddhars) and others should be taken into account.

Method of Administration of Kalpa Drugs

Kalpa drugs are given only after adhering to certain preliminary processes.

1. Vazhalal Kazhatral (Removal of mucous secretion from throat)

The juice of Karisalai (*Eclipta alba*) with ghee is painted over the uvula as well in and around the tonsil to enable the mucous to be vomited out. This process should be repeated once in the early morning for a period of 40 days for the complete removal of poisonous effects from the throat.

2. Malam Kazhatral (Cleaning of the bowels)

The juice of Katrazhai (*Aloe vera*) is mixed with castor oil and used as a lubricating enema a decoction of Kadukkai (*Terminalia chebula*) is a safe and gentle laxative without griping pain or other discomforts.

3. Milagu Karpam

After completion of the preliminary courses the next drug to be taken is Milagu (Piper nigrum). Pepper is to be taken daily 5 in numbers with the suitable medium. It should be increased at the rate of five per day up to a maximum of one hundred. Then it is gradually reduced at the rate of five per day. The time taken for the entire process is 40 days. This Karpam tones the entire body.

4. Karanthai Karpam (Basil plant)

There are 19 species of which Sphaeranthus Indicus, Sphaeranthus Hirtus and Sphaeranthus Zeylanicus are the most important species for therapeutic use.

Any of the above mentioned is dried, powdered and mixed with sugar at the ratio of 4:1.

This drug is taken daily 40 grains with honey, in two divided doses, on an empty stomach. It helps in eliminating the pathologic symptoms caused by the Pita and Vata.

5. Seenthi (*Tinospora cordiflora*)

Given for various kinds of fevers. It is a good tonic for strengthening the entire system.

6. Vembu (*Azadiracha indica*)

Riped bark tones up the nervous system improves the skin and connective tissues.

7. Karisalai (*Eclipta protrata*), Kuaimeni (*Acalypha indica*), Siruseruppadai (*Mollugo lotoides*)

These three drugs together reduce fat, tone up muscle, expel gas from stomach and 'kabam'' from respiratory system, and gives skin a healthy glow (Kabam is mucous).

8. Karuvumathai

Black datura is a rare species. The whole plant when bruised and taken, mixed with sugar is capable of rejuvenating the system.

9. Medicated Oil Bath

Milagu- *Piper nigrum*

Manjal-rhizomes of *Curcuma aromatica*

Kadukkai-fruits of *Terminalia chebula*

Nelli-fruits of *Emblica offinalis*

Veppam vithu- seeds of *Azadirachta indica*

A paste made out of the above ingredients in equal parts, cows milk and butter is used externally for taking bath twice a week while taking kalpa drugs.

Diet: The diet in the course of kalpa drugs is to be without common salt, tamarind, fish, meat and pungent foods. Sex should be minimized and/or avoided.

35

Spiritual Health

Here we'll explore the connection between the spiritual Self and physical, mental and emotional health and well-being.

Spirituality in this instance does not necessarily mean religious. Although some articles will lean toward one certain religion or another, most will be geared toward spirituality in terms of living out your passion, expressing the divine energy inside you as a means of balancing your Self. If a part of you is being neglected or overworked, it can manifest in your physical body as disease, and this is the basic premise behind Spirituality and Your Health.

Healing is not easy to define, though people have an innate sense of what they mean by it. For the majority of Westerners, their notions are likely to be based on early experiences with childhood diseases and illnesses. If granny's home remedies haven't worked in few hours or so, or if a fever or other symptoms develop that mom/dad feel are worrisome, the child is taken to the healer or physician. Many honest physicians will tell you that what they prescribe for these situations will produce a healthy state in a week and that the patient would take seven days to recover if they did nothing! In short, unless challenged very seriously, our bodies have remarkable capacities of self-healing. This point is not appreciated as much as it might be.

The overall effect of these early experiences in our Western lives leads us to think that Health is a state that only external influences cause us to lose and that we can be returned to Health only by the action of different external influences.

Some tribal peoples hold very different views. For them, disease is due to malign spirits, which need removing or placating in some way. Such cultures may believe that a malicious human has caused this spiritual effect and that it will take a powerful shaman to counteract the malign influences, both human and spiritual. Members of other cultures may regard such beliefs as superstitious nonsense but for those in the culture that has such belief structures, the experiences are real enough. Again, note that Health is seen as something that can be lost and which may need help to regain.

Between these two extreme views of the nature of Health *viz.*, due solely to either physical or spiritual influences, lie those therapies, often called "alternative therapies", that see Health as an outcome of interactions between one or more invisible bodies, and the physical body. These invisible bodies are often characterised by names such as Mental Body, Etheric Body, Emotional Body, and Spiritual Body. They are often represented as an interactive set of "energy fields. Here it is usually asserted that when the fields are "well balanced", the physical expression of those fields will be a healthy Physical Body.

This appears to be a very old idea. Certainly elements of it are to be found in Chinese medicine, shiatsu, acupuncture, naturopathy, and homeopathy, and it underlies the old Latin dictum "mens sane in corpore sano", "a healthy mind in a healthy body". It is implicit in much of psychotherapy where debilitating dysfunction in the physical may be traced to mental or emotional disorders. Because psychotherapeutic drugs may often ameliorate such disorders, there is certainly a link between biochemistry of the body and mental/emotional states in some circumstances.

To summarise, there is a wealth of evidence, some anecdotal, some quite hard core, of a connection between the biochemistry of the body, biochemical effects of drugs, emotional influences and the Health of human beings. But what are we to make of the "energy fields" and "energy bodies" that form a key part of the way alternative therapies are often discussed and practised?

The use of the word "energy" in these discussions opens the door to a lot of confusion and even to acrimonious debate. The word is not being used in its normal physical sense to mean the "E" of E=mc2, or the energy of an electrical current, or the energy in a rolling billiard ball, or the potential energy of water at the top of a water fall. It most definitely does not have the units of power, like the energy rating of electrical appliances.

The whole matter is made even more difficult by the fact that many speak of these energies as "positive or negative", that there can be "blockages in energy flow" and that such blockages can be highly detrimental to the Health of the physical body. It is further stated that these blockages can be felt, released, manipulated, *etc.* by therapists skilled in "energy adjustments".

It is widely agreed by holistic practitioners in these areas that negative emotions can influence one or more of the invisible bodies. In our Universe, all physical entities have an etheric construct whose structure is responsible for the physical manifestation of the entity. The invisible energy body is often called the etheric body, recognising that this does not correspond entirely with the beliefs of some other practitioners.

Most practitioners in this area believe that negative emotions are stored as "negative energy" in what I am calling the etheric body, causing it to be out of balance. They say this lack of balance leads to chronic disease states in the physical body. A strong supporter of this view is Louise Hay, whose books on the relationships between chronic disease states, negative emotions and the use of affirmations to help heal those negative emotions are well known. Homeopathic practitioners, however go further since they argue that all disease, even what Western medicine would label clearly as disease due to infectious agents, is due to the effects of emotions on/in the etheric body, or physical influences encountered in earlier stages of life.

They point out that even the most contagious of infectious agents does not infect everyone; there is always resistance to infection in any population, and they argue that this resistance is developed best in those whose etheric bodies have no emotional energy problems that open the door to infection. Such practitioners often state that it is specifically the immune system which is affected by "negative emotions". Infection is therefore seen as the result of a deficient immune system, caused by the impacts of the emotions on it, among other negative influences to which the immune system is sensitive.

There is a growing body of quite hard-core scientific evidence that links the emotions to the strength of the immune system. A whole new field, psychoneuroimmunology, has developed around research in this area. The immune system responds to infection by launching an inflammatory response, primarily mediated by three cytokines. Two of these cytokines produce changes in sleep patterns and all three induce a fever response and influence eating and social behaviour, including induction of behavioural depression. So the idea that the immune system is linked to the emotions is no longer untenable.

Specific examples may help to clarify these beliefs. A person with a long-repressed anger over childhood experiences is thought to store this anger in that area of the etheric body that controls the functioning of the stomach and duodenum. If long continued, the effect will be to produce ulcers. If the anger is released, then it is argued that the ulcer will self-repair.

Sadness, especially over someone much loved, can be stored in that part of the etheric body that manages the heart. If not processed and released, the emotional stress leads to "heart pain", which can be ameliorated by nicotine. But taking nicotine is palliative, not curative, and eventually produces its own side effects, one of which may well be a heart attack. If the heart pain is released, the need for nicotine can be reduced making withdrawal from smoking easier.

We need a lot of research to examine the clinical basis of what has led to the idea that "energy" is involved in Healing and in Health. We also need to examine in people the clinical symptoms which have led to concepts such as "energy blockages", and to what happens from the point of view of clients and therapists when these "blockages" are "released", as in body working or shiatsu.

It is my considered view that something which can be felt as warmth, tingling, or in a few cases as cold, both by the practitioner and the client, may be involved in "energy healing". I can produce and feel this effect myself and am aware of it in the hands of some others. I also know from personal experience that tissues can present with knots, cords, lumps, *etc.* in them which can be quite painful to the touch. They might well be called "fibrositis" or "muscles in spasm" in some cases if they were examined by conventional GP's, or a sports physiotherapist.

Such tissue lumps can be urged to "let go" and can disappear instantly, often with quite profound benefits to the client. Sometimes this "letting go" is accompanied by severe emotional side-effects. At first sight, the magnitude of these side effects seems to be out of all proportion to what has been done physically with the patient. Not only does this suggest that there is indeed a connection between the presence of the lumps and knots and the emotions, it means that such a treatment must be carried out in an environment that allows this potential outcome to be handled carefully and sensitively. The strength of the reaction shown by some patients to these very simple procedures can include nausea, vomiting and diarrhoea, severe tiredness and lassitude, and a total inability to function in their normal duties for several hours or even for a day or so. Emotional outbursts with floods of tears that may be hard to control are also possibilities. Similar patterns of emotional release sometimes accompany shiatsu treatment, rebirthing sessions, deep connective-tissue massage and certain kinds of body working.

So much for the down side; the up side is often very substantial improvement in the patients' sense of well-being. It would hardly be fair to call this process "healing" if that were not the case! In the specific case of "heart-nail removal", patients usually report a greater freedom to breathe deeply, reduction in need to smoke cigarettes and greater ease in giving up smoking

All masseurs, physiotherapists, and bodyworkers know that clients' bodies present with knots, cords, and sore spots. They also know that what is loosely called "tension" is associated with them and that particular people tend to "hold their tensions" in particular regions of the body. For some, perhaps a majority, it is the neck and shoulders, for others the back, or the scalp, or even the limbs. Traditional techniques of massage are directed at releasing these "tensions", a process often assisted by appropriate music and anything else that encourages relaxation. It is not at all uncommon to find that such a knotted area is intensely painful if pressed firmly and that it may release slowly or suddenly in the course of a massage. Shiatsu and related techniques often achieve rapid release of these "muscle tensions".

Unfortunately, it is a common experience of all such therapists that after the person has had their treatment and has been translated into a relaxed and less careworn state, they may well be back again next week with exactly the same pattern of tensions as the week before. Under these circumstances, the healing is temporary and it appears that environmental stress and/or return to traditional patterns of behaviour can reinstate the "energy blockages" quite quickly after they have "been removed." One is tempted to conclude that unless the emotional problem that may underlie the physical problem is addressed, it may be impossible to effect more than temporary change in the health of the client. If, as is believed by so many practitioners of alternative therapies, the real problem lies in the presence of "negative energies" in the etheric body, one might ask, "are there therapies that can free the etheric body of its negativity?" If so, one would expect that after such a treatment, the effect in the physical body would at least tend towards being long-lasting. One would hope to see genuine change in symptoms, altered behaviour, altered internal sense of well-being, and the eventual disappearance of that client from one's practice as they are "cured."

There do appear to be a host of therapies with the capacity to release "negative emotions" from the etheric body. Powerful therapies include bodyworking, rebirthing, reiki, shiatsu, acupuncture, deep connective-tissue massage, reflexology and all of these may be assisted by the right choice of aromatherapy products, homeopathy and naturopathy.

Any Western analysis of Healing using classical medical techniques is affected by our scientific and cultural attitudes towards the process. Thus, after it is established by an examination of the patient's vital signs that there is some deviation from the norm expected, an attempt will be made at diagnosis. Once a diagnosis has been made and hopefully confirmed, treatment will be recommended that should return the vital signs to normal. Sometimes this will be a straightforward process, but often it will involve rounds of treatment or treatment combinations, with a slow return to normality, or only a partial return to normality.

The whole process is concentrated on the physical body and the effects of external factors, including infectious agents, upon it. However, where drugs are used, one cannot eliminate the possibility that some of their effects are similar to those produced by naturopathy, homeopathy or aromatherapy. The diagnostic skills of medicos are known to vary widely. Some seem to have an uncanny sense of what is "wrong" with a patient, including the ability to sense when the problem is emotional or due to mental disease. Such gifted diagnosticians seem to use tests and technology only to confirm what they already sense. Other medicos who lack these clinical skills often find this ability in others infuriating.

The Western specialists who concentrate on particular forms of illness and disease. In their hands it becomes increasingly likely that the problem will be reduced to a problem in the specifics of a particular component or system of the body. Thus, if a woman presents with a breast lump, the issues are focussed immediately upon what should be done about the lump. Biopsies, mammograms and scans later, the lump will have had its character determined and its future made clear. Benign or a threat, to be removed or not, those will be the decisions to be made.

In that environment, it is hard to hold the view that it is the woman as a whole that has the lump, not just the breast. Should the lump get to be a metastatic cancer, over time it will become clearer that it is the whole person that has the disease, not just the breast, as cancer will turn up in the bones and eventually in other organs.

Even when we have a localised infection, say to the thumb of the right hand, it is a fact that the whole person has the infection, not just the thumb. The whole of the immune system is involved, not just that of the thumb, and the effects may be dealt with in part by the circulatory system, nervous system, spleen, liver, kidneys, just to mention the most significant. In the absence of antibiotics to which the infection is susceptible, the whole body could easily fall prey to septicaemia.

Where disease is less clearly associated with an infectious agent, the boundary between modern Western medicine and some of the alternative therapies is narrower. Someone suffering from depression, or anxiety and perhaps associated panic attacks, will be encouraged by a holistic practitioner to seek a longer recess to allow the time to get a feel for what has precipitated the situation. Physical exhaustion, acute mental and emotional stress, and poor sleep performance are often the underlying causes and dealing with them by this combination of approaches is reasonably successful.

36

Tantra Healing

Out of many of the tantrik worshipping, rituals and practices, spiritual healing is practiced for therapeutic purposes. Various sacred recitation from ancient texts are prescribed to attain self esteem and will force and for spiritual enlightenment. These practices are very

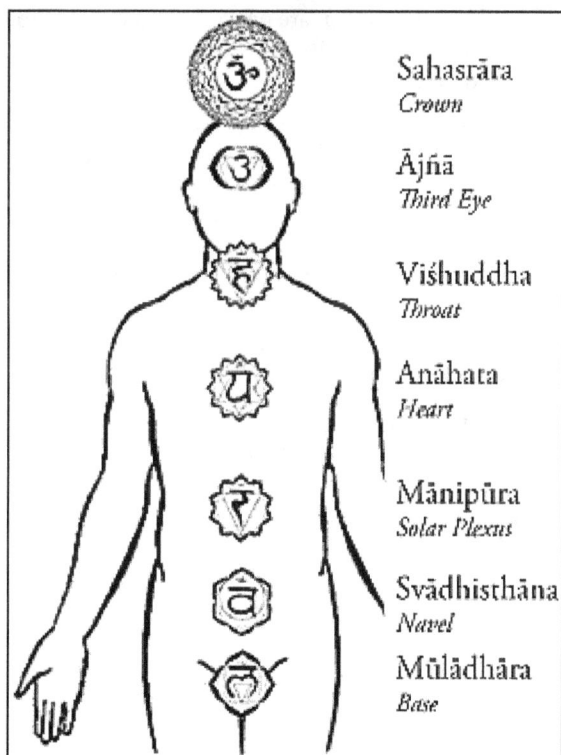

Sahasrāra
Crown

Ājñā
Third Eye

Viśhuddha
Throat

Anāhata
Heart

Mānipūra
Solar Plexus

Svādhisthāna
Navel

Mūlādhāra
Base

powerful but involves danger or risk and the procedures must be done under the guidance of a the spiritual teacher or guru only. Successful completion of tantrik worshipping and practices give blessings and a desirable self esteem. A tantrik guru who is positively inclined uses these benefits as a guide to motivate people for the benefit of mankind.

The biggest benefit can be achieved through Kundalini yoga which is a spiritual practice. Mantra is chanted (vibrated) to assure the purest inner guidance during the practice of Kundalini yoga. When we chant a mantra we are choosing to evoke the positive power contained in the mind, whether it is for prosperity, peace of mind, increasing intuition, or any other multitude of possible benefits inherent in mantras. The selected mantra corresponds to the area of different chakras or the spiritual centres within the body. Chanting of the selected mantra sets vibrations into motion which shall have an effect related to the individual chakra.

Many people want to know how to awaken the Kundalini. And as there are techniques for Kundalini Awakening, most of the techniques are kept secret by masters until the student is ready to receive them.

Here is a very powerful technique for Kundalini Awakening based on a Kriya Yoga. The first part of this technique for Kundalini awakening is to focus your breath starting at the base of your spine all the way up your spine and out of the top of your head. Do this for the inhale. As you inhale move your focus from the base of your spine, up the spine and out of the top of your head as though you are "breathing up" the energy.

You do not need to imagine anything. This is not about imagination but focus. Focus your breath to move up the spine and out the top of your head.

And the focus should be relaxed. You are not to force anything. You are simply gently suggesting the energy to move in this direction. So do focus the energy up but allow the Kundalini energy to move on it's own without force.

It may seem quite strange to focus your breath where your breath does not go but it is very effective. Breath is energy, "prana." And simply by focusing the breath in a certain area, the energy moves to that area.

After practicing this for a while, you will start to feel the energy rising. The sensations will be subtle at first but over time can become very blissful. It can almost feel like an orgasm, a rush of bliss moving up your spine.

The second part of this technique for Kundalini awakening is it to move the breath down your front as you exhale.

As you exhale, move the breath from the top of your head to the third eye, (located between your eyebrows) then to your throat and then to your heart chakra, located at the center of your chest.

Stop the breath at the center of your chest. And then again, inhale from the base of your spine, repeating the process above.

The third part of this technique for Kundalini awakening is to awaken the energy. Here one needs a spiritually enlightened teacher or Guru who will give you the sacred mantra for japan, it is to actually awaken your Kundalini by japan, chanting or simply by focusing their energy on you.

So to awaken this energy yourself, there is an easy technique but a very important one. And that is to silently repeat the name of deity that you feel connected to.

You can repeat the name as you inhale, and repeat the name as you exhale. The reason for this is that you take on the Energy of what you focus on. So simply by repeating the name of a deity, you take on his or her energy, their state of bliss. This technique is thousands of years old and is still used today because it is so effective. It has nothing to do with them as a personality or body but the energy that naturally radiates from them.

One should become perfectly desireless and should be full of Vairagya (no attachment to earthly attractions) before attempting to awaken Kundalini. It can be awakened only when a man rises above Kama (Sex), Krodha (Anger), Lobha (Lust), Moha (Desire), Mada (Alcoholic drinks or other narcotics) and other impurities. Kundalini can be awakened through rising above desires of the senses. In one word one should have full abstinence from above senses, only Bhakti (worshipping), Seva (service or offerings) and Batsalya (unconditional love similar to a mother's to her child) should be with him.

The Yogi, who has got a pure heart and a mind free from passions and desires will be benefited by awakening Kundalini. If a man with a lot of impurities in the mind awakens the power by sheer force through Asanas, Pranayamas and Mudras, he will break his legs and stumble down. He will not be able to ascend the Yogic ladder. This is the chief reason for people going out of the way or getting some bodily infirmities. There is nothing wrong in the Yoga. People must have purity first; then a thorough knowledge of the Sadhana, a proper guide, and a steady, gradual practice. When Kundalini is awakened there are many temptations on the way, and a practitioner without purity will not have the strength to resist.

A thorough knowledge of the theory is as essential as the practice. Some are of opinion that theory is not at all necessary. They bring one or two rare instances to prove that Kundalini has been awakened even in those who do not know anything about Nadis, Chakras and Kundalini. It might be due to the grace of a Guru or by mere chance. Everyone cannot expect this and neglect the theoretical side. If you look at the man in whom Kundalini has been awakened through the grace of a Guru, you will not at once begin to neglect the practical side and actually waste your time in passing from one Guru to the other. The man who has a clear knowledge of the theory and a steady practice, attains the desired goal quickly.

Awakening of Kundalini and its union with Shiva at the Sahasrara Chakra effect the state of Samadhi and Mukti. No Samadhi is possible without awakening the Kundalini.

For a selected few, any one of the above methods is quite sufficient to awaken the Kundalini. Many will have to combine different methods. This is according to the growth and position of the Sadhakas in the spiritual path. The Guru will find out the real position of the Sadhaka and will prescribe a proper method that will successfully awaken the Kundalini in a short period. This is something like the doctor prescribing a proper medicine to a patient to cure a particular disease. One kind of medicine will not cure the diseases of different patients. So also, one kind of Sadhana may not suit all.

There are many persons nowadays who foolishly imagine that they have attained purity, commit errors in selecting some methods and neglect many important items of Sadhana. They are poor, self-deluded souls. Self-assertive, Rajasic Sadhakas will select some exercises of their own fancy in an irregular manner and leave all the exercises when they get some serious troubles.

After Kundalini is awakened, Prana passes upwards through Brahma Nadi along with mind and Agni. You will have to take it up to Sahasrara Chakra through some special exercises such as Mahabheda, Sakti Chalana, *etc.* As soon as it is awakened, it pierces the Muladhara Chakra (Bheda). It should be taken to Sahasrara through various Chakras. When Kundalini is at one Chakra, intense heat is felt there and when it leaves that centre for another Chakra, the former Chakra becomes very cold and appears lifeless.

Freedom from Kama, Krodha, Raga and Dvesha and possession of balance of mind, cosmic love, astral vision, supreme fearlessness, desirelessness, Siddhis, divine intoxication and spiritual Ananda are the signs to denote the awakening of Kundalini. When it is at rest, a man has full consciousness of the world and its surroundings. When it is awakened he is dead to the world. He has no body-consciousness. He attains Samadhi state. When Kundalini travels from Chakra to Chakra, layer after layer of the mind becomes opened and the Yogi acquires psychic powers. He gets control over the five elements. When it reaches the Sahasrara Chakra, he is in the Chidakasa (knowledge space).

Awakening of the Kundalini Sakti, its union with Shiva, enjoying the nectar and other functions of the Kundalini Yoga that are described in the Yoga Sastras are misrepresented and taken in a literal sense by many. They think that they are Shiva and ladies to be Sakti and that mere sexual union is the aim of Kundalini Yoga. After having some wrong interpretation of the Tantric Yoga texts, they begin to offer flowers and worship their wives with lustful propensities. The term "Divine intoxication that is derived by drinking the nectar" is also misrepresented. They take a lot of wine and other intoxicating drinks and imagine to have enjoyed the Divine ecstasy. It is mere ignorance. They are utterly wrong. This sort of worship and sexual union is not at all Kundalini Yoga. They divert their concentration on sexual centres and ruin themselves.

It is easy to awaken the Kundalini, but it is very difficult to take it to Sahasrara Chakra through the different Chakras. It demands a great deal of patience, perseverance, purity and steady practice. The Yogi who has taken it to Sahasrara Chakra, is the real master of all forces. Generally Yogic students stop their Sadhana half-way on account of false Tushti (satisfaction). They imagine that they have reached the goal when they get some mystic experiences and psychic powers. They desire to demonstrate such powers to the public to get Khyati (reputation and fame) and to earn some money. This is a sad mistake. Full realisation alone can give the final liberation, perfect peace and Highest Bliss.

Rousing of Kundalini and its union with Siva at the Sahasrara Chakra effect the state of Samadhi and Mukti. Before awakening the Kundalini, you must have Deha Suddhi (purity of body), Nadi Suddhi (purification of Nadis), Manas Suddhi (purity of mind) and Buddhi Suddhi (purity of intellect). For the purification of the body, the following six exercises are prescribed: Dhauti, Basti, Neti, Nauli, Tratak and Kapalabhati. These are known as Shat-Karma or the six purificatory exercises in Hatha Yoga.

Pranayama for Awakening Kundalini

When you practise the following, concentrate on the Muladhara Chakra at the base of the spinal column, which is triangular in form and which is the seat of the Kundalini Shakti. Close the right nostril with your right thumb. Inhale through the left nostril till you count 3 Oms slowly. Imagine that you are drawing the Prana with the atmospheric air. Then close the left nostril with your little and ring fingers of the right hand. Then retain the breath for 12 Aums. Send the current down the spinal column straight into the triangular lotus, the Muladhara Chakra. Imagine that the nerve-current is striking against the lotus and awakening the Kundalini. Then slowly exhale through the right nostril counting 6 Oms. Repeat the process from the right nostril as stated above, using the same units, and having the same imagination and feeling. This Pranayama will awaken the Kundalini quickly. Do it 3 times in the morning and 3 times in the evening. Increase the number and time gradually and cautiously according to your strength and capacity. In this Pranayama, concentration on the Muladhara Chakra is the important thing. Kundalini will be awakened quickly if the degree of concentration is intense and if the Pranayama is practised regularly.

Kundalini Pranayama

In this Pranayama, the Bhavana is more important than the ratio between Puraka, Kumbhaka and Rechaka. Sit in Padma or Siddha Asana, facing the East or the North. After mentally prostrating to the lotus-feet of the Sat-guru and reciting Stotras in praise of God and Guru, commence doing this Pranayama which will easily lead to the awakening of the Kundalini. Inhale deeply, without making any sound. As you inhale, feel that the Kundalini lying dormant in the Muladhara Chakra is awakened and is going up from Chakra to Chakra. At the conclusion of the Puraka, have the Bhavana that the Kundalini has reached the Sahasrara. The more vivid the visualisation of Chakra after Chakra, the more rapid will be your progress in this Sadhana. Retain the breath for a short while. Repeat the Pranava or your Ishta Mantra. Concentrate on the Sahasrara Chakra. Feel that by the Grace of Mother Kundalini, the darkness of ignorance enveloping your soul has been dispelled. Feel that your whole being is pervaded by light, power and wisdom. Slowly exhale now. And, as you exhale feel that the Kundalini Shakti is gradually descending from the Sahasrara, and from Chakra to Chakra, to the Muladhara Chakra. Now begin the process again. It is impossible to extol this wonderful Pranayama adequately. It is the magic wand for attaining perfection very quickly. Even a few day's practice will convince you of its remarkable glory. Start from today, this very moment.

Experience of Awakening of Kundalini

During meditation you behold divine visions, experience divine smell, divine taste, divine touch, hear divine Anahata sounds. You receive instructions from God. These indicate that the Kundalini Shakti has been awakened. When there is throbbing in Muladhara, when hairs stand on their roots, when Uddiyana, Jalandhara and Mulabandha come involuntarily, know that Kundalini has awakened.

When the breath stops without any effort, when Kevala Kumbhaka comes by itself without any exertion, know that Kundalini Shakti has become active. When you feel currents of Prana rising up to the Sahasrara, when you experience bliss, when you repeat Om automatically, when there are no thoughts of the world in the mind, know that Kundalini Shakti has awakened. When, in your meditation, the eyes become fixed on Trikuti, the middle of the eyebrows, when the Shambhavi Mudra operates, know that Kundalini has become active. When you feel vibrations of Prana in different parts inside your body, when you experience jerks like the shocks of electricity, know that Kundalini has become active. During meditation when you feel as if there is no body, when your eyelids become closed and do not open in spite of your exertion, when electric-like currents flow up and down the nerves, know that Kundalini has awakened. When you meditate, when you get inspiration and insight, when the nature unfolds its secrets to you, all doubts disappear, you understand clearly the meaning of the Vedic texts, know that Kundalini has become active. When your body becomes light like air, when you have a balanced mind in perturbed condition, when you possess inexhaustible energy for work, know that Kundalini has become active. When you get divine intoxication, when you develop power of oration, know that Kundalini has awakened. When you involuntarily perform different Asanas or poses of Yoga without the least pain or fatigue, know that Kundalini has become active. When you compose beautiful sublime hymns and poetry involuntarily, know that Kundalini has become active.

In western culture, tantra is being described in such a way that it is supposed to mean uninhibited sex. It has been willfully misinterpreted. This is because books on tantra have been written by people who just want to sell books. They have never encountered real spiritual practitioners of the East and do not know real Tantra in any way.

The word "tantra" literally means a sacred methodology to achieve the union with the supreme power. This is an inner technology. These are subjective methods not objective methods. Sex is related to release of body fluids in genitals whereas Tantra Yoga is related to releasing sacred nectar in Sahasrara Chakra by awakning Kundalini Shakti.

37

Vastu

Vastu is an intrinsic energy-concept of science. We cannot see energy with our bare eyes but we can realize it and see its application in various forms. As every human aspect is governed with rules and regulations similarly, the nature also has certain key principles meant for smooth living of people. Vastu stands for 'nature law' and unawareness of law is not an excuse anywhere.

Here are some of the basic rules that one and all should follow:

☆ A good plot is the one wherein all 90 degrees corners have two sides with a front and back road.

☆ The open space in the North and the East should be more than that in the South and West.

☆ The South, West and South- West are the portions of the plot where maximum construction should be done.

☆ A pond or a water body should be positioned in Northeast corner of the house.

☆ A house should be designed keeping in mind the abundance inflow of sunlight and proper cross ventilation.

☆ The main door should open towards the North and East Side of the house.

☆ Trees or vegetation should not be grown in eastern, northern or northeast directions.

☆ Only small plants can be grown in the North, East, and North-east areas.

☆ The fireplace or oven must be placed in the eastern, southeast directions of the entire house. Placing it in such a way that the person cooking should face the East direction.

☆ Images of any warring scenes, demons, angry people *etc.* should not be placed in the house.

Vastu Tips

☆ While sleeping your head should be towards the south direction.

☆ The ceiling of the rooms in the south should never be low as compared to ceiling of the rooms in east and north side.

☆ Keeping shoes and slippers scattered around the house leads to feud.

☆ Dirty and washed clothes should be kept separately.

☆ Avoid sleeping under an exposed beam as it brings ill fortune.

☆ Don't keep a jug/glass of water near the telephone as the electric rays radiating from the telephone have a harmful effect on the water.

☆ Your face should be towards the east while cooking food.

☆ The cupboard containing food articles should be placed towards your right.

☆ Make sure that the ceiling of the pooja-room is lower than that of all the other rooms.

☆ Keep the hinges of doors and windows well-oiled, there should be no creaking sound while opening the doors and windows especially the ones at the entrance.

☆ Make sure that the vegetables are placed in a dish after cutting/chopping. Keeping them on the floor causes the nutrients to be absorbed by the floor.

☆ The house should always be kept spic and span. All the waste/garbage should be dumped or stored in the West direction of the house.

☆ Beauty aids such as combs should be kept tidily at the right place.

☆ Separate towels should be there for every family member and for the guests.

☆ One should never have 3 doors consecutively on a single wall, especially in the front of the house.

☆ One should never erect circular staircase in the house.

☆ It is considered very auspicious to view the rising sun after taking bath.

☆ One should always wear neat and clean clothes to work/while working.

☆ The safe or locker containing cash and jewels should be placed in such a way that it opens in the North direction.

Directions and their Respective Rules

☆ North-east (Eshan) is ruled by Sadasiva or God Himself. This direction is paramount, and so should be kept clean, open and highly receptive in a welcome "mode" always.

☆ East is ruled by Indra, the chief of the gods, giver of pleasures. The direction also represents the realm of the rising Sun projecting ultra-violet rays, which makes essential for health in several ways.

☆ South-east (Agneya) is the habitat of fire, the storehouse of energy.

☆ South is the abode of Yama, the god of death.

☆ South-west (Nairitya) is the abode of Putna, the demoness.

☆ West is the abode of Varuna, the god of ocean. It is also the direction of the setting Sun which gives infra-red radiation.

☆ North-west (Vayavva) is the abode of air/wind, the invisible, but the most effective blessing for all objects in need of motion/movement for their efficiency.

☆ North is the abode of Kuber, the god of wealth.

Vastu follows some prescribed norms that must be followed in a house to achieve good health. Here are enumerated some Vastu principles to keep your health in good condition:

☆ Always sleep with head towards South or East.

☆ Vastu Advice for HealthNever sit or sleep under the exposed beam.

☆ Face East or North while eating your meal.

☆ An ideal kitchen should be located in South-east.

☆ Avoid kitchen in North-east as it tends to create health problems in women.

☆ North-east is corner for flowing water, so keep some source of water here.

☆ South-west portion should be ideally provided to elderly people or main owner of house in order to keep them in good health.

☆ To have sound sleep, keep mobile phones and any other gadget away from bed. Also keeping iron stuff beneath the bed creates health problems for people.

☆ Growing bamboo plants in home is inauspicious, so avoid it.

☆ Do not keep exposed mirrors in home and cover all mirror including laptop screen and TV screen *etc.*

☆ Green colour is auspicious in North-east.

☆ Avoid constructing kitchen and toilet together and ensure that both the place have maximum distance from each other.

☆ Place under the staircase should be used for storage but avoid making any usable place here such as kitchen, toilet *etc.*

☆ There should be no room or any other functional place over or below kitchen and toilet.

☆ Keep the bed three inches away from the wall.

38

Yoga and Naturopathy

Naturopathy

Nature cure is a constructive method of treatment which aims at removing the basic cause of disease through the rational use of the elements freely available in nature. It is not only a system of healing, but also a way of life, in tune with the internal vital forces or natural elements comprising the human body. It is a complete revolution in the art and science of living.

Although the term 'naturopathy' is of relatively recent origin, the philosophical basis and several of the methods of nature cure treatments are ancient. It was practised in ancient Egypt, Greece and Rome. Hippocrates, the father of medicine (460-357 B.C.) strongly advocated it. India, it appears, was much further advanced in older days in natural healing system than other countries of the world. There are references in India's ancient sacred books about the extensive use of nature's excellent healing agents such as air, earth, water and sun. The Great Baths of the Indus Valley civilisation as discovered at Mohenjodaro in old Sind testifies to the use of water for curative purposes in ancient India.

The modern methods of nature cure originated in Germany in 1822, when Vincent Priessnitz established the first hydropathic establishment there. With his great success in water cure, the idea of drugless healing spread throughout the civilised world and many medical practitioners throughout the civilised world and many medical practitioners from America and other countries became his enthusiastic students and disciples. These students subsequently enlarged and developed the various methods of natural healing in their own way. The whole mass of knowledge was later collected under one name, Naturopathy. The credit for the name Naturopathy goes to Dr. Benedict Lust (1872 - 1945), and hence he is called the Father of Naturopathy.

Nature cure is based on the realisation that man is born healthy and strong and that he can stay as such as living in accordance with the laws of nature. Even if born with some inherited affliction, the individual can eliminate it by putting to the best use the natural agents of healing. Fresh air, sunshine, a proper diet, exercise, scientific relaxation, constructive thinking and the right mental attitude, along with prayer and meditation all play their part in keeping a sound mind in a sound body.

Nature cure believes that disease is an abnormal condition of the body resulting from the violation of the natural laws. Every such violation has repercussions on the human system in the shape of lowered vitality, irregularities of the blood and lymph and the accumulation of waste matter and toxins. Thus, through a faulty diet it is not the digestive system alone which is adversely affected. When toxins accumulate, other organs such as the bowels, kidneys, skin and lungs are overworked and cannot get rid of these harmful substances as quickly as they are produced.

Besides this, mental and emotional disturbances cause imbalances of the vital electric field within which cell metabolism takes place, producing toxins. When the soil of this electric filed is undisturbed, disease-causing germs can live in it without multiplying or producing toxins. It is only when it is disturbed or when the blood is polluted with toxic waste that the germs multiply and become harmful.

Materials in the healthy individual are removed from the system through the organs of elimination. But in the diseased person, they are steadily piling up in the body through years of faulty habits of living such as wrong feeding, improper care of the body and habits contributing to enervation and nervous exhaustion such as worry, overwork and excesses of all kinds. It follows from this basic principle that the only way to cure disease is to employ methods which will enable the system to throw off these toxic accumulations. All natural treatments are actually directed towards this end.

The second basic principle of nature cure is that all acute diseases such as fevers, colds, inflammations, digestive disturbances and skin eruptions are nothing more than self-initiated efforts on the part of the body to throw off the accumulated waste materials and that all chronic diseases such as heart disease, diabetes, rheumatism, asthma, kidney disorders, are the results of continued suppression of the acute diseases through harmful methods such as drugs, vaccines, narcotics and gland extracts.

The third principle of nature cure is that the body contains an eleborate healing mechanism which has the power to bring about a return to normal condition of health, provided right methods are employed to enable it to do so. In other words, the power to cure disease lies within the body itself and not in the hands of the doctor.

The nature cure system aims at the readjustment of the human system from abnormal to normal conditions and functions, and adopts methods of cure which are in conformity with the constructive principles of nature. Such methods remove from the system the accumulation of toxic matter and poisons without in any way injuring the vital organs of the body. They also stimulate the organs of elimination and purification to better functioning.

To cure disease, the first and foremost requirement is to regulate the diet. To get rid of accumulated toxins and restore the equilibrium of the system, it is desirable to completely exclude acid-forming foods, including proteins, starches and fats, for a week or more and to confine the diet to fresh fruits which will disinfect the stomach and alimentary canal. If the body is overloaded with morbid matter, as in acute disease, a complete fast for a few days may be necessary for the elimination of toxins. Fruit juice may, however, be taken during a fast. A simple rule is : do not eat when you are sick, stick to a light diet of fresh fruits. Wait for the return of the usual healthy appetite. Loss of appetite is Nature's warning that no burden should be placed on the digestive organs. Alkaline foods such as raw vegetables and sprouted whole grain cereals may be added after a week of a fruits-only diet.Another important factor in the cure of diseases by natural methods is to stimulate the vitality of the body. This can be achieved by using water in various ways and at varying temperatures in the form of packs or baths. The application of cold water, especially to the abdomen, the seat of most diseases, and to the sexual organs, through a cold sitting (hip) bath immediately lowers body heat and stimulates the nervous system. In the form of wet packs, hydrotherapy offers a simple natural method of abating fevers and reducing pain and inflammation without any harmful side-effects. Warm water applications, on the other hand, are relaxing.

Other natural methods useful in the cure of diseases are air and sunbaths, exercise and massage. Air and sunbaths revive dead skin and help maintain it in a normal condition. Exercise, especially yogic asanas,promotes inner health and harmony and helps eliminate all tension : physical, mental and emotional. Massage tones up the nervous system and quickens blood circulation and the metabolic process.Thus a well-balanced diet, sufficient physical exercise, the observation of the other laws of well-being such as fresh air, plenty of sunlight, pure drinking water,scrupulous cleanliness, adequate rest and right mental attitude can ensure proper health and prevent disease.

The word ' Vitamine' meaning a vital amine was proposed by a Polish Researcher, Dr. Cacimir Funk, in 1911 to designate a new food substance which cured beri-beri. Other terms were proposed as new factors were discovered. But the word vitamin, with the final 'e' dropped, met with popular favour.

Vitamins are potent organic compounds which are found in small concentrations in foods. They perform specific and vital functions in the body chemistry. They are like electric sparks which help to run human motors. Except for a few exceptions, they cannot be manufactured or synthesized by the organism and their absence or improper absorption results in specific deficiency disease. It is not possible to sustain life without all the essential vitamins. In their natural state they are found in minute quantities in organic foods. WE must obtain them from these foods or in dietary supplements.

Vitamins, which are of several kinds, differ from each other in physiological function, in chemical structure and in their distribution in food. They are broadly divided into two categories, namely, fat-soluble and water-soluble. Vitamins A, D, E and K are all soluble in fat and fat solvents and are therefore, known as fat-soluble. They are not easily lost by ordinary cooking methods and they can be stored in the body to some extent, mostly in the liver. They are measured in international units. Vitamin B Complex and C are water soluble. They are dissolved easily in cooking water. A portion of these vitamins may actually be destroyed by heating.

They cannot be stored in body and hence they have to be taken daily in foods. Any extra quantity taken in any one day is eliminated as waste. Their values are given in milligrams and micrograms, whichever is appropriate.

Vitamins, used therapeutically, can be of immense help in fighting disease and speeding recovery. They can be used in two ways, namely, correcting deficiencies and treating disease in place of drugs. Latest researches indicate that many vitamins taken in large doses far above the actual nutritional needs, can have a miraculous healing effect in a wide range of common complaints and illnesses. Vitamin therapy has a distinct advantage over drug therapy. While drugs are always toxic and have many undesirable side effects, vitamins, as a rule are non-toxic and safe.

Minerals

The term ' minerals ' refers to elements in their simple inorganic form. In nutrition they are commonly referred to as mineral elements or inorganic nutrients.

Minerals are vital to health. Like vitamins and amino acids, minerals are essential for regulating and building the trillions of living cells which make up the body. Body cells receive the essential food elements through the blood stream. They must, therefore, be properly nourished with an adequate supply of all the essential minerals for the efficient functioning of the body.

Minerals help maintain the volume of water necessary to life processes in the body. They help draw chemical substances into and out of the cells and they keep the blood and tissue fluid from becoming either too acidic or too alkaline. The importance of minerals, like vitamins, is illustrated by the fact that there are over 50,000 enzymes in the body which direct growth and energy and each enzyme has minerals and vitamins associated with it. Each of the essential food minerals does a specific job in the body and some of them do extra work, in teams, to keep body cells healthy. The mineral elements which are needed by the body in substantial amounts are calcium, phosphorous, iron, sulphur, magnesium, sodium, potassium and chlorine. In addition the body needs minute (trace) amounts of iodine, copper, cobalt, manganese, zinc, seleminum, silicon, flourine and some other trace elements.

Yoga

The word "YOGA" is derived from the Sanskrit (language) root "YUJ", which means join or union. The purpose of all Yoga is to unite man, the finite, with the infinite, with Cosmic Consciousness, truth, God, light or whatever other name one chooses to call the ultimate reality. Yoga, as they say in India, is a marriage of spirit and matter. Yoga was originated in India several thousand years ago. According to the research, it is over 6000 years old.

Yoga has several branches or divisions, but the goal, the aim of all of them is the same, the achievement of a union with supreme consciousness. In Karma Yoga, for instance, this is achieved through work and action, in Gnani Yoga, through knowledge and study, in Bhakti Yoga, through devotion and selfless love, in Mantra Yoga through repetitions of certain invocations and sounds. Raja Yoga (Royal Yoga) is the Yoga of consciousness, the highest form of Yoga. Its practice usually starts with Hatha Yoga which gives the body the necessary health and strength to endure the hardships of the more advanced stages of training. Hatha Yoga is the Yoga of physical well-being. It consist of several steps and is preceded by the YamaTiyama, there are some stages. The 1st stage is called Asana (or posture), the 2nd Pranayama (or breath control), the 3rd is Pratyahara (or nerve control), the 4th is Dharana (or mind control), the 5th is Dhyana (or meditation), and finally there is Samadhi, the state of ultimate bliss and spiritual enlightenment. Strictly speaking the last four stages of Hatha Yoga already merge into the realm of Raja Yoga.

Hatha Yoga : "Ha" stands for the sun and "Tha" for the moon. The correct translation of Hatha Yoga would be solar and lunar Yoga, since it deals with the solar and lunar qualities of breath and Prana.

"Prana" is a subtle life energy existing in the air in fluid form. Everything living, from men to amoeba, from plants to animals, is charged with Prana. Without Prana there is no life.

A Yogi can belong to any religion or to none at all. In this case, he usually forms his own relationship with the Ultimate Reality once he has come closer to it.

The Yogis regard the human body as a temper of the Living Spirit and believe that as such it should be brought to the highest state of perfection. Also the advanced practices of Yoga require great power of endurance. The body might not be able to stand the strain without special preparation.

Patanjali, lived roughly 2,500 years ago and is revered for his Yoga Sutras, 196 concise aphorisms that set forth the principles of yoga. he is called the Father of Yoga because he was the first to put into writing that had until that time been handed down only verbally from master or Guru, to pupil or Sishya.

The Yoga postures, breathing and relaxation exercises can be taken up by anyone who wants to improve his physical or mental condition. One need not go into the more advanced stages of the training. Person with lots of stress and anxiety, Yoga is very helpful.

Normally, one should not start yoga before the age of six and not after the age of sixty five, although many people do start later and still obtain good results. I have seen on TV, a lady from Singapore, aged 104 years old still can do any postures of Yoga exercises.

Yoga can not cure any ailment but it could be the part of healing. The healing work is done by nature. Yoga exercises can only help remove impurities and obstructions, so that nature may be given a chance to accomplish her task successfully.

Basic difference between Yoga and Gymnastics or Similar Exercises

Yoga Asanas are an art applied to the anatomy of the living body, whereas gymnastics are a form of engineering applied to the muscles of the body. The aim of Yoga postures is not merely the superficial development of muscles. These postures tend to normalize the functions of the entire organism, to regulate the involuntary processes of respiration, circulation, digestion, elimination, metabolism, *etc.* and to affect the working of all the glands and organs, as well as the nervous system and the mind. This result is placed in various postures. Each of these exercises creates a different totality in the functional relationship within the organism. Yoga is able to influence man physically, mentally, morally and spiritually. Yoga emphasizes the philosophy of exercise. Under its training one experiences a sense of awakening. All of one's capacities are heightened, and one achieves balance and stamina through these exercises, some of which are modeled after the movements of various animals. In Yoga, relaxation is taught as an art, breathing as a science, and mental control of the body as a means of harmonizing the body, mind and spirit.

A Few Yoga Postures

Ashtanga Yoga (Power Yoga)

Ashthnaga Yoga is light on meditation but heavy on stamina. It has little bit more difficult poses than other types of Yoga. Participants are encouraged to move quickly from one pose to another in an effort to build strength and flexibility. Ashthanga Yoga is suitable for anyone in good physical condition. Beginners can do same styles with moderate format. Adds up of Meditation with Ashtanga Yoga should be good for mental and physical fitness.

Iyengar Yoga

One of the soft styles of yoga, Iyengar is perfect for beginners and who haven't exercised in a while. In a lack of flexibility, you can use chairs and pillows to compensate if you have back or joint problems. Adds up of Meditation with Iyenger Yoga should be good for mental and physical fitness.

Vikram (Bikram) Yoga

This style of Yoga done in a hot room that is 38C or higher. Because to replicate the temperature environment of Yoga's birthplace in INDIA. This style of focuses on 26 postures that are meant to be performed in particular order. The exercises are intense and very physical, which, when combined with HEAT, makes out for a tough workout.

Hatha Yoga

This style of Yoga focuses on simple poses that flow from one to the other at very comfortable pace. Generally, participants are encouraged to go at their own pace, taking time will be better with lots of breathing and meditation into their workout. Hatha Yoga is ideal for those needing a stress release at the end of a tough day.

Some commonly practised Hatha Yoga postures are:

1. Standing Deep Breathing
2. Half Moon Pose
3. Awkward Pose
4. Eagle Pose
5. Standing Head To Knee Pose
6. Standing Bow Pulling Pose
7. Balancing Stick
8. Balancing Separate Leg Stretching
9. Triangle Pose
10. Standing Separate Leg Head To Knee Pose
11. Tree Pose
12. Toe Stand Pose

13. Dead Body Pose
14. Wind Removing Pose
15. Sit-up
16. Cobra Pose

Encyclopaedia of Medicinal Herbs and Shrubs

Agaricus Blazei (*Agaricus subrufescens*)

Actions: anti-tumor, anti-cancer effect

Indications: indigestion, diabetes, chronic and acute allergies, chronic fatigue, and disorders of the liver.

Known in Brazil as "God's Mushroom" because of its near miraculous curative properties to a wide range of disorders. Traditionally people have used the tea made from Agaricus Blazei to overcome numerous conditions and disorders related to the immune system, the cardiovascular system, digestion, diabetes, chronic and acute allergies, chronic fatigue, and disorders of the liver. Potent chemo preventive activity. Agaricus in now being used clinically across Asia for numerous conditions and disorders. It is also being consumed as a food and as a tonic herb as Agaricus is a gentle, edible mushroom that has no side effects. The active constituents have been identified as the Polysaccharide-protein complexes called Beta-1,6-D-glucan and an almost identical chemical, Beta-1,3-D-glucan. Agaricus Blazei is by far the richest natural source of Beta-1,6-D-glucan and Beta-1,3-D-glucan yet discovered.

These Polysaccharides are strong macrophage activators and work through the "glucan receptor" sites on the surface of these immensely important immune cells. Macrophages are our first line of defence and play the pivotal role in our immune response to any attack by intruders. Macrophages are the immune cells that recognize, engulf and destroy organisms, cells, and substances that don't belong in the body, including viruses, bacteria, yeast, heavy metals, pollutants, bits of dead tissue, mutated cells, and tumour cells. They take on any challenge and further stimulate the immune system to respond. Macrophages ignite intercellular communication by releasing chemical messengers called cytokines (interferon and interleukin). These cytokines are powerful proteins responsible for catalyzing and regulating several immune responses within the body. Agaricus Blazei extracts have been proven to stimulate the production of these cytokine. Macrophages travel from the site of their first contact with an intruder or toxin to the lymph system, where it communicates with T-cells, activating a specific immune response. Specific antibodies are then built to combat the intruder. Agaricus Blazei helps to stimulate and enhance this essential immune response. Beta-1,6-D-glucan and Beta-1,3-D-glucan have also been shown to have inhibitory effects on abnormal cell growths. It has been proposed that Beta-1,6-D-glucan and Beta-1,3-D-glucan may stimulate cell lysing (destroying) action by specialized immune system cells.

Anti tumor effect: Agaricus contains high level of effective elements such as polysaccharides, both Beta(1-3)D-glucan and Beta(1-6)D-glucan protein compounds, ribonucleic acid protein compounds, acid heteroglucan, xyloglucan, lectin, *etc.* When these elements are introduced into our white blood cells, it enhances the activity of macrophage, an antibody cell that destroys or delays the proliferation of cancer cells.

Anti cancer effect: Agaricus contains natural steroids, known for it's anti cancer effect. (It is different from the chemically produced steroids that enhance the body that is often said to be the cause of cancer). It is particularly effective in prevention of uteran cancer.

Preventative effect: Agaricus contains large amounts of non digestive dietary fibers that absorb cancerous materials in our body and discharge away from our system.

Blood glucose reducing effect: Effective elements mentioned above such as Beta(1-3) D-glucan and Beta(1-6)D-glucan protein compounds, ribonucleic acid protein compounds also has positive effects to reduce the blood glucose(sugar).

Blood pressure, Cholesterol and Arteriosclerosis reducing effect: The above mentioned dietary fiber and unsaturated fatty acid, such as linolin, contained in ABM, have effects to reduce blood pressure, cholesterol and arteriosclerosis.

Vitamin D_2 effect: Agaricus contains vitamins B_1 and B_2 but also contains large amounts of ergosterol, which would be converted to vitamin D_2 when it is dried in the sun or heated in a mechanical drying process. Vitamin D_2 has positive effects to lessen the danger of osteoclasis.

Agrimony (*Agrimonia eupatoria*)

Agrimony is grown throughout much of the United States and southern Canada. It is a perennial that reaches 2 to 3 feet tall, prefers full sun and average soils. Agrimony tolerates dry spells well.

An infusion of the leaves is used to treat jaundice and other liver ailments, and as a diuretic. It is also used in treating ulcers, diarrhea, and skin problems. Externally, a fomentation is used for athlete's foot, sores, slow-healing wounds, and insect bites.

Agrimony is used in Spiritual Healing in protection spells, and is used to banish negative energies and spirits. It is also used to reverse spells and send them back to the sender. It was believed that placing Agrimony under the head of a sleeping person will cause a deep sleep that will remain until it is removed.

Alfalfa (*Medicago sativa*)

Alfalfa is cultivated in many regions of the world. It is not picky as to soils, prefers full sun, and regular waterings, although it will tolerate dry spells. It is a perennial that grows to 1 to 3 feet tall, depending upon growing conditions.

Actions: alterative, diuretic, antipyretic, hemostatic, digestive tonic

Indications: ulcers, edema, arthritis, vitamin or mineral deficiency, chronic and acute cystitis, burning urine, prostatitis, peptic ulcers, rheumatic complaints, anemia, appetite stimulant, diabetes, fatigue, hemorrhages, pituitary problems, ulcers. Considered to be a good builder for those suffering from cancer, HIV, tuberculosis, and other wasting diseases. Alfalfa eliminates retained water, relieves urinary and bowel problems, helps in treating recuperation of narcotic and alcohol addiction. Used in treating anemia, fatigue, kidneys, peptic ulcers, pituitary problems, and for building general health.

Natural vitamin and mineral supplement. Contains organic minerals such as calcium, magnesium, phosphorus, and potassium and almost all known vitamins. Very high in chloropyll which is nearly identical to human hemoglobin. By itself it does not provide substance for building tissue. Its action is cleansing and detoxifying. Mild blood purifier, and a good tonic for Pitta and to a lesser degree Kapha-it can aggravate Vata in excess and cause emaciation. Restorative tonic. Traditionally used for chronic and acute digestive weaknesses. Aids in assimilation and metabolism of protein, carbohydrates, fats, iron and calcium. Helps build and regenerate normal strength and vitality. Aids acute and chronic inflammatory symptoms. Considered a good tonic for pregnancy. Used as a body cleanser, infection fighter and natural deodorizer. The enzymes have been shown to help to neutralize cancer. French scientists have shown that alfalfa can reduce tissue damage caused by radiotherapy or radiation. Helps decrease cholesterol levels and shrink arterial plaque. Has a high enzyme content and nuetralizes acids and toxins in the body. Also contains natural flouride and with the high levels of calcium supports healthy bones and teeth.

Allspice (*Pimenta dioica*)

Allspice comes from a tree that grows in Central and South America and the Caribbean, and prefers those climates.

Allspice is used as a paste to soothe and relieve toothache, and as a mouthwash to freshen the breath. Allspice encourages healing and is used in mixtures.

Aloe Vera (*Aloe vera, Aloe* spp.)

Best grown indoors in pots. Those living in the deep South, as in southern Texas or southern Florida, can grow aloe outdoors. Remember that Aloe is a succulent, not a cactus, so it needs water to keep the leaves fleshy and juicy.

The gel of the inner part of an aloe leaf is used to treat burns, skin rashes, and insect bites, as well as chafed nipples from breastfeeding, when applied to the affected area externally. Internally it can be used to keep the bowels functioning smoothly, or when there is an impaction, although it can cause intestinal cramping when taken internally, and there are other herbs that do this job better. It aids in healing wounds by drawing out infection, and preventing infection from starting. The fresh gel is best to use, rather than "stabilized" gels found in the stores. The fresh gel was used by Cleopatra to keep her skin soft and young.

Growing an aloe vera plant in the kitchen will help prevent burns and mishaps while cooking. It will also prevent household accidents, and guard against evil. It is opined that aloe was used to embalm the body of Christ.

Amaranth (*Amaranthus* spp.)

Amaranth is an annual, whose varieties grow from one to five feet tall. It does not transplant well, so sow it where you want it to grow. It is generally not picky about soil type, and tolerates heat and drought well.

Amaranth is used to battle stomach flu, diarrhoea, and gastroenteritis. It was used by Native Americans to stop menstruation and for contraception. Applied externally, it can reduce tissue swelling from sprains and tick bites. Not to be used by pregnant or lactating women.

It is believed to be associated with immortality, and is used to decorate images of gods and goddesses. It is sacred to the god Artemis. Woven into a wreath, it is said to render the wearer invisible. Also used in pagan burial ceremonies.

American Angelica (*Angelica archangelica*)

Angelica needs rich, moist garden soil in partial shade. It prefers wet bottomlands and swamps, and prefers the cooler northern regions to grow best. It is a perennial that can reach up to 6 feet tall.

Actions: tonic, emmenagogue, rejuvenative, diaphoretic, antispasmodic, analgesic

Indications: amenorrhea, dysmenorrhea, menstrual cramps, PMS, anaemia, headaches, colds, flus, arthritis, rheumatic pain, allergies, appetite stimulant, bronchial problems, exhaustion, gas, heartburn, menopausal symptoms, allergies, smooth muscle spasm.

One of the best tonics for women, nurturing the uterine organs and promoting their function and regulating the menstrual cycle. Improves circulation. Has warming properties. Should be avoided by diabetics as it increases blood sugar. Should be avoided during pregnancy. Contains phytoestrogens. Thought to reduce blood pressure by dilating blood vessels. Smooth muscle relaxer. Immune enhancer. Often used to balance female hormones and as an emmenagogue.

Angelica is a good herbal tea to take for colic, gas, indigestion, hepatitis, and heartburn. It is useful to add in remedies for afflictions of the respiratory system, as well as liver problems and digestive difficulties. Promotes circulation and energy in the body. It is often used to stimulate the circulation in the pelvic region and to stimulate suppressed menstruation. Angelica should not be used by pregnant women or diabetics. It's root is often used as a protective amulet, and has been used to banish evil by burning the leaves. It is also used to lengthen life, and is used in protection against diseases, as well as to ward off evil spirits. Adding it to a ritual bath will break spells and hexes. It has often been used to ward off evil spirits in the home.

American Ginseng (*Panax quinquefolius*)

Has similar properties to Panax Ginseng but is cooler in energy. Suitable as an overall tonic and promoter of vitality used in all conditions, such as weakness, heart problems, immune deficiency, hormone imbalances, and as an adaptogen. It is a better demulcent and tonic to the lungs, better for Pitta individuals, but more likely to aggravate Kapha. It is more calming to the brain than Oriental ginsengs, and even sedative to brain and nervous function. It is not inferior to the Korean ginseng and we should make better use of it.

Amla (*Phyllanthus emblica*)

Actions: carminative, appetizer, digestive, life promoting, tonic, nutritive

Indications: eye diseases

Excellent sustainer of youthful age, and provides strength to all sense organs. Helps in the growth of hair and it's luster, purifies the blood, stops the bleeding, relieves burning. Used in Ayurveda as a nutritive, rejuvenative and tonic. It is the highest natural source of Vitamin C. It is used to rebuild tissues after injury or illness, especially the blood, teeth, and gums, to improve eyesight, regulate blood sugar, and relieve allergies and inflammation. It is part of the revered longevity tonic Triphala.

Anise (*Pimpinellaa anisum*)

Anise likes warm, sunny areas with well-drained, rich sandy soils. It is suitable for all areas of North America. It is an annual, and grows 1-2 feet high. It needs 120 days to produce fully ripened seed heads.

A good herb for colic, gas, and indigestion. It can also be used in herbal remedies for coughing, as it aids in loosening phlegm. It is the mildest of the herbs used for these purposes.

Anise mixed with bay leaves provides an excellent bath additive prior to ritual. Using anise in potpourri around the house wards off evil, and anise in your sleeping pillow at night will chase away the nightmares. The essential oil is used in ritual baths prior to any divination attempts. It is believed that hanging an anise seed head on your bedpost will restore lost youth.

Apple (*Pyrus* spp.)

Apple trees grow over most of North America. They need a cool winter period, making them unsuitable for low desert or tropical regions. Check with your local nursery for varieties best suited to your area and growing conditions.

Apples are used to treat constipation. The pectin in fresh apples can help to lower cholesterol levels, an aid in treating heart disease. Crushed apple leaves can be rubbed on a fresh wound to prevent infection.

Apple blossoms are used in love and healing incenses. It is given as an offering to the dead, since it is a symbol of immortality. Apple wood is used to make magical wands. Pouring apple cider on the ground in your garden before you plant gives the earth life.

Arjuna (*Terminalia arjuna*)

Actions: cardiac tonic, diuretic, anti-ischemic and cardioprotective

Indications: stress, nervousness, high blood pressure

Helps maintain a healthy heart and reduces the effects of stress and nervousness. Arjuna promotes effective cardiac functioning and regulates blood pressure. Arjuna has been the herb of choice in Ayurveda for cardiovascular health. Arjuna's ability to suppress the blood's absorption of lipids indicates that it has cholesterol-regulating properties. Its principle constituents are 1È2-sitosterol, ellagic acid and arjunic acid. Anti-ischemic and cardioprotective agent in hypertension and in ischemic heart disease, especially in disturbed cardiac rhythm, angina or myocardial infarction. The bark powder possesses diuretic and a general tonic effect in cases of cirrhosis of the liver, in addition to prostaglandin enhancing and coronary risk factor modulating properties. It induces a drug-dependent decrease in blood pressure and heart rate. Heals all types of tissues, from fractures and contusions to damage on the heart from surgery or disease. This may be due in part to Arjuna's high calcium content.

Recent studies have investigated the mechanism of this activity and have shown a dose-dependent regulation of blood pressure and heart rate. There was also a slight increase in the HDL-to-total cholesterol ratio and an overall improvement in the cardiovascular profile. The bark of Arjuna inhibits the oxidation of LDL and accelerates the turnover of LDL-cholesterol in liver. This enhances the elimination of cholesterol from the body. The suppression of hepatic cholesterol biosynthesis by Terminalia arjuna is the mechanism responsible for a significant lowering of beta-lipoprotein lipids and the recovery of HDL components in hyperlipidemia. In a study on the efficacy of the bark powder in treating congestive cardiac failure (CCF), over 40 per cent of the cases showed marked improvement. CCF due to congenital anomaly of heart and valve disease was also brought under control. 4 out of 9 cases of CCF due to chronic bronchitis were also relieved by the treatment. Arjuna is also known to relieve symptomatic complaints of essential hypertension *viz.* giddiness, insomnia, lassitude, headache and the inability to concentrate. Oral administration of an aqueous suspension of the bark powder reduces coagulation, bleeding and prothrombin time.

Arjuna also helps to strengthen the spiritual heart.

Arrowroot (*Maranta arundinacea*)

It is a large, perennial herb found in rainforest habitats.

It is believed to have digestive and antacid properties, also stimulates the appetite. It's normally found in the spice rack at grocery stores, and a little is generally all it takes.

Ashitaba (*Angelica keiske*)

Actions: strenghtening tonic, anti-bacterial, anti-aging.

Indications: pre menstrual syndrome, menopause, in diabetes to normalize blood sugar, upset stomach, indigestion, bloating of abdomen.

Asian herb belonging to the same genus as Angelica Sinensis, "Angelica" comes from the Latin name for angel and was given to this herb because of its godly effects, namely its extraordinary ability to slow the aging process, The "longevity herb" is one of nature's most dynamic nutritional food sources, and has a vast spectrum of other beneficial properties which augment a number of systems in the body, especially the immune system and the circulatory system. Gives one a beautiful skin complexion. Contains Chalcones which are rarely found anywhere in the natural world, but are the key factor in ashitaba. Research has shown that the unique properties of ashitaba are at least partly due to these remarkable compounds. The chalcones that are in ashitaba are known as Xanthoangelol, Xanthoangelol-E and 4-Hydrooxyderricin. These organic compounds are flavonoids and they give the plant its characteristic yellow sap. This differentiates ashitaba from all other strains of angelica. The antioxidant activity of flavonoids is due to their molecular structure, and these structural characteristics of certain flavonoids found in ashitaba confer surprisingly potent antioxidant activity exceeding that of red wine, green tea, or soy.

Astragalus (*Astragalus membranaceous*)

Astragalus strengthens metabolism and digestion, raises metabolism, aids in strengthening the immune system, and is used in the healing of wounds and injuries. It is often cooked with broths, rice, or beans for a boost to the healing energies during those illnesses that prevent one from eating normally.

Actions: Deep immune activator, diuretic, tonic, hypotensive, vasodilator, anti-stress.

A digestive aid containing polysaccharides which enhance immune activity and T-cell function. In Traditional Chinese Medicine astragalus is considered a deep immune tonic that increases the "bone marrow reserve", increasing the body's ability to produce more immune effector cells (such as T-cells), protecting us from "pathogens". Also used as a daily tonic when one is not feeling well; it has the ability to build the energy reserves in the body and exhibits several anti-stress properties. Astragalus root has also been indicated as an aid in the side effects of chemotherapy as well as having the ability to inhibit tumour growth. This herb strengthens metabolism and digestion, aids in strengthening the immune system and is used in the healing of wounds and injuries.

Ashwagandha (*Withania somnifera*)

Actions: tonic, rejuvenative, aphrodisiac, nervine, sedative, astringent.

Indications: general debility, sexual debility, nerve exhaustion, convalescence, problems of old age, emaciation of children, loss of memory, loss of muscular energy, spermatorrhea, overwork, tissue deficiency, insomnia, paralysis, multiple sclerosis, weak eyes, rheumatism, skin afflictions, cough, difficult breathing, anemia, fatigue, infertility, glandular swelling, hypothyroidism.

Best rejuvenative herb for the muscles, marrow, and semen and for Vata constitution. Used in all conditions of weakness and tissue deficiency in children, the elderly, those debilitated by chronic diseases, those suffering from overwork, lack of sleep, or nervous exhaustion. Inhibits aging and catalyzes the anabolic processes of the body. Sattvic in quality it is one of the best herbs for the mind, upon which it is nurturing and clarifying. It is calming and promotes deep, dreamless sleep. Is a good food for weak, pregnant women-it helps stabilize the fetus. Also regenerates the hormonal system, promotes healing of tissues, and can be used externally on wounds, sores, *etc.* Ashwagandha was shown to slow the circulation of adrenaline in the bloodstream, making it useful for all types of anxiety, panic, and hypertension. Stimulates thyroid function.

Basil (*Ocimum basilicum*)

Basil will grow in any well-drained, fairly rich soil, and full sun. It can be grown throughout most of North America. It is an annual, which reaches 2-3 feet tall. Pinch off the tips to promote bushiness and flower buds to maintain growth.

Actions: diaphoretic, febrifuge, nervine, antispasmodic, anti-bacterial, antiseptic

Indications: colds, cough, sinus congestion, headaches, arthritis, rheumatism, fevers, abdominal distention.

The name Basil comes from the Greek word for King and was thought of as the Herb of Kings. Like other members of the mint family, basil has been used medicinally for digestive troubles. It has a mild, aromatic odor and a warm, sweet flavor with a slight licorice taste.

Perhaps the most sacred plant of India-its quality is pure sattva. Opens the heart and mind, bestowing the energies of love and devotion. Strenghtens faith compassion and clarity. Gives the protection of the divine by clearing the aura and strenghtening the immune system and nervous system. Contains natural mercury which gives the seed of pure awareness. Removes excess Kapha from the lungs and nasal passages, increasing prana and sensory acuity. Removes excess Vata from the colon, improves absorption, and strenghtens the nerve tissue, increasing memory. In astral travel or projection, Tulsi creates a purified sphere of protection, free from negative energies. Basil is used to treat stomach cramps, vomiting, fevers, colds, flu, headaches, whooping cough, and menstrual pains. It is also used to reduce stomach acid, making it a valuable part of any treatment for ulcers, and a valuable addition to any recipe using tomatoes for those with sensitive stomachs. Externally, it can be used for insect bites, to draw out the poisons. It has been used in other countries to eliminate worms from the intestines, and the oil from basil leaves is applied directly to the skin to treat acne.

Basil protects from evil and negativity, and aids in attracting and keeping love. It is used for purification baths, and in wealth and prosperity rituals. Carrying a basil leaf in your pocket brings wealth, and if powdered basil is sprinkled over your mate while they sleep, it is supposed to eliminate infidelity from your marriage.

Bayberry (*Myrica cerifera*)

Bayberry, taken in small doses, increases the vitality of your total body systems, improving circulation. It can also be used as a poultice over varicose veins to strengthen the blood vessels. A douche made of the tea is used for vaginal infections. Tea made of Bayberry is a good gargle for sore throat and tonsillitis.

It is believed that the oil of Bayberry brings prosperity and luck.

Bay Laurel (*Laurus nobilis*)

The leaves are burned to enhance psychic powers and to produce visions. Worn in an amulet, it will provide protection from evil and negativity. The leaves are used as decorations during the Yule season, and placed in your window it will protect against lightning striking your house. Write a wish on a bay leaf and then burn it if you want the wish to come true. Sprinkling the crushed leaves in your cupboards will keep out cockroaches and other insect pests.

Bergamot (Monarda citriodora)

Bergamot grows to 2 feet tall, and is a member of the mint family, so grow it as you would a mint.

Also known as Oswego tea and Bee Balm. It is used to treat nausea and vomiting, and cold and flu relief. The essential oil is used to treat acne, coughs, fevers, tension, stress, and depression.

Used in money and success spells and rituals in folk medicine.

Bilberry (*Vaccinium myrtillus*)

Actions: anti-oxidant, anti-inflammatory.

Indications: cataracts, vascular disorders, peptic ulcers, diabetes, mitral valve prolapse, diabetic retinopathy, macular degeneration, glaucoma, varicose veins, Reynard's Syndrome, diarrhoea. Possesses significant collagen-stabilizing action. Has the ability to cross-link collagen fibers. Reduces free radical damage as an antioxidant. Inhibits enzymatic cleavage of collagen by enzymes secreted during inflammation. Hinders the release of compounds that promote inflammation. Decreases capillary fragility and permeability. Decreases the permeability of the blood-brain barrier. Anti-aggregation effect on platelets. Reduces hyperglycemia. Smooth muscle relaxer. Anti-ulcer activity. Used for most disorders of the eye. Bilberry contains anthocyanins, bioflavonoids which strengthen the capillaries throughout the body. In addition to the protective and healing effects on the eyes and their diseases, such as glaucoma and retina disorders, Bilberry is used wherever the capillaries are not functioning well, as in poor peripheral circulation and varicose veins. Finally, Bilberries are antioxidant to all the tissues which they affect.

Bistort (*Polygonum bistorta*)

Bistort prefers damp soils, such as in cultivated fields. It is native to Europe, but has been grown in Nova Scotia and as far south as Massachusetts. It is a perennial that reaches up to 30 inches tall.

Bistort root, when ground and mixed with echinacea, myrrh, and goldenseal, is a great dressing for cuts and other wounds. It is also a powerful astringent, used by mixing a teaspoon in a cup of boiled water, and drunk several times a day, as a treatment for diarrhoea and dysentery. The same mixture can be used as a gargle for sore throats. Bistort is good to drive out infectious disease, and is effective for all internal and external bleeding.

An amulet fashioned of the root of Bistort is carried when one wishes to conceive. Sprinkle an infusion of bistort around your home to keep out unwanted visitors of the mischievous variety, such as poltergeists, sprites, *etc.*

Blackberry (*Rubus villosus*)

Blackberries are perennial vines that grow in many areas, depending on the variety. They require full sun, very good air circulation, fertile soil that is kept moist, not soggy. Do not grow where you have grown other fruits or vegetables, to avoid transferring diseases to the young vines. Some varieties need pollinators, so check with your local nurseries to find a variety best suited to your needs and climate. A syrup made from the root is used to treat diarrhoea and upset stomach (good for treating children). An infusion of the leaves is good for treating diarrhoea and sore throat.

Blackberry leaves are used in money spells, as are the berries.

Avoid transferring diseases to the young vines. Some varieties need pollinators, so check with your local nurseries to find a variety best suited to your needs and climate.

Black Cohosh (*Cimcifuga racemosa*)

Black Cohosh grows in open woody areas. It needs good soil and partial to mostly shade to do well. It has been grown as far south as Georgia, and as far west as Missouri. It is a perennial which reaches 3 - 8 feet tall.

Black Cohosh is useful in all conditions dealing with arthritis. It improves blood circulation, and is used in treating delayed and painful menstruation, and is often used in conjunction with other herbs in treating menopause symptoms. It should not be used during pregnancy. Black Cohosh can be poisonous in large doses. It contains a chemical much like estrogen, so those advised by their doctor's not to take the Pill shoud avoid using this herb.

Black Cohosh leaves laid around a room is said to drive away bugs, and to drive away negativity.

Brahmi (*Bacopa monniera*)

Actions: sedative, tonic.

Indications: memory loss, epilepsy, insomnia, learning disorders, stress, nervous tension.

Bacopa is an Ayurvedic herb which shares the name Bramhi with Gotu Kola. Used in India for memory enhancement, epilepsy, insomnia, and as a mild sedative.

Bacopa is also used to improve the sharpness of the sense organs, and to increase learning speed and retention in children and adults. It strengthens the nervous system in a calming and centering way.

Traditional application suggests that Bacopin has a direct effect on improving brain functions, increasing concentration, and in promoting memory functions. Bacosides are believed to play a protective role in the synaptic functions of the nerves in the hippocampus, the seat of memory. Nerve impulses are transmitted across the synapses and their degeneration is believed to contribute to impaired memory and cognition. Bacopa may improve higher order cognitive processes that are critically dependent on the input of information from our environment such as learning and memory. Research indicated that Bacopa heals the brain tissue by augmenting the protein the body uses to build new neurons to replace damaged ones. This action was also found to increase the efficiency of the transmission of nerve impulses in the brain. In other modern studies, Bacopa was found to increase protective antioxidants in the brain and nervous tissue, acting in a way comparable to Vitamin E.

Bladderwrack (*Fucus vesiculosus*)

It is a seaweed that grows in the ocean.

It is used to strengthen and promote the glands. It controls the thyroid and regulates metabolism. It is a sustainer to the nervous system and the brain, and is a terrific boost for pregnancy and the developing child. It contains over 30 minerals and vitamins.

Blessed Thistle (*Cnicus benedictus*)

Blessed Thistle is generally found along roadsides and in wastelands. It is an annual, and reaches to 2 feet tall. Most folks consider this a pesky weed, so cultivation is not common.

Blessed Thistle is used to strengthen the heart, and is useful in all remedies for lung, kidney, and liver problems. It is also used as a brain food for stimulating the memory. It is used in remedies for menopause and for menstrual cramping. Often used by lactating women to stimulate blood flow to the mammary glands and increases the flow of milk.

Blue Cohosh (*Caulophylum thalictroides*)

Blue Cohosh grows best in deep, loamy, moist woodlands. The berry of this plant is poisonous, and the plant itself can irritate the skin. The root is the part used in herbal medicine. It has a range from southern Canada, as far south as the Carolinas, and as far west as Missouri. This herb is best purchased from the stores, rather than cultivated.

Blue Cohosh is used to regulate the menstrual flow. It is also used for suppressed menstruation. Native Americans used this herb during childbirth to ease the pain and difficulty that accompany birthing, as well as to induce labor. This herb should not be taken during pregnancy, and should be taken in very small amounts in conjunction with other herbs, such as Black Cohosh

Blue Green Algae (Cyanobacteria)

A balanced nutritive tonic, containing all the amino acids needed by the body, in the correct proportions, high amounts of B vitamins, 17 different beta-carotene type nutrients, 2000 enzymes, and more gamma-linoleic acid than any other plant source, including the better-known Evening Primrose oil. By entering and nourishing the cells at such a concentrated level, Blue-Green Algae is helpful for a myriad of health problems, since good nutrition and cellular structure are necessary, regardless of any passing illness or other influence. This algae is especially renown to powerfully increase stamina and physical endurance. It is also reported to improve memory, mental energy, attitude and digestion, and reduce stress. There has been research done indicating that it supports the healing of tissue and boosts the immune system, maybe partially due to its high (alkaline) pH. Has very strong life force energy-chi.

Seems to increase group mind with others who are taking it, and also increase intelligence.

Boneset (*Eupatorium perfoliatum*)

Boneset prefers damp to moist rich soils. It is a North American native perennial that reaches 2 to 4 feet high, and grows in partial sun.

Used for treating severe fevers, as well as flu and catarrh conditions. One to two tablespoons of the tincture in hot water is used for sweat therapy to break fevers.

An infusion sprinked around the house will drive away evil spirits and negativity.

Borage (*Borago officinalis*)

Borage was once widely planted in gardens throughout Europe. It was brought to the United States, and now grows wild in much of the eastern half of the nation. It is an annual that grows in most soils, tolerates dry spells, and prefers full sun, reaching to 2 feet in height.

Used for treating bronchitis, rashes, and to increase mother's milk. The infusion is used as an eyewash.

Carrying the fresh blossoms brings courage. The tea will induce your psychic powers.

Butcher's Broom (*Ruscus aculeatus*)

Actions: anti-inflammatory

Indications: varicose veins, thrombosis, hemorrhoids and atherosclerosis

Strengthening effect on vessel walls. Constricting action on veins. Useful in post-operative conditions. Improves peripheral circulation. Increases circulation to brain, legs and arms. Effective for arthritic and rheumatic pains.

Buchu (*Barosma betulina, Agathosma betulina*)

Actions: diuretic, diaphoretic, stimulant, astringent.

Indications: bladder, kidney problems, nephritis, prostate problems, urethritis.

Used for genito-urinary tract. Absorbs excessive uric acid, reduces bladder irritations. Astringent and disinfectant to mucus membranes. Great for inflammation of the urethra and cystitis. Useful in early stages of diabetes.

Burdock (*Arctium lappa*)

Actions: alterative, diaphoretic, diuretic, antipyretic

Indications: inflammatory skin conditions, rashes, cold with fever and sore throat, toxins in the blood, lymphatic clogging, nephritis, edema, kidney inflammation, hypertension, arthritis, eczema, gout, lungs, rheumatism.

Burdock Root is used to treat skin diseases, boils, fevers, inflammations, hepatitis, swollen glands, some cancers, and fluid retention. It is an excellent blood purifier. A tea made of the leaves of Burdock is also used for indigestion. Very useful for building the systems of young women. Helps clear persistent teenage acne if taken for three to four weeks. Used with dandelion root for a very effective liver cleanser and stimulator. Strong action in cleansing the blood and lymphatics. Clears congestion, reduces swelling, dispels toxins. Tonic and rejuvenative for pitta, claers pittagenic emotions like anger, aggression and ambition. Used in most ama conditions like arthritis, and toxic fever. Promotes kidney function. Nourishes hypothalamus and pituitary glands. Aids in the metabolism of carbohydrates (contains inulin). Abundance of minerals. Excellent for skin diseases. Hypoglycemic. Possible anti-tumor, anti-biotic, anti-fungal properties.

Used to ward off all sorts of negativity, making it invaluable for protective amulets and sachets. Add to potpourri in the house.

Calamus (*Acorus calamus*)

It is a perennial herb growing to 1 m by 1 m at a medium rate in light (sandy), medium (loamy) and heavy (clay) soils. Suitable pH: acid, neutral and basic (alkaline) soils. It cannot grow in the shade. It prefers wet soil and can grow in water.

The flowers are hermaphrodite and are pollinated by Insects.

Indications: colds, cough, asthma, sinus headaches, sinusitis, arthritis, epilepsy, shock, coma, loss of memory, deafness, hysteria, neuralgia, pain, tension.

Used in Ayurveda for millennia as the supreme tonic for the mind and spiritual self, it is one of the most renowned herbs of the ancient Vedic seers. Rejuvenative for Vata- for the brain and the nervous system, which it purifies and revitalizes. Clears the subtle channels of toxins and obstructions. Promotes cerebral circulation, increases sensitivity, sharpens memory and enhances awareness. It is sattvic and one of the best herbs for the mind along with gotu kola for which its purposes can be combined to facilitate meditation. Helps trasmute sexual energy and feeds Kundalini. In large doses it is an emetic. Taken nasaly it directly revitalizes prana. Its Vedic name, Vacha, means "speaking" or the power of the word, communication, and intelligence that this herb supports.

Calamus is a purifier and rejuvenative to the brain and nervous system, and increases perception and awareness, especially of the Sattvic, or highest spiritual plane of being. Calamus is considered clearing to the subtle channels, and its North American relative was used in much the same way- to clear and cleanse the body on all levels. Calamus is also used in Ayurveda to balance the effects of marijuana use, and to aid in cleansing and recovery from overuse of drugs.

Bitter Grass (*Calea zacatechichi*)

It is an herbaceous perennial small shrub which can grow up to 3 meters tall, but in normal conditions most specimens tend to be half that size. Each branch produces many small oval leaves, which have serrated edges and curl under. The younger leaves are brilliant green on top and violet underneath. In the right lighting this plant is luminescent and will dramatically standout from its surroundings. C. zacatechichi may be grown from germinated seeds. The dried husks should be removed before planted. The plant likes rich top soil and lots of water. Calea is called "the dream herb", and is used to promote dreams of divination. as an appetite stimulant, cleansing agent for deep wounds and minor burns, to treat diarrhea, reduce fevers, as a application to heal skin rashes and swollen scalps, and most notably to relieve headache pains. The Aztecs once used the plant to treat "cold stomach". The plant is still used in Mexican folk medicine as a laxative and treatment for fever. A tea made from the leaves is particularly good for the stomach and disorders of the digestive system. It is also used for menstrual complaints. In clinical trials, Calea has been proven to increase the duration of rapid eye movement, or dreaming, sleep. Calea creates an interesting sleep pattern, in which users report quick onset of deep sleep with many dreams, followed by immediate awakening after the dreams. Coming out of sleep right after the dream often helps with dream recall, rather than sleeping on until morning and losing the dream's details.

California Spikenard (*Aralia californica*)

It is a perennial, herbaceous plant, 3-5 feet tall; its thick, fleshy rootstock features long, thick roots and produces one or more branched stems growing up to 6 feet high. Stems are smooth, dark green or reddish. The leaves are alternate and usually ternate, 6-21 toothed, pointed leaves, weakly heart-shaped, with doubly serrate margins. Its tiny, greenish-white flowers grow in panicled umbels during July and August.

Actions: demulcent, expectorant, tonic, alterative, adaptogenic.

Indications: Atonic states, cough and irritation of the broncho-pulmonary tract; catarrhal affections; chronic laryngitis with excess, abundant mucus; chronic pharyngitis with thick tenacious mucus; chronic bronchitis with profuse secretions and debility; acute cough with faucial irritability, wheezing, dry mucus; influenza; adrenal cortex hypofunctions; primipara, with irritability; distress last trimester; thin, subanemic blood with hypersensitivities.

Adaptogenic mood elevator for the winter blues, end of relationships, family reunions and other stressful conditions.

Soothing expectorant for chronic moist lung problems, some types of bronchitis, mild emphysema.

Caraway (*Carum carvi*)

Caraway can be found in meadows, woods, and rocky areas. It prefers a rich soil. Native to Europe, Asia, and Africa, it also grows wild here in North America. It is a biennial that reaches 1 1/2 - 2 feet high.

Caraway aids digestion, can help promote menses, can increase a mother's milk, and is good to add to cough remedies as an expectorant.

Carry Caraway in an amulet for protection. Carrying caraway seeds promotes the memory. It can also guard against theft. It is said to promote lust when baked into breads, cookies, or cakes.

Cascara Sagrada (*Rhamnus purshiana*)

Cascara Sagrada is a tree that is native to the Pacific Northwest regions of North America.

Used in treating chronic constipation, and is a stimulant to the whole digestive system. It is a safe laxative, and is useful for treating intestinal gas, liver and gall bladder complaints, and enlarged liver.

Sprinkled around the home before going to court, it will help you to win your case. It is ued in money spells and in repelling evil and hexes.

Catnip (*Nepeta cataria*)

Catnip will grow in most soils, and tends to enjoy a bit of the dry spells once it is established. It grows throughout North America, and is a perennial reaching to 3 feet high.

Catnip is effective alone or in herbal remedies for colds, flu, fevers, upset stomach, and insomnia. Particularly good for children with upset stomachs in a very mild infusion.

The large well dried leaves are used to mark pages in magical books. Used in conjunction with rose petals in love sachets. It will also create a psychic bond between you and your cat. Grow near your home to attract luck and good spirits.

Catuaba (*Erythroxylum vacciniifolium*)

Actions: aphrodisiac, nervine, tonic.

It is a low shrubby tree with yellow and orange flowers that grows in the forests of northern Brazil.

Indications: fatigue, impotence, insomnia, failing memory.

The historic traditional use of this herbal tonic is legendary. Many say it is the most famous of all the Brazilian aphrodisiac plants. It is a strong tonic and nervous system fortifier. It is known for its general capability of giving strength and relieving fatigue. The effect is pronounced in men, especially as a libido enlivener. Many songs praising its wonders have been sung by the Brazilian Indians. It is often used in combination with Muira Puama for tonifying the male sexual and urinary organs, specifically as an aphrodisiac and for impotence. It has reported benefits in relieving insomnia from hypertension, restless sleeping patterns, and even in helping to arrest failing memory." It is said that regular consumption of Catuaba bark over a period of time leads first to erotic dreams, which are later followed by increased libido. Catuaba is considered an "innocent" aphrodisiac, meaning no adverse side effects have been reported in its use.

Cat's Claw (*Uncaria tomentosa*)

It is native to the rain forest regions of Central and South America, especially in the upper Amazon region of Peru and neighboring countries. Maranhao, Brazil is the most eastern area where it has been found growing wild. It favors mountain slopes in the organic soils of the primary (old growth) rainforest, between 250 and 900 meters above sea level. It can also be found in disturbed forest but rarely in secondary forest.

Used in treatment of arthritis, gastritis, tumors, dysentery, female hormonal imbalances, viral infections. It is effective in aiding treatment of the immune system, the intestinal system, and the cardivoascular system. This herb should not be taken by those who have received an organ transplant, nor by pregnant or nursing women. Considered a good herb for cleansing. Traditionally used for cancer and tumors. Traditionally used by Peruvian native women for birth control. Research shows that it helps stimulate immune system. Research as an adjunct to chemotherapy has positive results. Current research for use in AIDS and cancer treatments.

Cayenne (*Capsicum* spp.)

Cayenne pepper plants like a good, rich soil, plenty of water, and full sun. The peppers are dried after ripening. For herbal use, the peppers are usually ground into a powder and mixed with other powdered herbs in capsules.

Actions: stimulant, diaphoretic, expectorant, carminative, alterative, hemostatic, anthelmentic, astringent, antispasmodic, antibiotic, antifungal.

Indications: indigestion, ama, poor absorption, abdominal distension, worms, sinus congestion, chronic chill, poor circulation, antioxidant support, atherosclerosis, pain disorders, diabetic neuropathy, cluster headaches, arthritis, psoriasis. is very effective added to liniments for all sorts of arthritis and muscle aches. Internally it benefits the heart and circulation when taken alone or added to other remedies. It is also used to stimulate the action of other herbs. Capsicum is also used to normalize blood pressure. It will stop bleeding both externally and internally, making it excellent for use with ulcers. It is used in antibiotic combinations, for menstrual cramps, and as a part of treatment for depression. Helpful in cases of heart attack and heart weakness for revival purposes. Burns toxins from the colon. May cause disturbance in the mind if used to much and also increase pitta. Warms the body. Increases heart action but not blood pressure. Used for flu and colds. Helps with ulcers. Stimulates the body. Thins the blood. Helps prevent heart attacks. Used in tincture over inflamed joints and arthritis. Lowers cholesterol. Useful for cluster headaches. Topically, useful for psoriasis, shingles, trigeminal neuralgia and pain due to mastectomy and chemotherapy or radiation.

Cayenne pepper scattered around your house will break bad spells. Adding it to love powders will ensure that your love will be spicy, and can inflame the loved one with passion.

Cedar (*Cedrus deodara*)

There are many types of cedars that grow throughout the world. Cedars are evergreen perennial coniferous tree reaching 40–50 m tall, exceptionally 60 m, that are attractive in any landscape.

It is an astringent, antiseptic, diuretic, expectorant and sedative, useful in phthisis, bronchitis, blennorrhagia, and also for eruptions on the skin.

In folk medicine, it is a symbol of power and longevity. Hung in the home it will protect against lightning. Cedar chips used in rituals or burnt attracts money, and is also used in purification and healing.

Celastrus (*Celastrus orbiculatus*)

It is a deciduous Climber growing to 12 m at a fast rate. It can grow in light (sandy), medium (loamy) and heavy (clay) soils. Suitable pH: acid, neutral and basic (alkaline) soils. full shade (deep woodland) semi-shade (light woodland) or no shade. It prefers moist soil and dappled shade; shady edge; not deep shade. It is in flower from Jul to August, and the seeds ripen from Nov to February. The flowers are dioecious and pollinated by Bees.

Actions: acrid, bitter, thermogenic, emollient, stimulant, intellect promoting, digestive, laxative, emetic, expectorant, appetiser, aphrodisiac, cardiotonic, anti-inflammatory, diuretic, emmenagogue, diaphoretic, febrifuge and tonic.

Indications: abdominal disorders, leprosy, pruritis, skin diseases, paralysis, cephalalgia, arthralgia, asthma, leucoderma, cardiac debility, inflammation, stranguary, neuropathy, amenorrhoea, dysmenorrhoea and fever, stimulating the intellect and sharpening the memory, vitamin B-deficiency.

Powerful cognitive enhancer, it increases both memory and one's ability to learn. It induces a sense of well-being, and a feeling of being in control. It also acts as a central nervous system stimulant. Celastrus is used as an aphrodisiac and as a powerful brain tonic to stimulate intellect, sharpen memory, and increase intelligence. The Ayurvedic name of this plant, Magzsudhi, means "brain clearer". In Ayurvedic medicine, Celastrus is used to improve the capabilities mentally disabled children. It is also useful as a tonic for the healthy brain, with many antioxidant properties.

Celastrus is helpful with spiritual and meditative work that requires mental acuity and recall, such as Qabalah, sacred geometry, or complex chanting. Celastrus is also helpful for meditations on the flow of consciousness and thought.

Chamomile [*Matricaria chamomilla* (German chamomile), *Anthemus nobilis* (Roman chamomile)]

Chamomile is an annual that adapts to most soils, likes lots of water and full sun. It grows up to 20 inches tall.

The tea is used for nerves and menstrual cramps. The tea is also useful for babies and small children with colds and stomach troubles. Also used to calm the body for inducing sleep in insomniac conditions. It is also a good wash for sore eyes and open sores.

Chamomile is used in prosperity charms to attract money. Added to incense, it will produce a relaxed state for better meditation. Burned alone it will induce sleep. Added to a ritual bath, it will attract love. Sprinkle it around your property to remove curses and bad spells.

Chickweed (*Stellaria media*)

This annual spreading plant is usually hated as an obnoxious weed by the typical gardener. It is found throughout temperate areas of North America and of Europe, the plant's native homeland. It prefers full sun, average to poor soils, and infrequent watering.

Chickweed is an excellent source of many B vitamins and various minerals. It is used to treat bronchitis, pleurisy, coughs, colds, and as a blood builder. Externally it is good for skin diseases, and the tea added to the bath is good for soothing skin irritations and rashes.

Chickweed is used in spells to attract love and to maintain a relationship.

Chastetree (Vitex) (*Agnus castus*)

is a deciduous Shrub growing to 3 m at a medium rate. It grows in light (sandy) and medium (loamy) soils, prefers well-drained soil and can grow in nutritionally poor soil. Suitable pH: acid, neutral and basic (alkaline) soils. It cannot grow in the shade. It prefers dry or moist soil. The flowers are hermaphrodite and pollinated by Insects.

Actions: stimulating, normalizing.

Indications: dysmenorrhea, pre menstrual syndrome, acne in teens, excessive bleeding, irregular cycles, fibroids and other uterine or ovarian growths and inflammations, excessive cramps, and amenorrhea.

It has the effect of stimulation and normalizing pituitary gland functions, especially its progesterone function. Chaste Tree may be called an amphoteric remedy, as it can produce apparently opposite effects (though in truth it is simply normalizing). The greatest use of Chaste Tree lies in normalizing the activity of female sex hormones, and it is thus indicated for dysmenorrhoea, premenstrual stress and other disorders related to hormone function. Because Vitex contains progesterone, it is most helpful for problems which occur during the second half of the cycle, from ovulation to menstruation. This is the phase during which the symptom package known as PMS occurs, and Vitex can help with all PMS-related problems- emotional upset, cramps, migraines, acne, swollen feet, *etc.* Although it works primarily on these second-half problems, Vitex should be taken throughout the entire cycle, supplemented by estrogenic herbs during the first half of the cycle. It is especially beneficial during menopausal changes. And in a similar way, it may be used to aid the body to regain a natural balance after the use of the birth control pill. Recent findings confirm that Chaste Tree helps restore a normal estrogen-to-progesterone balance. It can not only ease, but with time, actually cure premenstrual syndrome, which has been linked to abnormally high levels of estrogen, especially if symptoms tend to disappear when menstruation begins. European herbalists also use it today to treat fibroid tumors and other female complaints. As the baby-boom generation passes into menopause, this herb will be rediscovered by women looking for a natural alternative to estrogen replacement therapy. For most purposes, it is necessary to take Vitex daily as a long-term therapy. Improvement is usually seen after the third cycle from beginning, but it is necessary to continue for many months to effect long-term re-balancing.

Cinnamon (*Cinnamomum zeylanicum*)

This shrub requires a subtropical to tropical area with full tropical sunshine and moist soil. It has been known to survive short mild frosts but nothing too prolonged, or below 32° Fahrenheit. Cinnamon is among many other species since the tropical rainforest it natively grows in are abundant in the amount of different species.

It is a stimulant to other herbs and the body, enabling herbal remedies to work faster. It is also a blood purifier, an infection preventer, and a digestive aid.

Caution: Do not ingest cinnamon oil.

In folk medicine, it is burned in incense, cinnamon will promote high spirituality. It is also used to stimulate the passions of the male. It should also be burned in incenses used for healing. The essential oil is used in aromatherapy.

Clover (*Trifolium pratense*)

Grow clover as you would lawn grasses. Clover is an excellent cover crop, planted in fallow areas and turned under in the fall, it makes an excellent fertilizer for poor soils. Clover is used as a nerve tonic and as a sedative for exhaustion. It is used to strengthen those children with weak systems, and is used with children for coughs, bronchitis, wheezing, as it is mild to their systems. It is often used in combination with many other drugs in the treatment of cancer. It is also used for skin eruptions (acne).

Clover brings luck, prosperity, and health. Carrying a three-leaf clover gives you protection. Worn over the right breast it will bring you success in all undertakings.

Cloves (*Eugenia caryophyllata*)

A bushy, evergreen tree with a medium-sized crown, growing 8-20 metres tall. Several parts of the tree are aromatic, including the flower buds, leaves and bark. Clove grows best on tropical mountain slopes at lower elevations as part of a mixed forest and grows in small clusters.

Clove oil will stop a toothache when it is applied directly to the cavity. It is very warm and stimulating to the system, and is very useful with people who have cold extremities. Cloves will promote sweating with fevers, colds, and flu. It is often used in remedies for whooping cough. Cloves are also safe and effective for relieving vomiting during pregnancy.

In folk medicine, cloves are worn in an amulet to drive away negativity and hostility, and stop gossip. It is often carried to stimulate the memory, and can be added to attraction sachets. Clove oil is also worn as an aphrodisiac, and the buds when eaten are said to stir up bodily lusts. It is placed in sachets with mint and rose to chase away melancholy and to help one sleep soundly. Carried, it can also bring comfort to the bereaved and mourning.

Codonopsis (*Codonopsis pilosula, Codonopsis lanceolata*)

Actions: stomachic, nutritive tonic, hypotensive, stimulant.

Indications: high blood pressure, indigestion, chronic coughs, shortness of breath.

Chinese Codonopsis is considered a nourishing tonic, it has been used to promote digestion, absorption, metabolism and tone and strengthen the stomach and spleen. The herb stimulates the nervous system, increases body resistance and has been shown to increase the number of red blood cells and hemoglobin content. Animal studies have shown that it dilates peripheral blood vessels and inhibits adrenal cortex activity, causing a lowering of blood pressure. This herb also benefits the lungs and is helpful in treating chronic coughs and shortness of breath. Chinese traditional medicine practitioners often use Codonopsis as a Ginseng substitute.

Coltsfoot (*Tussilago farfara*)

Coltsfoot is a perennial that prefers damp, clay soils. It grows 5 to 18 inches high, and likes full to partial sun.

Used to treat respiratory problems, and is soothing to the stomach and intestines. Combine with horehound, ginger, and liquorice root for a soothing cough syrup.

Coltsfoot is added to love sachets and is used in spells of peace and tranquility.

Comfrey (*Symphytum officinale*)

Comfrey prefers well-drained soils and partial shade. It grows from Canada to Georgia, as far west as Missouri, in the wild. It is a perennial that grows to 3 feet high. It can be started from seed, but you will be more successful with cuttings. Once established, it will spread vigorously. Harvest leaves when the flowers bud, and roots in the autumn after the first frost.

Actions: nutritive tonic, demulcent, expectorant, emollient, vulnerary, astringent, haemostatic.

Indications: cough, lung infections, coughing blood, lung haemorrhage, gastrointestinal ulcers, blood in urine, diarrhoea, dysentery, sprains, fractures, wounds, sores, boils.

It is a powerful tonic and vulnerary. Nutritive and rejuvenative tonic to the lungs and the mucous membranes. Can be used in most conditions where the membranes are inflamed, bleeding, or wasting away. Promotes tissue growth, externally and internally, and healing throughout the body when it has been afflicted by disease or traumatic injuries.

A poultice of comfrey heals wounds, burns, sores, and bruises. It is a powerful remedy for coughs, ulcers, healing broken bones and sprains, and is used in treating asthma. Large amounts or dosages can cause liver damage, but there are no problems with using it externally. Used internally, it is best and safest to use a tea, rather than capsules.

Carrying comfrey during travel will ensure your safety. Put some in your luggage to prevent it being lost or stolen. It will also bring luck to the carrier.

Cornflower (*Centaurea cyanus*)

Cornflower is adaptable to many soils and conditions. It is an annual that grows 1 - 2 feet tall.

Juices from the stems of this plant are used externally to treat wounds and cuts.

Cornflower is used to promote and enhance phsychic sight, as well as normal eyesight.

Cramp Bark (*Viburnum opulus*)

Cramp Bark is one of the best female regulators in the herb world. It is a uterine sedative, aiding in menstrual cramps and afterbirth. It helps to prevent miscarriage, as well as internal haemorrhage.

Cordyceps (*Cordyceps sinensis*)

Actions: adaptogen, tonic, immune stimulant, restorative.

Indications: exhaustion, impotence, backache, chronic cough, asthma, debility, anaemia.

Similar to Ginseng, being used to strengthen the body after exhaustion or long term illness. It has traditionally been used for impotence, backache, to increase sperm production and to increase blood production. In China, Cordyceps is used medically to regulate and support the gonads and as a lung and kidney tonic. It is used specifically for excess tiredness, chronic cough and asthma, impotence, debility, anaemia, to build the bone marrow and reduce excess phlegm. The clinical studies done on Cordyceps in China have been numerous and remarkable, indicating the fungus can improve liver functions, reduce cholesterol, adjust protein metabolism, improve immune functions, inhibit lung carcinoma and has a therapeutic value in the treatment of aging disorders including loss of sexual drive. These are just some of the reported benefits attributed to this remarkable fungus. Not recommended if nursing or pregnant.

Damiana (*Turnera diffusa*)

Actions: stimulant, aphrodisiac, tonic.

Indications: bronchitis, emphysema, hormone balancer, hot flashes, menopause, Parkinson's disease, sexual stimulant.

Damiana is used to regulate the female cycles. It is also used to stimulate the sexual appetite. It is good for urinary problems and nervousness, as well as hypertension.

Increases sperm count in males. Aids in ovulation in female. Useful as a hormone balancer for both sexes. Beneficial for exhausted state of the body. Increases sexual prowess. Mild aphrodisiac. Digestive aid. Useful in removal of mucus congestion. Contains testosterone precursors and like action. Has tonic action on central nervous system. Damiana helps the general health of the sexual organs, and is antiseptic to the urinary tract. Damiana is also very protective to the tissues of the reproductive organs, and used to heal "honeymoon cystitis", the pain and inflammation from frequent lovemaking. Damiana is stimulating and rejuvenating to the reproductive organs, as well as to the nervous system, especially increasing the tactile sensation of the skin. It is also used as a general women's tonic and is believed to increase fertility in both men and women. It can be over-stimulating to the nervous system, and is best taken in small, frequent doses, for finite periods of time.

Damiana is used in infusions to incite lust, and is burned to produce visions.

Dandelion (*Taraxum officinale*)

Dandelion is a common yard, garden, and roadside weed. Do not gather where chemicals have been used, and don't gather those near roadsides, as they have been contaminated from exhausts.

Dandelion benefits all functions of the liver. It clears obstructions (such as stones) and detoxifies poisons that gather in the liver, spleen, and gall bladder. It will also promote healthy circulation. The juice from a broken leaf stem can be applied to warts and allowed to dry; used for 3 days or so it will dry up the warts. It is also used to treat premenstrual syndrome, as it is a diuretic. It has been shown to reduce cholesterol and uric acid. Dandelion also helps clear skin eruptions when used both internally and externally. Actions: alterative, diuretic, lithotriptic, laxative, bitter tonic.

Indications: liver problems, jaundice, gall stones, congested lymphatics, breast sores, breast cancer, hepatitis, diabetes, edema, ulcers, blood cleanser and weight loss aid.

Primarily a detoxifying herb for Pitta and ama conditions. Specific for problems of the breast and mammary glands, breast sores, tumors, cysts, suppression of lactation, swollen lymph glands. It clears and cleanses the liver and gall bladder and dispels accumulated and stagnated Pitta and bile. Good for detoxification from a meat diet and over eating of fried or fatty foods. Enhances flow of bile. High choline content. Diuretic, especially water retention due to heart problems. Possible anti-tumor agent. Hypoglycemic. High inulin content. Useful in pre menstrual syndrome. Used to treat jaundice, hepatitis, and gallbladder disease. Strengthens weak arteries. Aids in removal of toxins from body. Useful in muscular rheumatism. Helps clear obstructions in the biliary ducts. Useful for the liver, gallbladder, kidneys, pancreas, spleen and circulatory system.

It is a rich source of potassium, and contains more vitamin A than carrots.

It is a sign of rain when the down from a ripened dandelion head falls without wind helping it to do so. To blow the seeds off a ripened head is to carry your thoughts to a loved one, near or far.

Dill (*Anethum graveolens*)

Dill grows in most regions of North America. It needs sun and a well-drained soil, and frequent waterings. It is a hardy annual, biennial in the deep southern regions, that reaches 2 - 3 feet tall. Dill matures quickly, and self-sows for the following year. Plant in six week intervals for a season-long supply of fresh dill.

Dill is used to treat colic, gas, and indigestion.

Dill is used in love and protection sachets. The dried seed heads hung in the home, over doorways, and above cradles provides protection. Add dill to your bath to make you irresistible to your lover.

Dong Quai (*Angelica sinensis*)

Actions: antispasmodic.

Indications: treat PMS, menopause, and menstrual irregularities, anemia, internal bleeding, stroke, blood purifier, brain nourisher, female glands, hot flashes, menopause, menstruation, nervousness.

Helps dissolve blood clots and remove morbid matter from body. Increases circulation especially to the pelvic area. Used to treat female gynecological ailments. Should not be used during pregnancy nor menses. Helps relieve pelvic congestion. Useful for women who do not have enough energy. A powerful herb for the female reproductive system. It is a powerful uterine tonic and hormonal regulator. Used in premenstrual syndrome formulas as well as menopausal formulas.

Caution: Weak estrogenic activity. Do not use in cases of bleeding uterine fibroids or breast cancer. Do not use when menstruating as it thins the blood and brings blood to the pelvic region.

Dragon's Blood (*Daemonorops draco*)

The resin of Dragon's Blood is used externally as a wash to promote healing and stop bleeding. Internally it is used for chest pains, post-partum bleeding, internal traumas, and menstrual irregularities.

Added to love incenses and sachets, it increases the potency of other herbs used. A piece of the plant is often used under the mattress as a cure for impotency. It is also used in spells to bring back a loved one. A pinch added to other herbs for magical purposes will increase their potency.

Devil's Club (*Oplopanax horridus*)

Actions: diuretic, laxative, tonic.

Indications: colds, tuberculosis, diabetes, stomach pain, cancer, sore throat and hoarseness, arthritis, rheumatism, headache relief, hangovers, hyperthyroid, gallstones, blood poisoning, staphylococcus or streptococcus infections, regulating blood sugar.

Used to relieve viral cold symptoms, combat tuberculosis, combat and cure adult-onset diabetes, resolve "stomach pain", treat cancer, as a topical wash for staphylococcus or streptococcus infections, re-establish menstrual flow after giving birth, used internally (strong infusions were consumed exclusively for several days, excluding all food and water) and externally(as a topical wash) for rheumatism and arthritis, consumed regularly as a general health tonic, gargled for sore throat and hoarseness, used to quell fevers.

Contemporary coastal peoples use the tea of Devil's Club for: headache relief, fresh or dried bark tea for rheumatism, respiratory ailments, stomach ulcers, hangover cure, to treat and prevent blood poisoning, for gallstone remediation, to cure hyperactive thyroid (part of the lethargy attributed to consumption of devil's club tea may be due to thyro suppressive action). For those patients with sugar blues, it is not only helpful in modulating blood sugar release and utilization, but there is often a distinct mood enhancement and improved sense of well-being, and decreased cravings for sweet foods. Ability to stop breast milk flow and to initiate menstruation post partum strongly suggests that this herb effects estrogen metabolism; the anti hyper thyroid activity suggests further endocrine involvement as does the resolution of type II diabetes. I suspect that the weight loss action is affected by thyroid suppression in either the pituitary or hypothalamus, as a progressive state of minimal hypothyroidism develops. The post-consumption lethargy experienced by many may be thyro suppression. Certainly the eventual return of lost weight and further weight gain support the hypothyroid idea. The long term use of small amounts of Devil's Club may act as both an adaptogenic and endocrine tonic.

Echinacea (*Echinacea purpurea*)

Echinacea, also known as Purple Coneflower likes the prairies and other open, dry places. It adapts to most soils, in full sun, except wet ones. It grows over most of North America. It is a perennial, and reaches to about 2 feet tall. The root is used ground, and the leaves are used for teas. Actions: alterative, diaphoretic, antibacterial, antiviral, antiseptic, analgesic, immunostimulant.

Indications: toxic conditions of the blood, blood poisoning, gangrene, eczema, poisonous bites or stings, venereal diseases, prostatitis, infections, wounds, abcesses, prostate problems and cancer.

It is a natural antibiotic and immune system stimulator, helping to build resistance to colds, flus, and infections. It increases the production of white blood cells, and improves the lymph glands. The tea from this herb should be used in all infections, and has been used in treating skin cancers and other cancers. Please note that if you suffer from any auto-immune disorder, you should use Echinacea, or any other immune stimulant herb, only under the guidance of a professional, such as a naturopathic doctor, TCM practitioner, *etc.*

Best detoxifying agent in western herbalism. Natural antibiotic that counters effect of most toxins in the body. Cleanses blood and lymph systems, catalyzes the action of white blood cells to area of infection and increases their amount, helps arrest pus formation and tissue putrefaction. Destroys ama. May be combined with liquorice or marshmellow to prevent dizziness or ungroundedness. Save for sickness because the body will adapt to it. Don't take with meals.

Echinacea is used as an offering to the spirits or gods and goddesses to strengthen a spell or ritual.

Elder (*Sambucus nigra*)

Elder is a tree or shrub, growing to 30 feet tall. It prefers moist areas throughout North America. The leaves, bark, and roots of the American varieties generally contain poisonous alkaloids and should not be used internally.

Actions: diaphoretic, diuretic, alterative, antiseptic.

Indications: flu, colds, acne, inflammation and fever.

This herb is an internal cleanser when fighting flu and colds. Acne can also be treated with this herb when taken as a tea Elder flower oil is a remedy for chapped skin and is also used to cleanse the body, build the blood, treat inflammation, fever and soothe the respiratory system. Being an antiseptic it heals external wounds and is also used as an insect repellant.

Elder flowers, mixed with mint and yarrow blossoms, are excellent internal cleansers when fighting flu and colds. A tea of the elder flowers and sassafras is a remedy for acne. Elder flower oil is a remedy for chapped skin. Elder is used to cleanse the body, build the blood, treat inflammation, fever, and soothes the respiratory system. The leaves can be used as an antiseptic poultice for external wounds, and as an insect repellant. The Greeks used a tea from the root as a laxative. Elder can be toxic, especially if fresh, most notably the stems as they contain cyanide. This herb should not be used internally by pregnant or lactating women.

The branches of the sacred elder are used to make magical wands for ritual. Scattering the leaves in the four winds will bring protection. Elderberry wine, made from the berries, is used in rituals. In Denmark, it is believed to be unlucky to have furniture made of elder wood. Grown near your home, elder will offer protection to the dwellers. It is used at weddings to bring good luck to the newlyweds. Flutes made form the branches are used to bring forth spirits.

Elecampane (*Inula helenium*)

Elecampane enjoys roadsides and damp fields and pastures. Plant it in full sun in a damp, but not soggy, location. It is a perennial that grows 3 - 6 feet tall. The root is most commonly used.

Elecampane is used for intestinal worms, water retention, and to lessen tooth decay and firm the gums. It gives relief to respiratory ailments. It is usually used in combination with other herbs. Externally it is used as a wash for wounds and itching rashes. It is burned to repel insects.

Add this herb to love charms and amulets of all kinds. Used with mistletoe and vervain, it is especially powerful. Use when scrying for better results.

Eucalyptus (*Eucalyptus globulus*)

Eucalyptus reigns among the tallest trees in the world, capable of reaching heights of over 250 feet tall. It thrives only in areas where the average temperature remains above 60 degrees, and is adaptable to several soil conditions.

Eucalyptus oil is a powerful antiseptic, and is used to treat pyorrhoea (gum disease), and is used on burns to prevent infections. The oil breathed in will help clear the sinuses, as will the steam from boiling the leaves. When mixed with water or vegetable oils, it makes a good insect repellant. A small drop on the tongue eases nausea.

Healing energies come from the leaves. A branch or wreath over the bed of a sick person will help spread the healing energies. The oil is added to healing baths, and for purifications.

Eucommia (*Eucommia ulmoides*)

It is a deciduous tree grows at a fast rate up to 12 m by 8 m in light (sandy), medium (loamy) and heavy (clay) soils and prefers well-drained soil. It cannot grow in the shade. It prefers moist soil.

Actions: tonic, strengthening, anti-inflammatory, sedative and diuretic.

Eucommia bark is a superb Yang Jing tonic, used to strengthen the back (especially the lower back), skeleton, and joints (especially the knees and ankles). It is believed to confer strength and flexibility to the ligaments and tendons. It is often used by athletes to strengthen the entire body. Eucommia is considered to be one of the great longevity tonics of Chinese tonic herbalism and is also believed in China to be a safe and effective sex tonic for men and women. It helps regulate blood pressure. It is especially beneficial to those with high blood pressure. It is very safe. Primarily used for its value in regulating blood pressure Eucommia is also regarded highly for use as a longevity herb second only to Ginseng. This is attributed to the tonic effect it has on the liver and kidneys. In Japan Eucommia is currently being marketed for its actions as an anti-inflammatory and as a sexual tonic. Many people use it for lumbago in Asia. The new popularity it is enjoying in Japan however is connected to the weight loss claims many users experience. It reportedly stifles the craving for food.

Evening Primrose (*Oenothera biennis*)

The American variety is found throughout North America. It enjoys dry soils and full sun. It is a biennial, and grows 3 - 6 feet tall. The seed oil is the most commonly used portion of the plant. Some nurseries sell evening primrose, but they are actually a small, showy hybrid of the perennial Missouri Primrose, and does not have the same medicinal uses, so be sure you are buying the plant you really want.

Evening Primrose oil stimulates to help with liver and spleen conditions. In Europe, it has been used to treat Multiple Sclerosis. It lowers blood pressure, and eases the pain of angina by opening up the blood vessels. It has been found to help slow the production of cholesterol, and has been found to lower cholesterol levels. Used with Dong Quai and Vitex, it is a valuable part of an herbal remedy for treating the symptoms of pre-menstrual syndrome (PMS) and menstrual cramping.

Eyebright (*Euphrasia officinalis*)

Eyebright is adaptable to many soil types in full sun. It is a small annual, growing 2 - 8 inches high. It attaches itself by underground suckers to the roots of neighboring grass plants and takes its nutrients from them. To be cultivated, it must be given nurse plants on whose roots it can feed.

Eyebright stimulates the liver to remove toxins from the body. It has been used internally and externally to treat eye infections and afflictions, such as pink-eye. The herb strengthens the eye, and helps to repair damage.

Eyebright is used to make a simple tea to rub on the eyelids to induce and enhance clairvoyant visions.

False Unicorn (*Chamaelirium luteum*)

False Unicorn grows primarily in very moist areas, in partially shady areas, throughout North America.

False Unicorn is very soothing for a delicate stomach. It also stimulates the reproductive organs in women and men. This herb is very important for use during menopause, due to its positive effects on uterine disorders, headaches, and depression.

Fennel (*Foeniculum vulgare*)

Fennel prefers dry, sunny areas. It is a perennial that can reach 4 - 6 feet high, and grows in most average to poor soils.

Fennel helps to take away the appetite. It is often used as a sedative for small children. It improves digestion, and is very helpful with coughs. It is also used for cancer patients after radiation and chemotherapy treatments. Enriches and increases the flow of milk for lactating women.

It is used for scenting soaps and perfumes to ward off negativity and evil and grown near the home for the same purpose.

Fenugreek (*Trigonella foenum-graecum*)

Fenugreek likes dry, moderately fertile soil in a sunny location. It is an annual, and grows to 1 - 3 feet tall.

Actions: stimulant, tonic, expectorant, rejuvenative, aphrodisiac, diuretic, antiseptic.

Indications: dysentary, dyspepsia, chronic cough, allergies, bronchitis, influenza, convalescence, dropsy, toothache, neurasthenia, sciatica, arthritis, diabetes, bronchial catarrh, lowers cholesterol, lung infections, mucus, stomach irritations, weight loss, sinus congestion and appetite suppressant.

Good herbal food for convalescence and debility, particularly that of the nervous, respiratory, and reproductive systems. Promotes digestion and liver function. Softens and dissolves hardened accumulated mucus. Helps the lymphatic system cleanse. Contains lecithin. Lowers cholesterol. Inhibits intestinal absorption of cholesterol. Contains lipotropic (fat dissolving) substances. Helps prevent water retention. Used in treatment of diabetes. Stimulates pancreatic secretions. Contains considerable quantities of diosgenin and tigogenin (substances from which sex hormones are made). Stimulates lactation. The seeds are spermicidal. Contains large amounts of amino acids: lysine, tryptophan, leucine, histidine and arginine. Contains mucilaginous fiber with high viscosity.

Fenugreek is used to soften and expel mucous. It has antiseptic properties and will kill infections in the lungs. Used with lemon and honey, it will help reduce a fever and will soothe and nourish the body during illness. It has been used to relax the uterus, and for this reason should not be taken by pregnant women.

It is believed that adding a few fenugreek seeds to the mop water used to clean household floors will bring money into the household.

Feverfew (*Tanacetum parthenium*)

Feverfew bears a resemblance to chamomile. It prefers dry places, will tolerate poor soil, and is a hardy biennial or perennial, growing to 2 1/2 feet. It prefers full sun. Feverfew is used to treat colds, fevers, flu, and digestive problems. It is often used to end migraines and other headaches.

In folk medicine, Feverfew is carried for protection against illnesses involving fever, as well as for preventing accidents.

Flax (*Linum usitatissimum*)

Flax grows in a wide range through North America. It is an annual that grows to 3 feet high, and is adaptable to many soils and conditions.

Flax is used as an aid to achieving cardiovascular health, to help in menopause, and as a mild laxative. The seed and the seed oil are being studied as a possible cure for cancer. The oil helps slow the kidney disease that accompanies lupus.

In folk medicine it is used to attract money and wealth, and is used in healing spells and rituals.

Fo Ti (*Polygonum multiflorum*)

It is a perennial climber grows upto 4.5 m (14ft 9in) in light (sandy), medium (loamy) and heavy (clay) soils. Suitable pH: acid, neutral and basic (alkaline) soils. It can grow in semi-shade (light woodland) or no shade. It prefers moist soil.

Actions: tonic, rejuvenative, aphrodisiac, astringent and nervine.

Indications: anaemia, neurasthenia, impotence, lower back pain, premature greying or falling out of hair, enlarged lymph glands, excessive menstruation, leucorrhoea, arteriosclerosis and diabetes.

Chinese rejuvenative herb that builds the blood and sperm and strengthens the muscles, tendons, ligaments, and bones. Strengthens the kidneys, the liver and the nervous system and is famous restorative for the hair. Often combined with gotu kola, which is similar in its effects.

Frankincense (*Boswellia carterii*)

It is a small deciduous tree, which reaches a height of 2 to 8 m growing on rocky limestone slopes and gullies, and in the 'fog oasis' woodlands of the southern coastal mountains of the desert-woodland.

Frankincense relieves menstrual pains, and treats rheumatic aches and pains. Externally it is used for liniments and for its antiseptic properties.

In religious practices it is burned to raise vibrations, purify, and exorcise. It will aid meditations and visions. The essential oil is used to anoint magical tools, altars, *etc.*

Garlic (*Allium sativum*)

Garlic is a perennial herb that likes moderate soil and lots of sun and warmth. The plant grows to 2 feet tall. The bulb is the most common used portion, although the greens are often used in salads.

Actions: stimulant, carminative, expectorant, alterative, antibiotic, anti fungal, anti-bacterial, antispasmodic, aphrodisiac, disinfectant, anthelmentic and rejuvenative.

Indications: colds, cough, asthma, heart disease, hypertension, arteriosclerosis, palpitation, skin diseases, parasitic infections, rheumatism, hemorrhoids, edema, impotence, hysteria, cancer prevention, diabetes, high blood pressure, high cholesterol, infection. It is a powerful rejuvenative and detoxifier. Good for chronic fevers. Cleanses ama and kapha from the blood and lymphatics. Increases groundedness. It increases semen but can have an irritating effect on the reproductive organs. Blood thinner. Health building herb. Dissolves cholesterol. Stimulates lymphatic system. Kills parasites. Does not destroy the body's normal flora. Anti-biotic component is known as allicin (sulfur compound) which is estimated to equal 15 standard units of penicillin. Used in China to treat cryptococcal meningitis. Protective against influenza. Lowers cholesterol. Inhibits platelet aggregation. Garlic is a powerful natural antibiotic. It can stimulate cell growth and activity. It reduces blood pressure in hypertensive conditions. A main advantage to using garlic for its antibiotic properties is that it does not destroy the body's natural flora. It is excellent for use in all colds and infections of the body. When ingesting the raw cloves, a sprig of parsley chewed immediately after will freshen the breath.

In folk medicine peeled garlic cloves placed in room to ward off diseases. It is hung in new homes to dispel negativity and evil. It is a strong protective herb. Often placed a clove under the pillow of sleeping children to protect them.

Can increase dullness of the mind. Good medicine but not good for common usage for those practicing yoga.

German Chamomile (*Matricaria chamomilla*)

Actions: diaphoretic, carminative, nervine, antispasmodic, analgesic, emmenogogue, emetic, harmonizing, anti-inflammatory and sedative.

Indications: headaches, indigestion, digestive and nervous problems of children, colic, eye inflammations, jaundice, dysmenorrhea, amenorrhea, insomnia, appetite stimulant, bronchitis, menstrual cramps, fevers, hysteria, menstrual suppressant, nervousness, tension and irritability.

In moderation it is good for all constitutions, and is particularly good for Pitta. Helps relieve bilious, digestive headaches, relieves congestion of the blood and promotes menstruation. It is a sattvic herb and is balancing to the emotions. It sedates nerve pain and strengthens the eyes. Mixed with a little ginger, chamomile makes a completely balanced beverage. Useful for nerves. Helps promote thyroxine. Gentle for children. Reduces tension. Aids digestion. Soothes the stomach. Helps reduce menstrual cramps. Helps with infant colic.

Ginger (*Zingiber officinale*)

Ginger grows through most of North America. It reaches to 6 inches high, and is a perennial. The ground root is the part used for healing.

Ginger is an excellent herb to use for strengthening and healing the respiratory system, as well as for fighting off colds and flu. It removes congestion, soothes sore throats, and relieves headaches and body aches. Combined with other herbs, it enhances their effectiveness. It is also very effective in combatting motion sickness. Actions: stimulant, carminative, expectorant, anthelmintic, analgesic, anti-inflammatory, anti-ulcer effects and antioxidant.

Indications: colds, flus, indigestion, vomiting, belching, abdominal pain, laryngitis, arthritis, haemorrhoids, headaches, heart disease, nausea and vomiting, motion sickness, migraine headaches, gas and flatulence.

Perhaps the best and most sattvic of the spices, it was called the universal medicine. Useful in digestive and respiratory diseases and in arthritic conditions, and it is tonic to the heart. It relieves cramps in the abdomen, including menstrual cramps, soothe the digestive system. Ginger is believed to aid in relieving the symptoms of motion sickness. Removes congestion, relieves headaches. Excellent for upset stomach and indigestion. Helps remove mucus from the body. Used in the treatment of headaches. Excellent remedy for nausea. Speeds food through the large intestines. Cleansing agent for bowels. Relieves wind. Useful for motion sickness and morning sickness. Helps symptom of vertigo and dizziness. Ginger tea is excellent for that common nausea and diarrhoea associated with the flu or 24 hour virus. Inhibits prostaglandins. Inhibits platelet aggregation. Accelerates calcium uptake by the heart muscle. This reduces high blood pressure. Recommended during pregnancy for treating morning sickness and digestive problems, as well as safe to use during pregnancy for colds and sore throats.

In folk medicine it is used in passion spells, to "heat up" the relationship. It is used in success spells, and to ensure the success of spells.

Maidenhair (*Gingko biloba*)

It grows 50 to 70 feet tall. It prefers temperate areas with moist soils, and needs full sun and high humidity.

Actions: antioxidant and anti-depressant.

Indications: Raynaud's disease, arthritic and rheumatic problems, arteriosclerosis, anxiety, tension, asthma, and bronchial congestion, cerebral vascular insufficiency, dementia, depression, impotence, inner ear dysfunction, multiple sclerosis, neuralgia and neuropathy, peripheral vascular insufficiency, premenstrual syndrome, retinopathy and vascular fragility.

Increases blood flow to brain. Improves memory. Used in Europe for early stages of Alzheimer's Disease. Increases circulation of blood to retina. Used to treat muscular degeneration. Also used to treat ear problems such as tinnitus. Dilates blood vessels. Inhibits clumping of platelets. Stabilizes cell membranes. Free radical-scavenging effects. Increases cellular uptake of glucose, thus restoring energy production. Promotes increased nerve transmission rate. Tones blood vessels. Inhibition of platelet aggregation. Improves recovery of acute cochlear deafness. Used to treat impotence. Has anti-allergic properties. Gingko Biloba is used to treat memory loss and difficulties, and is used to treat head injuries. It is also used to treat tinnitus, circulatory problems, strengthening the cardiac system, impotence, asthma, allergies that affect breathing, and Alzheimer's disease, in its early stages. Its properties enable the opening of the smaller veins, helping to improve circulation to all organs and especially the heart.

In folk medicine it is held or carried to help improve the memory. A mild tea prior to bedtime will help to remember dreams during sleep.

Ginseng [*Panax quinquefolia* (American ginseng), *Eleutherococcus senticosus* (Siberian ginseng), *Panax* spp.]

Ginseng can be very difficult to grow. Germination of disinfected seeds can take up to a year or more. Plant in early autumn in raised beds of very humus-rich soil. Plants must be shaded at all times. Roots are not harvested until the plants are at least 6 years old. Take care during harvesting and drying not to break off any of the "arms" of the root. Dry for one month before use.

Actions: tonic, rejuvenative, stimulant, aphrodisiac, demulcent and nervine.

Indications: old age, senility, debility, emaciation, fatigue, impotence, convalesence, to improve energy, childhood diseases, circulation, colds, colic, fevers, flu, gas pains, headache, indigestion, morning sickness, settles stomach, toothache and whooping cough.

One of the best tonic and rejuvenative herbs, promoting growth and revitalization of the body and mind. Particularly good for Vata disorders of tissue deficiency found in old age. Those who are not weak may find it a stimulant. It is excellent for promoting weight and tissue growth in the body including nerve tissue. Beneficial effect on heart and circulation. Reduces cholesterol. Has estrogenic activity. Used to treat high blood pressure. Helps reduce blood sugar levels. Do not use where there is inflammation, high fever and burning sensations. Increases resistance to stress by affecting the adrenal-pituitary axis. Thought to be an aphrodisiac. Central nervous system stimulant. Whole-body tonic. Not suggested for teenage girls. Considered an adaptogen or an aid in helping the body adapt to stress. Ginseng stimulates the body to overcome all forms of illness, physical and mental. It is used to lower blood pressure, increase endurance, aid in relieving depression, and is a sexual stimulant. The dried root is used for healing purposes. It has been used throughout ancient times to the present day for use in conjunction with most herbs in treating all sorts of illnesses, including cancers, digestive troubles, and memory. It is used to tone the body during stress and to overcome fatigue. During menopause it aids in rejuvenating the system and balances hormones, as well as aids in regulating hot flashes.

In folk medicine it is carried to guard your health and to attract love. It will also ensure sexual potency. Ginseng is an effective substitute for mandrake in all spells.

Goldenseal (*Hydrastis canadensis*)

Goldenseal prefers rich soils in partial shade. It is a perennial herb that grows 6 - 18 inches high. The dried ground root is the part most often used, although the dried leaves are used in teas. It is difficult to grow successfully, and the plants need to be at least 6 years old before harvesting.

Actions: bitter tonic, antipyretic, alterative, antibiotic, antibacterial, antiseptic and laxative.

Indications: jaundice, hepatitis, diabetes, obesity, ulcers, infectious fever, malaria, swollen glands and lymphatics, haemorrhoids, eczema, pyorrhoea, menorrhagia and leucorrhoea.

Destroys yeast and bacteria in the GI tract and clears the flora. Sedates and regulates liver and spleen function, along with sugar and fat metabolism, reducing toxins and excess tissue from the body. Purifies the mucuos membranes and is good for all catarrhal conditions. Has a negative impact on good intestinal flora-it is contraindicated in most deficiency conditions. For deep seated fevers it can be combined with hot herbs like ginger or black pepper. It is one of the strongest anti-Pitta herb. Goldenseal is another natural, powerful antibiotic. It should not be used by pregnant women. The herb goes straight to the bloodstream and eliminates infection in the body. It enables the liver to recover. When taken in combination with other herbs, it will boost the properties for the accompanying herbs.

In folk medicine it is used in prosperity spells, as well as healing spells and rituals.

Goathead (Gokshura) (*Tribulus terrestris*)

It is a tap-rooted flowering plant that grows on light textured soils but will grow over a wide range of soil types, found wild throughout China, India, western parts of Asia and southern parts of Europe and Africa.

Actions: diuretic, lithotriptic, tonic, rejuvenative, aphrodisiac, nervine and analgesic.

Indications: difficult of painful urination, edema, kidney or bladder stones, chronic cystitis, nephritis, hematuria, gout, rheumatism, lumbago, sciatica, impotence, infertility, seminal debility, venereal diseases, cough, dyspnea, hemorrhoids and diabetes.

Effective in most urinary tract disorders because it promotes the flow of urine, cools and soothes the membranes of the urinary tract, and aids in the discharge of stones. Stops bleeding, strengthens kidney function and nourishes the kidneys. Strengthens the reproductive system by increasing semen. It is invigorating to postpartum women. Rejuvenative to Pitta, and calms Vata with a sedative effect on the nervous system. Free of the side effects of most diuretics. Sattvic in nature and promotes clarity. Taken with milk it is a strong aphrodisiac. With ginger it relieves back and neck pain. With ashwaganda it is a powerful revitalizer.

Gotu Kola (*Centella asiatica*)

It is an evergreen perennial growing to 0.2 m (0ft 8in) by 1 m (3ft 3in) on old stone walls and rocky sunny places in lowland hills and especially by the coast in central and southern Japan. Shady, damp and wet places such as paddy fields, and in grass thickets.

Actions: nervine, rejuvenative, alterative, febrifuge and diuretic.

Indications: nervous disorders, epilepsy, senility, premature aging, hair loss, chronic and obstinate skin conditions, veneral diseases, fevers, cellulite, wound healing, varicose veins and scleroderma.

Main revitalizing herb for the nerves and brain cells. Increases intelligence, memory, longevity, and decreases senility and aging. It fortifies the immune system, both cleansing and feeding it, and strenghtens the adrenals. Powerful blood purifier and is specific for skin diseases including leprosy, syphilis, eczema, psoriasis, cervicitis, vaginitis, and blisters. Helps rebuild energy reserves. Combats stress. Improves reflexes. Nourishes brain cells. Reduces fevers. Neural tonic. Used for diseases of the blood, and nervous system. Does not contain caffeine. Contains asiaticoside which stimulates hair and nail growth, increases development of blood vessels into connective tissue, increases tensile integrity of the dermis, exerts a balancing effect on connective tissue. Speeds healing time of wounds. Has been used clinically for anal fissure, bladder ulcers, burns, cellulite, cirrhosis, dermatitis, fibrocystic breast, hemorrhoids, keloids, leprosy, lupus, mental retardation, peptic ulcer, retinal detachment scleroderma, surgical wounds, tuberculosis, and venous disorders. Gotu Kola is and excellent mental stimulant. It is often used after mental breakdowns, and used regularly, can prevent nervous breakdown, as it is a brain cell stimulant. It relieves mental fatigue and senility, and aids the body in defending itself against toxins.

In religious practices it is used in meditation incenses.

Green Tea (*Camellia sinensis*, Various spp.)

Grown in wide range of temperate climates around the world.

Green tea has recently come into prominence as an effective anti-oxidant. It has been shown to reduce the risk of many forms of cancer, and it has the ability to stabilize blood lipids, making it part of an overall cardiac care regimen. It aids in treating high cholesterol, hypertension, and stimulates immune functions. This herb eases mental fatigue, and may lower the risks for arteriosclerosis. It can also help to prevent plaque buildup on the teeth. People who are sensitive to, or cautioned to reduce or avoid, caffeine, can still use the decaffeinated form of Green Tea, which is still shown to have the same medicinal properties and qualities.

In folk medicine it is used for prosperity rituals, and to honor deities around the world.

Guggul (*Commiphora wightii*)

Guggul is a much-branched, spiny, deciduous shrub growing up to 6 metres tall on rocky and open hilly areas or rough terrain and sandy tracts in warm and semiarid to arid areas at elevations from 250 - 1,800 metres.

Actions: rejuvenative, stimulant, alterative, nervine, antispasmodic, analgesic, expectorant, astringent and antiseptic.

Indications: arthritis, rheumatism, gout, lumbago, nervous disorders, neurasthenia, debility, diabetes, obesity, bronchitis, whooping cough, dyspepsia, haemorroids, pyorrhoea, skin diseases, sores and ulcers, cystitis, endometrisis, leucorrhoea, tumours, liver disorders, cardiovascular disease, bone fracture, inflammation, obesity, elevated cholesterol and triglycerides and atherosclerosis.

Strong purifying and rejuvenative powers. Rejenative for vata and kapha. Increases white blood cell count, disinfects secretions including sweat, mucus and urination. Increases appetite, clears the lungs, helps heal the skin and mucous membranes. Regulates menstruation, catalyzes tissue regeneration, particularly nerve tissue. Reduces fat, toxins, tumors, and necrotic tissue. Best medicine for arthritic conditions. Research indicates excellent for obesity; significantly lower serum cholesterol. Protects against atheriosclerosis. Reduced body weight of animals in research studies. Reduced serum cholesterol levels by 11 per cent after four weeks use in research. Considered safe for long-term use. No known interactions.

Jiaogulan (*Gynostemma pentaphyllum*)

Actions: anti-oxidant, anti-aging, tonic, adaptogenic

It is a climbing, perennial vine native to China, Japan, and parts of southeast Asia. The plant is dioecious growing to 8 m in light (sandy), medium (loamy) and heavy (clay) soils and prefers well-drained soil. It cannot grow in the shade. It prefers moist soil.

Indications: arteriosclerosis, broncial asthma, hepatitis, low energy, sexual dysfunction, immune disorders, emotional imbalances, inflammations, cancer and blood sugar imbalances.

Its common name in its native China translates as "Magical Grass", and it is highly revered as a tonic to all systems and bodily functions. It has a unique position as both a tonic and specific cure, and so is helpful in a wide variety of complaints. Its broad usefulness comes from its excellent adaptogenic properties, through which it helps the body to respond to many stresses and adverse conditions. It has been used to combat fatigue and low energy, increase oxygen efficiency and metabolism, improve digestion and appetite, and balance nervous system function. Its effects on the nervous system are especially interesting, as it can be calming or stimulating as needed. It can also work in both directions regarding weight, increasing or decreasing according to the direction of imbalance.

In China, the herb is known as 'herb of immortality' as it is believed to delay aging and increase endurance. Its usages include prevention of growth of cancer and of high blood fat and arteriosclerosis, cure of bronchial asthma and hepatitis, strengthening of the body and prevention of senility. Its taste is sweet and aromatic. A rare exception of a herb that combines a set of uses that keep you physiologically young. It regulates blood pressure, supports a healthy cholesterol balance, supports cardiovascular function, supports the body's natural adrenal process, supports the body's healthy immune system, and regulates hormones. As one of the main adaptogenic herbs it helps to support a number of body functions and helps to support the body in times of stress so it can be beneficial for a range of people. It is particularly recommended for those who want to try and improve their energy levels, who would like to support and maintain their blood pressure, support maintaining a healthy cholesterol balance, support with a healthy immune system, assist the circulation and aid a bright and clear mind. It is also used to enhance athletic performance.

Hawthorn (*Crataegus oxycantha*)

Hawthorn is a deciduous tree or shrub, that can reach 40 feet tall. It grows throughout North America. It is tolerant of most soils, but prefers alkaline, rich, moist loam. Consult a nursery for the best species to use in your area. The fruit is the part used in healing.

Hawthorn is effective for curing insomnia. Hawthorn is used to prevent miscarriage and for treating nervousness. Hawthorn has been used for centuries in treating heart disease, as regular use strengthens the heart muscles, and to prevent arteriosclerosis, angina, and poor heart action. Actions: stimulant, carminative, vasodilator, antispasmodic and diuretic.

Indications: heart weakness, artheriosclerosis, valvular insufficiency, hypertension, heart palpitations, blood clots, insomnia, food stagnation, abdominal tumors, angina pectoris, enlarged heart, angina, congestive heart failure and high blood pressure.

Special action on the heart, streghtening the heart muscle and promoting longevity. Good for Vata heart conditions like nervous palpitations or the heart problems of old age like cholesterol and arteriosclerosis. In promoting digestion it helps move accumulated food masses, or even tumors, in the GI tract. Helps increase weight in the body. Used in poultice for drawing thorns, splinters, *etc.* Fruit is used for nervousness. Used to prevent miscarriage. Strengthens the heart, lowers blood pressure, lowers cholesterol. Mild dilation of coronary vessels, increased enzyme metabolism in heart muscle, increased oxygen utilization by the heart. Flavonoid components possess collagen-stabilizing action, increase intracellular vitamin C levels, cross-link collagen fibers.

Religious: The leaves are used to make protection sachets. They are also carried to ensure good fishing. In Europe, Hawthorn was used to repel witchcraft spells. Bringing branches of it into the home is supposed to portend death. It is incorporated into spells and rituals for fertility. It will protect the home from damaging storms.

Hazel (*Corylus avellana*)

Hazelnut trees do best when planted in a well-drained, fertile, slightly acid soil. They do best where the winter temps are above -10.

Hazelnuts (raw, roasted, ground or paste) are used as a dietary source of protein and minerals. They are a source of vitamin E, unsaturated fatty acids, and linolenic acid. Hazelnut oil also is used in the cosmetic industry as well as in confectionery.

Approximately 100 g of nuts delivers 15 mg of vitamin E.

Caution: Use with caution in patients with known allergy to peanuts.

In religious practices it's forked branches are used for divining, and the wood makes wonderful wands. Hazel nuts hung in the house will bring luck, and can be carried to cause fertility. Eaten, the nuts bring wisdom.

Heather or Ling (*Calluna vulgaris*)

Heather prefers rocky or sandy soils and full sun. It is an evergreen shrub that grows up to 0.6 m (2ft) by 0.5 m (1ft 8in). It is in leaf 12-Jan It is in flower from Jul to October, and the seeds ripen from Oct to November. The flowers are hermaphrodite (have both male and female organs) and are pollinated by Bees, flies, lepidoptera, wind.

It is a good urinary antiseptic and diuretic, disinfecting the urinary tract and mildly increasing urine production. The flowering shoots are antiseptic, astringent, cholagogue, depurative, diaphoretic, diuretic, expectorant, mildly sedative and vasoconstrictor. The plant is often macerated and made into a liniment for treating rheumatism and arthritis, whilst a hot poultice is a traditional remedy for chilblains. An infusion of the flowering shoots is used in the treatment of coughs, colds, bladder and kidney disorders, cystitis *etc*. A cleansing and detoxifying plant, it has been used in the treatment of rheumatism, arthritis and gout. The flowering stems are harvested in the autumn and dried for later use. The plant is used in Bach flower remedies - the keywords for prescribing it are 'Self-centredness' and 'Self-concern'. A homeopathic remedy is made from the fresh branches. It is used in the treatment of rheumatism, arthritis and insomnia.

In folk medicine is carried as a guard against rape and violent crime. In potpourri, it adds protection. When burned with fern, it will bring rain. It is noted for attracting wildlife.

Henbane (*Hyosycamus niger*)

Henbane grows wild throughout temperate North America. The plant can grow up to 25-100 cm tall. It has bright green, hairy leaves that are oval or elongated with toothed edges. Biennial plants bloom in May and June, while annual plants often bloom from July to September. The yellow flowers sit in long, one-sided tassels and a single plant can produce up to 8,000 seeds. All parts of the herb are extremely toxic.

Henbane stops pain, and lessens perspiration. Henbane is very toxic, so it should not be used by pregnant women or the weak or children, and should be used in only extremely small amounts for external use only, and not on a regular basis. A poultice of leaves or it's juice mixed with olive oil can be used externally to treat ear ache, or applied to the skin to relieve pain from neuralgia, sciatica, arthritis and rheumatic conditions. A poultice of leaves is used briefly to remove pain from wounds. Henbane leaf is used for spasms of the digestive tract. It's leaf oil applied directly to the skin for treating scar tissue.

In folk medicine it is thrown into the water to bring rain. In olden times, it had many more uses, but is seldom used today due to its poisonous nature.

Holly (*Ilex aquifolium, Ilex verticiallata, Ilex vomitoria*)

Holly likes slightly acid soils, and can tolerate poor, sandy soil. It needs full sun, and grows to about 4 feet tall.

The Ilex aquifolium variety is a good astringent, diuretic, expectorant, and febrifuge agent. On the other hand, Ilex verticiallata's bark makes up for a great bitter tonic, astringent, and febrifuge, while its berries are cathartic and vermifuge. Ilex vomitoria is an efficient emetic and stimulant.

In folk medicine it is used as a protective plant, and used as decoration during the Yule season. Planted outside the home, it will also afford protection. Sprinkle holly water on newborn babies to protect them.

Hops (*Humulus lupulus*)

Hops prefers full sun, and will adapt to many soils. It is a perennial vine that reaches to 30 feet in height. The portion of the plant used in healing are the dried flowers. Well-known as an ingredient in beer, with an equally long history as a healing herb as well. Hops contains strongly estrogenic compounds, also useful for regulating and balancing the menstrual cycle, and is widely used for Menopausal imbalances as well.

Hops' other popular use is as a calmative and anti-stress herb, whether sewn into a dream pillow or taken internally. Hops gives most users a clear state of calm, useful for still meditation or to stop nervousness but still remain alert. Hops can also be used for dreamlike astral travel or lucid dreaming and meditation.

Hops is a sedative. Therefore, it is useful in treating insomnia and nervous tension. It is mild and safe. It is used in brewing beer and ales. Hops is also used for treating coughs, bladder ailments, and liver ailments. Externally it is used to treat itching skin rashes and hives. It also removes poisons from the body.

In folk medicine, it is used in healing incenses. Sleep pillows often include hops to induce sleep and pleasant dreams.

Horehound (*Marrubium vulgare*)

Horehound likes dry sandy soils and full sun. It is a perennial (except in very cold climates) that reaches to 3 feet tall. It is a vigorous grower and can become a pest if not carefully controlled. It needs little water, tolerates poor soils, and does best in full sun. It blooms during its second year.

It is used in children's cough remedies, as it is a gentle but effective expectorant. It acts as a tonic for the respiratory system and stomach. In large doses it acts as a laxative.

In folk medicine, it is used in incenses for protection. It is also used in exorcisms.

Epimedium (Horny Goat or Barrenwor)

It is a perennial growing to 0.4 m (1ft 4in) by 0.3 m (1ft) at a medium rate in light (sandy), medium (loamy) and heavy (clay) soils. It prefers dry or moist soil.

A time-tested aphrodisiac that increases libido in men and women in a balanced way, increasing Yang (male virility and drive) without damaging Yin (female receptivity and peace). This herb is also an overall tonic and restorative, and brings warm feelings of euphoria and increases nerve sensitivity in the genital areas. It improves erectile function in men. Restore sexual fire, boost erectile function, allay fatigue and alleviate menopausal discomfort.

Horseradish (*Armoracia lapathifolia*)

Horseradish is a perennial plant that is cultivated throughout the world for its long, tasty root. It will grow two to three foot tall by its second year.

Grind some of the fresh root, combine it with a carrier oil, and use it to massage away muscular aches, and help loosen chest congestion. It can be used to warm a cold body, and to clear up drippy sinuses.

Horseradish is part of the Jewish Passover ritual. It also repels evil around the home and property.

Horsetail (*Equisetum arvense*)

Actions: diuretic, lithotriptic, diaphoretic, alterative, haemostatic and anti-inflammatory, astringent.

Indications: edema, nephritis, burning urethra, kidney stones, gall and bladder stones, stomach ulcers, broken bones, wounds, menorrhagia, venereal diseases, mouth sores, canker sores, dysentery and diarrhoea, enlarged prostate, lung problems, asthma, tuberculosis, cystitis, hair, skin, and nails and haemorrhoids.

Horsetail contains large amounts of minerals, especially silica, and so is used in malnutrition-related disorders which manifest as poor hair, skin, or fingernail condition. Effective diuretic and blood cleanser. Good for Pitta conditions and has a strong stone removing action for kidney, bladder and gall stones. Somewhat irritant-should not be taken for long periods of time. Promotes the healing of broken bones and supplies nutrients to the bone tissue. Helps clear and brighten the eyes and removes toxicity from the blood. Good for infectious fevers and flus. Used for urinary tract disorders and chronic cystitis. Good for hair, nails, and skin. Considered strengthening to connective tissue. Stops bleeding and is useful for wounds.Horsetail is used in treating urinary tract infections. It aids in coagulation and decreases bleeding. It will also help broken bones heal faster, and will help brittle nails and hair, due to its high silica content. It has also been used as part of a treatment for rheumatoid arthritis. The plant alone, boiled in water, makes an effective foot soak for tired feet, or for the treatment of athlete's foot. Do not use if pregnant or nursing.

In folk medicine, whistles made from the stalks of Horsetail are used to call the spirits.

Horse Chestnut (*Aesculus hippocastanum*)

Actions: tonic, narcotic and febrifuge.

Indications: intermittent fevers, rheumatism and neuralgia, circulation problems, hemorrhoids, prostrate enlargement and varicose veins.

A restorative to the veins and capillaries, reducing their fragility and permeability and helping them to perform their job of moving blood more efficiently. These effects are most pronounced in the intestines, uterus, and prostate. This balancing action can also be applied to conditions such as varicose veins, edema of the legs, nighttime cramps, especially in the legs, headache and skin reactions due to liver stagnation, and menstrual imbalances.

Hyssop (*Hysoppus officinalis*)

Hyssop prefers dry conditions, tolerates most soils, and full sun. It is a member of the mint family. It is a perennial shrubby plant growing to 3 feet tall.

Hyssop is used in treating lung ailments. The leaves have been applied to wounds to aid in healing. The tea is also used to soothe sore throats. It has been used to inhibit the growth of the herpes simplex virus.

In folk medicine, it is used in purification baths and rituals, and used to cleanse persons and objects.

Iceland Moss (*Cetraria islandica*)

Iceland Moss grows in cold, humid mountain areas and wooded areas. It grows to 4 inches tall.

Iceland Moss, a lichen, has been used for centuries to treat all kinds of chest ailments. It is used to nourish the weak, elderly, and weakly children.

Ivy (*Hedera helix*)

Ivy has many different varieties, and most will adapt to many different soil and growing conditions. It grows throughout North America.

The leaves have been used externally as a poultice to treat sores, ulcers, and other skin eruptions.

In folk medicine, it is grown to grow up the outside of the home to act as a guardian and protector. It is worn by brides to bring luck to the marriage.

Jasmine (*Jasminum officinale*)

Jasmine is best grown indoors in pots. It is an evergreen vine. It likes bright light, but no direct sun, some support such as a trellis, lots of water, and occasional fertilizing. Jasmine tea is drunk for its calming affect, especially after dinner, as well as for its aphrodisiacal qualities. Jasmine oil used in massage is soothing to the skin, and reported to be an aphrodisiac. It is used in aromatherapy to treat depression and nerve conditions, and as a massage oil for menstrual cramps. A drop of the essential oil in almond oil, massaged into the skin, is said to overcome frigidity.

In folk medicine, it is used in love sachets and incenses. It is used to attract spiritual love. Carrying, burning, or wearing the flowers attracts wealth and money. If burned in the bedroom, Jasmine will bring prophetic dreams.

Jatoba (*Hymenaea courbaril*)

Jatoba is a huge canopy tree, growing to 30 m in height, and is indigenous to the Amazon rainforest and parts of tropical Central America.

Actions: tonic, energizer, anti-fungal and decongestant.

Indications: asthma, laryngitis, bronchitis, lung weakness and chronic coughs as well as in the treatment of yeast and fungal infections, cystitis, bladder infections, arthritis, prostatitis, bursitis and hemorrhage

In the Brazilian rainforest, Jatoba bark is considered a tonic and energizer and is especially popular amongst Brazilian lumerjacks to increase strength, vigor and productivity. Not recommended if nursing or pregnant.

Jojoba (*Simmondsia chinensis*)

Jojoba is a native shrub of the Sonoran Desert and typically grows upto 1–2 meters tall, with a broad, dense crown, but there have been reports of plants as tall as 3 meters.

Jojoba oil from the seed has been used to promote hair growth and relieve skin problems for centuries. It is effective in treating dandruff, psoriasis, dry and chapped skin.

Juniper (*Juniperus communis*)

Junipers of all species are adaptable to many growing conditions. They are low-maintenance plants. Choose a species suited to your landscape needs, to avoid problems later, as some folks plant them with no regard for their eventual size, and sometime find they have a nuisance on their hands as the plant matures. If you want berries, you must plant a male and a female juniper.

Juniper has been used to clear uric acid from the body. It is high in natural insulin, and has the ability to heal the pancreas where there has been no permanent damage. It is useful for all urinary infections and for water retention problems. Juniper is used externally as a compress to treat acne, athlete's foot, and dandruff.

In folk medicine, it is used to protect from accidents and theft. Grown at your doorstep, it will offer your home protection. It is used in incenses for protection.

Kava Kava (*Piper methysticum*)

Kava Kava grows on many South Pacific tropical islands. It grows 8 to 20 feet tall. Try growing it as a houseplant, with frequent misting for humidity, or placing on humidity trays. Seeds and plants are available from several mail-order sources.

It is used to treat insomnia and nervousness. Relieves stress after injury. Used as a tea for pains associated with nerve and skin diseases. Large doses can cause a build up of toxic substances in the liver.

In folk medicine, a tea of kava kava is drunk to offer protection against evil and to invite good luck. Sprinkle the tea around the home and property for the same uses.

Kelp (*Macrocystis pyrifera*)

It is an edible brown seaweed that is rich in vitamins and minerals especially iodine. There are about 30 different species. Kelp grows in "underwater forests" in shallow oceans.

It is rich in Iodine besides vitamins B_1, B_2, and B_{12} and minerals such as potassium, calcium and iron and used in treating dry, scaly skin and scalp, constipation, fatigue, dysfunctional thyroid operation, infertility, increased rate of still births and growth abnormalities.

It gives protection against radiation poisoning. Iodine is also important for the thyroid, immune system, and female hormone regulation. It has anti oxidant, anti cancer, anti inflammatory properties, useful in treatment of neurodegenerative diseases that are accompanied by microglial activation.

Kudzu (*Pueraria lobata*)

It is a perennial climber growing to 10 m (32ft 10in) at a fast rate. It grows on light (sandy), medium (loamy) and heavy (clay) soils and prefers well-drained soil. It cannot grow in the shade. It prefers moist soil and can tolerate drought.

Actions: tonic, diaphoretic, diuretic, antispasmodic and relaxing.

Indications: muscular tension, fevers, dysentery, thirst, headache, allergies, diarrhoea, angina, intestinal cramping and hypertension.

It may relieve muscular tension and spasms, fevers associated with flu and colds, colitis, sudden nerve deafness, angina pectoris, and offers support for alcohol withdrawal. Increases blood flow to the brain, and lowers blood pressure.

Kratom (*Mitragyna speciosa*)

It is a tropical deciduous and evergreen tree in the coffee family native to Southeast Asia in the Indochina and Malaysia floristic regions.

Kratom is one of the most effective and pleasurable psychoactive herbs available. It produces euphoria, alleviates pain (physical and emotional), reduces anxiety and emotional stress, increases mental focus, improves stamina, and suppresses coughs. Many people enjoy the sensual effects of kratom and like to combine it with sex. It is also an effective remedy for premature ejaculation. Many people report that kratom is an effective treatment for restless legs syndrome (RLS), arthritis, and fibromyalgia.

These effects are caused by alkaloids that bind to opioid receptors (the same receptors responsible for the effects of opioid drugs, such as morphine). Kratom has been used in South-East Asia for conditions such as intestinal infections, muscle pain, to reduce coughing and diarrhoea. Malay and Thai natives use Kratom for its opium and coca-like effects to enhance tolerance towards for hard work in extreme heat. Due to its analgesic (pain reliving) properties, Kratom has also been used by some to treat pain. Some research has been done with regards to analgesic potential, but neither Kratom nor its alkaloids have been utilized in modern medicine. Mitragynine has less analgesic effects than morphine, though it is still potentially useful in treating minor to moderate pain. Generally there is a high amount of mitragynine in the Kratom plant, making it practical to extract. 7-hydrohydroxymitragynine has been shown to have much stronger analgesic effects than, but it is less abundant in the Kratom plant. It might not be practical to extract this alkaloid from the Kratom plant, but if mitragynine can be converted into 7-hydrohydroxymitragynine, it might emerge as a replacement for modern medication. Kratom is usually taken orally as a tea, by swallowing powdered leaves, or chewed fresh. Coarsely ground leaves are best for preparing a tea. The effects last about 6 hours. At low doses it induces mild euphoria, improves mental focus, increases stamina, and reduces fatigue. Low doses do not interfere with most ordinary activities. At strong doses the effects are profoundly euphoric and immensely pleasurable. Typically people describe the effects as dreamy, ecstatic, and blissful. Many people experience dream-like closed-eye visuals.

It is used in folk medicine as a stimulant (at low doses), sedative (at high doses), recreational drug, pain killer, medicine for diarrhea, and treatment for opiate addiction.

Lavender (*Lavandula officinalis*)

Lavender likes light sandy soil and full sun. It grows to 18 inches. It should be mulched in colder climates for winter protection for this perennial.

Actions: carminative, diuretic and antispasmodic.

Indications: depression, nervous tension, stress and headaches.

Lavender tea made from the blossoms is used as an antidepressant. It is used in combination with other herbs for a remedy for depression and nervous tension and stress. It is also used as a headache remedy. Lavender tea made from the blossoms is used as an antidepressant. It is used in combination with other herbs for a remedy for depression and nervous tension and stress. It is also used as a headache remedy.

In folk medicine, it is used in purification baths and rituals. It is used in healing incenses and sachets. Carrying the herb will enable the carrier to see ghosts. The essential oil will heighten sexual desire in men. Lavender water sprinkled on the head is helpful in keeping your chastity. The flowers are burned to induce sleep, and scattered throughout the home to maintain peaceful harmony within. Carrying lavender brings strength and courage.

Lemon Balm (*Melissa officinalis*)

Lemon Balm is a perennial that can reach up to 3 feet high. It needs full sun and rich soil with regular watering. Plant where you can enjoy the lemon scent of the leaves from a porch, deck, or open window.

Actions: diaphoretic, carminative, nervine and anti-spasmodic.

Indications: fever, flu, colds, headaches, depression, menstrual cramps, insomnia, nervous stomach, hyperthyroidism, herpes simplex, indigestion, tension and ear infection.

Considered by Paracelsus to be a cure-all, and was the herb he revered most. Appears to work as a sedative in tension cases, but at the same time as a mood elevator and antidepressant, and so can be useful for relieving tension in people who might have overly depressive reactions to heavier, downer-type sedatives. Lemon Balm is used to treat children with fever, flu, and colds, and ear infections. Lemon Balm is used in spells to bring success, and in healing spells. It is often used in spells to find love and friendship. It is elevating and cheering.

In adults, it treats colds, headaches, depression, menstrual cramps, insomnia, and nervous stomachs. It has also been used to treat hyperthyroidism, herpes simplex, and indigestion. The crushed leaves are applied to wounds and insect bites to aid in healing. The essential oil, also known as Melissa, is used in aromatherapy to ease depression.

In folk medicine, it is used in spells to bring success, and in healing spells. It is often used in spells to find love and friendship.

Liquorice (*Glycyrrhiza glabra*)

Licorice is a perennial that reaches 3 to 7 feet tall. Hard freezes will kill it, so it grows best in warm sunny climates.

Actions: demulcent, expectorant, tonic, rejuvenative, laxative, sedative and emetic.

Indications: cough, colds, bronchitis, sore throat, laryngitis, ulcers, hyperacidity, painful urination, abdominal pain, general debility, Addison's disease, drug withdrawal, female complaints, hypoglycemia, to increase energy, viral infections, inflammation, menstrual and menopausal disorders, peptic ulcers and canker sores.

Liquefies mucus and facilitates its dicharge from the body. In large doses it is a good emetic for cleansing the lungs and stomach of Kapha. Mild laxative which soothes and tones the mucuos membranes, reliving muscle spasms and reducing inflammation. Helps harmonize the qualities of other herbs, countering heat and dryness and reducing toxicity. For colds and respiratory affliction it combines well with ginger. With ginger and cardamom it is a tonic for the teeth. It is a restorative and rejuvenative food. Sattvic in quality it calms the mind and nurtures the spirit. It nourishes the brain and increases cranial and cerebrospinal fluid, promoting contentment and harmony. It improves the voice, vision, hair, and complexion and gives strength. Contains phytoestrogens. Stimulates adrenal glands. Supports endocrine system. Used to treat ulcers. Shown to have anti-viral, anti-tumor, anti-inflammatory properties. Used to treat hypoglycemia. Important herb when recovering from illness. Shown to be useful in the treatment of stomach cancer. Helps remove fluid from the lungs. Helps remove old matter from the small intestines. Increases T cell counts.

Liquorice Root is a great source of the female hormone estrogen. It is used for coughs and chest ailments. It is an important herb to use when recovering from an illness, as it supplies needed energy to the system. Used as a remedy for stomach and heart problems, indigestion, and most respiratory ailments. Helps to normalize and regulate hormone production.

Topical applications: herpes, eczema, psoriasis.

Caution: Do not use if you have high blood pressure or retain water easily. Do not use large doses for long periods of time. Liquorice can increase blood pressure. Should not be used by pregnant women as it can sometimes lead to high blood pressure with prolonged use.

In folk medicine, it's root was buried in tombs and caskets to help the soul pass easily into the Summerland. Chewing on a piece of the root will make you passionate. It is added to love sachets, and an ingredient in spells to ensure fidelity.

Lobelia (*Lobelia inflata*)

Actions: antispasmodic, emetic, expectorant, diaphoretic, stimulant (small doses) and sedative (large doses).

Indications: arthritis, colds, congestion, bronchitis, asthma, convulsions, cough, ear infections, epilepsy, food poisoning, lung problems, nervousness, pain, pneumonia, worms, headaches, cramps, toothaches and earaches.

The most common use for Lobelia is as an aid to quitting smoking, it seems to mimic the effects of nicotine in the body, thus helping to reduce cigarette cravings, while at the same time opening the lungs to remove phlegm and toxic buildup, and promoting healthy blood flow to the lungs.

Removes obstructions. Relaxant. Removes congestion. Relieves bronchial spasms. Allows blood and oxygen to flow freely throughout the body. Used as a respiratory stimulant, anti-asthmatic. Lobelia stimulates the vomiting center at the base of the brain. Expels secretions from the lungs and bronchiole. Stimulates the adrenal gland to release hormones that cause the bronchial muscles to relax. Considered one of the most valuable herbs in the plant kingdom. Useful immediately during an asthma attack.

Caution: toxic in large amounts.

Lovage (*Levisticum officinale*)

Lovage is a perennial that grows 3 - 7 feet tall. It is adaptable to many conditions, and does best in full sun.

Lovage root eases bloating and flatulence. It is also used with other herbs to counteract colds and flu.

In folk medicine, it is added to baths to clean the pysychic portion of the mind. Added to baths with rose petals will make you attractive to the opposite sex.

Mandrake (*Podophyllum peltatum*)

Flowering plant with large umbrella like leaves, usually grows in shades. The mature fruits are edible and quite tasty. However, they are poisonous when unripe and green. Do not try to eat them until they are yellow and soft.

Mandrake is a very strong gland stimulant. It is used to treat skin problems, digestion, and chronic liver diseases. It is most often combined with other herbs. It is very powerful and should be used with caution, as well as in very small small dosages. Pregnant women should not use this herb. It is potentially very toxic to anyone if improperly used.

In folk medicine, it is used in the home as a powerful protection. It is carried to promote conception, and men carry it to promote fertility and cure impotency.

Marigold (*Calendula officinalis*)

Marigold is an annual plant that comes in many sizes and colors. It is adaptable to many soils. Give plenty of water and full sun.

Actions: vulnerary, antispasmodic, alterative, anti-inflammatory, diaphoretic, astringent, antifungal, cholagogue, emmenagogue and febrifuge.

Indications: fungal infections, digestive inflammation, ulcers, indigestion and painful menstruation.

Anti-fungal activity and may be used both internally and externally to combat fungal infections. Internally it is used to reduce digestive inflammation and is therefore helpful in treating gastric and duodenal ulcers. Also useful for relieving indigestion and gall bladder problems, and can help normalize delayed or painful menstruation. Marigold is a great first aid remedy. It relieves headaches, earaches, and reduces fevers. It is excellent for the heart and for the circulation. It is also used externally to heal wounds and bruises.

In folk medicine, marigolds are kept in any room to heighten the energy within. Placed under the pillow before bed, it induces clairvoyance. Planted in rows with tomatoes, it will keep pests from them and other vegetables. Planted near the porch/deck, it will keep mosquitoes away. It is also used to attract and see the fairies. Scattered under the bed, they protect during sleep. Add to bath water to win the respect of everyone you meet.

Marjoram (*Origanum majorana*)

Marjoram is a perennial herb growing 1 - 3 feet tall. It likes all kinds of soils, and prefers full sun and rich soil. It is grown as an annual or wintered indoors in cold regions.

It is useful for treating asthma, coughs, and is used to strengthen the stomach and intestines, as well as used with other herbs for headaches.

In folk medicine, it is added to all love charms and sachets. A bit in each room will aid in protection of the home. If given to a grieving or depressed person, it will bring them happiness.

Marshmallow (*Althea officinalis*)

Marshmallow needs marshes and swamps to grow. It is a perennial growing to 4 feet tall.

Marshmallow aids in the expectoration of difficult mucous and phlegm. It helps to relax and soothe the bronchial tubes, making it valuable for all lung ailments. It is an anti-irritant and anti-inflammatory for joints and the digestive system. It is often used externally with cayenne to treat blood poisoning, burns, and gangrene.

Meadowsweet (*Filipendula ulmaria*)

Meadowsweet is a perennial that prefers wet soils and marshes. It grows to 6 feet high, and prefers partial to full sun.

Meadowsweet is used to treat headaches, fever, arthritis, rheumatism, menstrual cramps, and flu, as well as diarrhoea in children. Use it in place of aspirin, or white willow. It is used to rebuild the digestive system during recovery from drug and alcohol abuse.

In folk medicine, it is used in love spells, and blossoms placed in the home will bring peace and tranquillity to those who live there.

Milk thistle (*Silybum marianum*)

The Milk Thistle is an annual or biennial plant growing up to 1.5 metres in height. Milk Thistle prefers sunny, dry stone cliffs, the edges of paths, waste dumps, and pasture land.

Actions: antioxidant.

Indications: liver disorders, hepatitis, cirrhosis of the liver, gallstones and psoriasis.

As a liver rejuvenative it restores impaired liver function. Used for the treatment of liver diseases and amanita mushroom poisoning. Increases milk production in nursing mothers. Helps prevent plaque buildup and hardening of the arteries. Stimulates protein synthesis. Lowers fat deposits in liver. Increases secretion and flow of bile from liver to gallbladder. Increases production of liver cells to replace damaged old ones. Used as a long-term treatment for addiction recovery after the patient has stopped taking drugs or consuming alcohol, and in acute, emergency care for poisonings, whether chemical or drug, it is also useful for neutralizing industrial poisons and reduce damage to liver if taken within 72 hours of a large toxic exposure. Protects against liver damage due to chemotherapy, chemicals, alcohol, hepatitis. Milk Thistle accomplishes these effects by scavenging the free radicals that the liver creates as a by-product of trying to process toxins, as well as stimulating the production of proteins and new cells in the liver. These effects can also be used as part of a protective, regenerative, cleansing program during and after illness or chemotherapy.

Mugwort (*Artemisia vulgaris*)

Mugwort likes dry areas in full sun. It is a perennial shrubby plant that grows 1 - 6 feet tall, depending upon growing conditions.

Mugwort is used in all conditions dealing with nervousness, shaking, and insomnia. It is used to help induce menses, especially combined with cramp bark. Often used to stimulate the liver and as a digestive aid. It should not be used by pregant women. Fresh juice form the plant is used to treat poison ivy.

In folk medicine, it is added to divination incenses. It is carried to prevent poisoning and stroke. An infusion made of mugwort is used to cleanse crystals and scary mirrors. Placed beneath your pillow, or in a dream pillow, it will promote astral travel and good dreams.

Muira puama (*Ptychopetalum olacoides*)

Muira Puama or Liriosma, is a tree that grows in the rain forests of Brazil.

Actions: aphrodisiac, stimulant, astringent and nervous system tonic.

Indications: nervous problems, low energy, impotency, frigidity, cold hands or feet, menstrual cramps, PMS, stress, trauma, nervous tension, rhuematism, paralysis and intestinal problems (dysentary and parasites)

Commonly referred to as Potency Wood or Raiz del Macho (Root of the Male), these names give a good indication of the aphrodisiac qualities this root is said to possess. Used as a mild tonic for depressed basal metabolism, it often exhibits distinct sexual stimulation in both sexes. This herb is influential in treating the symptoms of nervous problems and disorders, such as neurasthenia, neuralgia and nervous depression. It makes an excellent tonic for the nervous system. Also indicated for those with low energy levels but no particular organic or emotional problems, impotency, frigidity, general lassitude, poor sacral stimulation, cold hands and feet, menstrual cramps and PMS and is thought to fortify the stomach and intestines. This root combines well with Brazilian Catuaba bark. Not recommended for nursing or pregnant women.

Mullein (*Verbascum* spp.)

Mullein is adaptable to many soils. It prefers full sun. It is a biennial plant growing to 8 feet tall. It is a prolific self-sower.

Actions: expectorant, astringent, vulnerary, antispasmodic, analgesic, sedative and anti-bacterial.

Indications: colds, bronchitis, asthma, hay fever, dyspnoea, sinusitis, cough, lung haemorrhage, swollen glands, earache, mumps, nerve pain, insomnia, diarrhoea, dysentery, pain, pleurisy, sinus congestion, tuberculosis, remove lymphatic congestion, sore throat and earaches.

Powerful herb for dipelling heat and congestion from the lungs and nasal passages. It is mucilaginous and soothing to the mucus membranes. Dispels accumulated Kapha, cleansing the bronchii and the lymphatics. Specific for mumps, earaches, and glandular swellings. Relieves inflammation of the nerve tissue and allays irritation. Good pain killer. Helps induce sleep. Calms irritated nerves. Loosens and thins mucus. Cleans lymphatic system. Tones mucus membranes of respiratory tract. Inhibits absorption of allergens through membranes. Reduces pain and swelling.Mullein is a terrific narcotic herb that is not addictive or poisonous. It is used as a pain killer and to bring on sleep. It loosens mucous, making it useful for treating lung ailments. It strengthens the lymphatic system.

In folk medicine, it is worn to give the carrier courage. The leaves are also carried to prevent animal attacks and accidents when in the wilderness. In a sleeping pillow it will guard against nightmares. Use as a substitution in old spells for "grave dust".

Maca (*Lepidium meyenii*)

It is an annual or biennial plant growing up to 1.5 metres in height, which grows in the Andes of Peru.

Actions: aphrodisiac, nutritive and specific tonic.

Indications: low libido, low energy, depression, stress, menopause, erectile dysfunction, infertility and menstrual disorders.

It is very high in many vitamins and minerals, especially vitamins B, C, and E, and calcium, zinc, iron, iodine, and magnesium. It is also rich in sugars, carbohydrates, amino acids, and essential fatty acids. It is used to generally support the whole body, and specifically give more energy, mental clarity, and muscle tone. It is also considered an aphrodisiac, and has hormone-balancing effects.

For women it enhances libido, increases energy, is a sexual stimulant, helps overcome depression, produces a general state of well-being, reduces stress, fights menopausal symptoms such as hot flashes, balances hormones, improves athletic performance, allieviates PMS. For men, it increases energy, stamina and endurance, improves male potency, increases levels of DHEA, reduces stress, increases seminal volume, improves athletic performance, increases testosterone, and enhances fertility.

Mugwort (*Artemisia vulgaris*)

It is a perennial plant growing up to 1.2 m by 0.7 m at a fast rate in light (sandy), medium (loamy) and heavy (clay) soils and prefers well-drained soil. It can grow in semi-shade (light woodland) or no shade. It prefers dry or moist soil and can tolerate drought.

Actions: antispasmodic, diaphoretic, emmenagogue, haemostatic, anthelmintic and antiseptic.

Indications: dysmenorrhea, menorrhagia, infertility, preventive for miscarriage, sciatica, convulsions, hysteria, epilepsy, depression, mental exhaustion, insomnia, gout, rheumatism and fungal infections.

Used for stomach disorders, especially those acid in nature, as it reduces acid secretions and may even lessen the damage from chronic over-acidic conditions, such as gastric ulcers, acid reflux, *etc.* Mugwort is also becoming popular as an antioxidant to regulate fat metabolism, especially in the liver. Too much fat in the blood causes blood to be sluggish, thick, and have fat cell pile-ups and strain on the capillaries, resulting in dyspeptic headache, inflamed areas and congestion of the digestive system. Mugwort seems to help these fats move along and out more smoothly. It can also help with frontal or allergy-related headaches. Warms lower abdomen, strengthens uterus and the fetus, regulates menstruation, relieves menstrual cramping and headaches, relieves pain.

Mugwort is a strong emmenagogue, and should be avoided in pregnancy.

Myrrh (*Commiphora molmol, Commiphora myrrha*)

It is normally found in open Acacia, Commiphora bushland on shallow soil, chiefly over limestone at an altitude of 250-1300 m having annual rainfall 230-300 mm, it prefers shallow soil, chiefly grows on limestone.

Actions: alterative, emmenogogue, astringent, expectorant, antispasmodic, rejuvenative, analgesic, antiseptic, antibiotic and anti-viral.

Indications: amenorrhea, dysmenorrhea, menopause, cough, asthma, bronchitis, arthritis, rheumatism, traumatic injuries, ulcerated surfaces, anaemia, pyorrhoea, bad breath, colon, gums, haemorrhoids, lung diseases, mouth sores, skin sores, sore throat, menstrual congestion, pains, gas, mouthwash, hypothyroidism. Myrrh is a powerful antiseptic, being a remedy second only to echinacea. It is a strong cleaning and healing agent, soothing the body and speeding the healing process. It is often used with goldenseal. It is most often used in mouthwashes, gargles, and toothpastes for fighting and preventing gum disease.

Used for preventing decay, reversing the aging process, and rejuvenating body and mind. Works specifically on the blood and female reproductive system. Helps dispel old and stagnant blood from the uterus, and aids in new tissue growth. Catalyzes healing of wounds and sores, while stopping pain. Helps dispel repressed emotions, as its purifying action extends to the subtle body. Possesses true tonic and rejuvenative powers along with strong detoxifying effects. Does not weaken the body with long term usage as do Golden Seal and other bitter detoxifiers. Yet it is not as effective as these in acute conditions. Antiseptic on mucous membranes. Disinfectant to wounds. Cleansing and healing agent of stomach and colon. Good gargle for sore throat, spongy gums, pyorrhoea, and canker sores. Helps in waste elimination. Inhibits bacterial growth. Stimulates capillary activity. Increases digestion. Promotes absorption of nutrients. Increases number of white blood cells. Good for delayed menstruation. In combination with ashwagandha has a synergistic effect that boosts thyroid function.

Caution: It is very rough on the kidneys. Do not use internally for longer than two weeks at a time. Do not use during pregnancy.

In folk medicine, it is burned to purify and protect. It is used to consecrate and purify ritual tools and objects needing to be blessed.

Myrtle (*Myrtus communis*)

Myrtle is an evergreen plant that prefers warm climates. It has small pointed leaves, and grows to about 12 feet high. Its blossoms are small, white, and in clusters. The leaves are gathered and dried for use in August.

Myrtle is used to treat bronchitis, bruises, bad breath, wounds, colds, sinusitis, and coughs.

In folk medicine, it is burned as an incense to bring beauty, to honor Diana and to Venus, and is a symbol of glory and happy love. Myrtle tea will make you look beautiful to your loved one. A distillation of the leaves and flowers combined will make a wonderful beauty wash for the face, and is known as "angel water". It is used in spells to keep love alive and exciting. Grow on each side of the house to preserve and protect the love within.

Neem (*Azadirachta indica*)

Neem is a tropical evergreen, native to India and Burma, and growing in southeast Asia and western Africa. It can grow to reach 50 feet tall in a desirable climate, and tolerates drought and poor soils. It can live up to 200 years. Neem bears fruit at 3 to 5 years of age. It cannot take freezing temperatures, so those in northern climates can grow it as a houseplant. Summer it outdoors, and place near a bright or sunny window in the winter. It needs organic fertilizers, well drained soil, and as large a pot as possible. Ten gallon sized pots are recommended as the minimum. Be careful not to overwater. This plant, known as "the village pharmacy" in India, has been used for at least 4000 years for its medicinal qualities. All parts of the plant are used. Neem has been used to treat a wide range of ailments, including wounds, burns, sprains, bruises, earache, headache, fever, sore throat, food poisoning, shingles, colds, flu, hepatitis, mononucleosis, fungal infections, yeast infections, sexually transmitted diseases, acne, skin diseases, heart diseases, blood disorders, kidney problems, digestive problems, ulcers, periodontal diseases, nerve disorders, malaria, fatigue, and a host of others. It is being closely studied for use in battling AIDS, cancer, diabetes, allergies, and as birth control for both men and women. Neem should not be used for more than two weeks at a time. For chronic ailments, it should be used on a schedule of two weeks on, one week off, or as directed under the guidance of a health practitioner. It is anti-bacterial, anti-viral, anti-septic, and strengthens the body's overall immune responses. It should not be used by internally by pregnant women.

In traditional uses it's leaves were traditionally strewn on the floor of temples at weddings, to purify and bless the area and the couple, and the air was fanned with neem branches during the ceremony. The bark was burned to make a red ash for religious decoration of the body in adulthood. Neem branches were used to cover the body at death, and the wood used to burn the funeral pyre. Neem is considered to be the Goddess Neemari Devi.

Nettle (*Urtica urens*)

It is an erect, rhizomatous, dioecious, perennial; unbranched plant, it can grow up to 6 ft. in height, naturally occurs in Coastal Plain, Piedmont and Mountain provinces. It inhabits damp, rich, disturbed areas, primarily in calcareous soils, in floodplains and moist open forests, and can tolerate dry soils and some shade. As nettle is considered a bothersome weed, the spines on a nettle plant can cause painful stinging, so it is not a good idea to include it in your herbal garden. Very nutritious.

Actions: alterative, astringent, haemostatic, diuretic and anti-inflammatory.

Indications: allergies, asthma, urinary complaints, hives, urinary stones, nephritis and cystitis, to stop bleeding both internally and externally, blood purifier, high blood pressure, auto-immune diseases such as rheumatoid arthritis and prostate.

A popular spring tonic, Nettles contain large amounts of nutrients, especially iron, calcium, silicon and potassium, and are opening and cleansing to the body, stimulating all types of elimination. Nettles are also reputed to be alkalinizing to the blood, and to both remove uric acid and prevent its crystallization in the kidneys and joints, possibly helping with arthritis. Strengthen and support whole body. Natural anti-histamine. Stops bleeding. Good for allergies. The plant is used for treating high blood pressure, gout, PMS, rheumatism, and ending diarrhoea, scurvy, liver and prostate problems. Externally it is used as a compress to treat neuralgia and arthritis. It is a very high source of digestible iron. It also treats anaemia, fatigue, edema, menstrual difficulties, eczema, enlarged prostate (especially when combined with Saw Palmetto), urinary tract problems, hay fever and allergies. It is very supportive during pregnancy. Use the infusion as a hair rinse to treat dandruff and to stimulate hair growth.

In folk medicine, it is sprinkled around the room to protect it. It is also burned during ceremonies for exorcism. Stuffed in a poppet and sent back to the sender of a curse or bad spell, it will end the negativity. Nettles gathered before sunrise and fed to cattle is said to drive evil spirits from them.

Nutmeg *(Myristica fragrans)*

It is moderately tall (15 meters or 50 feet), tropical, evergreen tree, with smooth, gray-brown bark and dark green leaved evergreen tree, Myristica fragans, that is cultivated for two spices derived from its fruit, "nutmeg" and "mace." Nutmeg is produced from the dried, ripe, inner seed and mace from the seed coat (arillus) that separates the seed from its outer husk.

A small amount of nutmeg, about the size of a pea, can be taken once daily over a long period (6 months to a year) to relieve chronic nervous problems, as well as heart problems stemming from poor circulation. Added to milk, and baked fruits and desserts, it aids in digestion, and relieves nausea. Large doses can be poisonous, and may cause miscarriage for pregnant women.

In folk medicine, it is believed that nutmeg will help to get relieve from clairvoyance, and ward off rheumatism. It is included in prosperity mixtures. Nutmegs are carried as good luck charms.

Oak *(Quercus robur)*

It is an evergreen tree or shrub growing to 25 m (82ft) by 20 m (65ft) at a slow rate.

Oak bark is used to treat diarrhoea, dysentery, and bleeding. For external use the bark and/or leaves are boiled and then applied to bruises, swollen tissues, wounds that are bleeding, and varicose veins.

The oak is considered the most sacred of all the trees in folk medicine. The most powerful mistletoe grows in oaks. The leaves are burned for purification, and the branches make powerful wands. The acorn is a fertility nut. It is carried to promote conception, ease sexual problems, and increase sexual attractiveness. The leaves and bark are used in binding spells. Planting an acorn in the dark of the Moon will bring you money. Oak wood carried will protect from harm, and hung in the home it will protect the home and all within.

Oats *(Avena sativa)*

Oats are an annual grass that grows up to 4 feet tall. Easiest to purchase from a health food store, as much is needed to be beneficial, and takes up more room than the average gardener has available. It does make a pretty ornamental grass in the garden and around foundations.

Oats are a traditional food for those recovering from an illness. It also supplies necessary fibre in the diet. Oats made into packs and pastes clear up many skin disorders, such as acne.

In folk medicine, it is used in prosperity and money spells, and in rituals to the harvest.

Olive *(Olea europaea)*

It is an evergreen tree or shrub native to the Mediterranean, Asia and Africa. It is short and squat, and rarely exceeds 8–15 m (26–49 ft) in height. Actions: antibiotic, anti-bacterial, anti-viral and anti-fungal.

Indications: weakened immune system, all types of infection(viral, bacterial, fungal, parasitic), colds, flu, HIV, chronic fatigue, yeast infections, chicken pox, high blood pressure and cholesterol, arthritis, and fibromyalgia.

Olive leaf helps your immune system fight dangerous Viruses, Bacteria and other Invaders without building Antibiotic Resistance! It acts like an antibiotic by fighting the "bad" microbes we are exposed to, it destroys only the bad bacteria and protects the good. Antibiotics kill invaders but don't strengthen the immune system. HIV/AIDS Scientific/ Documented Research: Oleuropein the active ingredient of the olive leaf maintains acclaimed powers against a virus, retrovirus, bacterium and protozoan. Oleuropein brings some highly important claims: Oleuropein interferes with critical amino acid production essential for viruses, inactivates viruses and prevents a virus from spreading, budding or assembling, directly penetrates infected cells stopping viral replication, neutralizes reverse transcriptase and protease and can stimulate phagocytosis where cells ingest harmful microorganisms.

Onion (*Allium cepa*)

Onion is a perennial herb that grows from a bulb. It prefers rich garden soils and plenty of water. The greens above ground can be used alone, and the bulb harvested by pulling from the ground, and allowing the tops to dry before storing in a dry location, with temperatures between 35 and 50 degrees F.

Onion leaves and immature bulbs are consumed as vegetable.

It contains vitamins B and C and minerals Calcium and Iron.

It is used in ear-ache, colic pain *etc.* Onion is used externally as an antiseptic. Internally, it can alleviate gas pains, reduce hypertension, and reduce cholesterol.

In folk medicine, it has been used as a charm against evil spirits. It is believed that halved or quartered onions placed in the home absorb negativity.

Orange (*Citrus aurantium*)

Oranges prefer a rich, sandy soil, and warm year-round temperatures. For most of us, that means growing them indoors as house plants.

Oranges are rich in Vitamin C and Potassium. One orange supplies around 116.2 per cent of the daily value for vitamin C. It's peel contains citral, an aldehyde that hassle the action of vitamin A. In addition, it is a rich source of Vitamins A, B1,B2, Calcium and Iron. It has antioxidant and immune booster properties.

It is protective as it contains long-acting liminoids which has the ability to promote optimal health. Compounds in orange peel lower the Cholesterol as effectively as statin drugs. Rich fibre in the fruit helps to prevent atherosclerosis. It gives protection against rheumatoid arthritis. It helps in preventing ulcers and reduces the risk for stomach cancer.

In folkmedicine, the dried peel is added to love charms. The fresh or dried orange flowers added to the bath makes you attractive. The fruit itself hinders or banishes lust. Orange juice is used in rituals in place of wine.

Oregano (*Origanum vulgare*)

Oregano is a perennial that prefers well-drained, slightly alkaline soil and full sun. It is propagated by seed, root division, or cuttings.

Oregano is used to promote perspiration as a treatment for colds, flu, and fevers. A tea of oregano is often used to bring on menses and relieve associated menstrual discomfort. It is also used in baths and inhalations, as well drinking the infusion, to clear lungs and bronchial passages. Internally and externally it can help alleviate dry itching skin. The essential oil is used to treat viral infections, respiratory ailments, and muscle aches. Pregnant women should not ingest large amounts of oregano.

In folk medicine, it is used to help forget and let go of a former loved one, such as a former spouse, boyfriend, girlfriend, *etc.* Burn in incenses or drink the infusion to aid in spells for letting go.

Oregon Grape (*Mahonia aquifolium*)

Oregon Grape is an evergreen shrub which can grow to a height of 7'-10' and up to 5' wide. It often forms scraggly clumps. Leaves with the spines on the edges are stiff and leathery in texture, with a dark green upper surface during the summer and a bright red-purplish color in the winter. In April yellow flowers form followed by clusters of blue, grape-like fruit. While the berries are edible, they are very sour.

Actions: alterative, antipyretic, laxative, antiseptic and antibacterial.

Indications: skin conditions, liver problems, gastro-intestinal problems, psoriasis, eczema, mucus membrane infections, acne, jaundice, infections and constipation.

Excellent blood purifier that stimulates the liver and gall bladder. It has been shown to release iron which is stored in the liver into the blood stream, which can be useful in anaemia. In studies it has been found that it block the adhesion of streptococcus, E. Coli, and salmonella to various mucus membranes. Its antiseptic effects act in particular on the kidneys and urinary system. Aids assimilation of nutrients. Increases appetite.

Caution: Stimulates the thyroid. This can be useful for short periods but depleting over long periods of time.

Osha (*Ligusticum porteri*)

Osha is an herbaceous perennial growing from 50 to 100 cm tall or more. In winter, the above-ground parts die back to a thick, woody and very aromatic rootstock. The plant has deeply incised, elliptic or lance-shaped leaf segments that are 5 to 40 mm in width with larger basal leaves. The white flowers appear during late summer, and are approximately 2 to 5 mm in diameter with five petals. They are grouped in flat-topped, compound umbels and are followed by reddish, oblong, ribbed fruits 5 to 8 mm in length. Osha often grows in rich, moist soils in wooded habitats—from pine-oak woodland to spruce-fir forest—but it is also found on slopes and in meadows with drier, rocky soils from 1,500 to 3,505 meters.

Actions: stimulant, antibacterial, expectorant, antiseptic, purifying tonic, anti inflammatory, anti-viral, antirheumatic, antispasmodic, diaphoretic, digestive febrifuge and stomachic.

Indications: colds, low immunity, allergic reactions, sore throats, respiratory problems, viral infections, eruptive fevers, digestive complaints, toothache, painful menstruation, retained placenta, tuberculosis and headaches.

Stimulates the circulation, kidneys and uterus. To soothe sore throats and irritation of the gums, helps loosen phlegm and is an effective treatment for viral colds and flu. Osha Root is arguably the best American herb for lung and throat infections. It stimulates the macrophages or resident white blood cells of the lungs, numbs sore throats, bronchio-dilates the lungs to assist in expectoration, warms the lungs and helps one to breathe more deeply. Osha can be used as a preventative for those prone to sore throats and lung congestion or who get secondary infections from allergies. As Osha Root brings more blood into the lungs, it assists in dilation of the lungs when constricted. Therefore, it is helpful for emphysema, pneumonia, allergies, smokers cough, and athletically induced asthma. It is more effective than Echinacea and Goldenseal when one is already acute and congested. It increases oxygen utilization and uptake into the body, which aids in motion and air sickness.

Oat Straw (*Avena sativa*)

Oat straw is an annual grass which grows in moderate temperatures. It grows well in dry wastelands, cultivated ground and meadows alike. It can be grown in heavier soils although it normally prefers sandy or loamy soils. It requires good drainage but can grow in a high acid soil. It does require sun, and cannot grow in the shade. Oat straw tolerates drought well.

Actions: nervine, antispasmodic, tonic and sedative.

Indications: stress, insomnia, anxiety, depression, PMS, colitis, ulcers, digestive ailments and quitting smoking.

Great for stress, emotional challenges and a burnt out nervous system. Oatstraw helps with insomnia, a chattering mind, anxiety, nervous exhaustion, nerve pain, recovery from substance abuse, emotional trauma, tension, depression, pre menstrual syndrome, and nervous headaches. It helps to regenerate and nourish the nervous system, it contains alkaloids which may help rebuild the myelin sheath, the protective wrapper around nerves, and so is often useful for long term nerve problems such as chronic pain from injury or illness. It is gentle, calming and soothing. Oatstraw is also well known to be useful for all stages of a woman's life. It can help with anxiety, insomnia, pain, and nervousness during menstruation and menopause. The ability of oatstraw to help regenerate the nervous system can have a very beneficial effect on a woman's hormone cycle. The nervous system plays an important role in communicating hormonal messages in the body. If the nervous system is weak or exhausted this can have an effect on a woman's hormonal balance. Oatstraw helps to balance the nervous system and the endocrine system. It is also gentle enough to use during pregnancy and is found in many tonic pregnancy tea formulas. Regardless of menstruation, menopause, or pregnancy, most women experience varying levels of stress and tension. Oatstraw is one of my favourite herbal remedies for all levels of depression, emotional challenges, and stress and it actually has a very nice taste. It contains high amounts of bone building materials.

Passionflower (*Passiflora incarnata*)

Passionflower grows best below zone 7 outdoors, so grow indoors in the north. It prefers partial shade and a fertile soil with good drainage. It grows to 25-30 feet as an outdoor vine. Prune old growth in the winter or very early spring to encourage flowering.

Actions: nervine, sedative, diuretic and anodyne.

Indications: insomnia, nervous disorders, hyperactivity, restlessness, pain, fevers, muscles aches, flu, insomnia, menopause, ADD (attention deficit disorder), Parkinson's, epilepsy, shingles, anxiety and hypertension.

Relaxes muscles. Relieves nervousness, agitation and exhaustion. It does not have intense sedative or painkilling action, but has a combination of effects which increases respiration, but at the same time slows the heart rate, dilates the blood vessels and lowers blood pressure. Helpful in controlling convulsions in children. Used in Italy to treat hyperactive children. Used for various neurological disorders. Used as a sedative to calm nerves during menopause. Has pain-killing effect. Eye wash used for eye infections, inflamed eyes, and dimness of vision. It can also be helpful in conditions of muscle spasm, from injury, menstrual problems, or even stomach cramps. It has a very balanced action in cases of insomnia, and most people are calmed enough by it to sleep deeply, yet still wake refreshed and alert.

It is used to treat hyperactivity, insomnia, Parkinson's disease, nervous tension, and the infusion is used to treat eye infections and eye strain. Native American tribes used it as a poultice for bruises and injuries, as well as for an overall tonic.

Passionflower will calm a troubled household when placed inside the home according to religious interpretation. Spanish missionaries believed the flowers were a symbol of Christ's crucifixion, and the crown of thorns of Christ's passion, giving this plant its name.

Pau d'arco or Trumpet Bush (*Tabebuia avellanedae*)

It is a large canopy tree indigenous to tropical regions of South America, it grows up to 30 meters tall and 3 meters wide. Pau d'arco trees only grow in tropical regions.

Actions: alterative, antipyretic, anti-biotic, anti-viral and anti-fungal.

Indications: ulcers, diabetes, rheumatism, yeast, cystitis, lupus, Parkinson's disease, and anaemia, slows and inhibits growth of cancers and tumors, skin diseases, candida yeast growth, acne, hepatitis, allergies, asthma, poisons attacking the liver, flu and herpes.

Strong immune stimulant that grows in South America, where it is used for many toxic blood related conditions. It is a powerful herb with antibiotic and virus-killing properties. It gives the body the energy needed to defend itself and to help resist diseases. It is used in South America to battle cancer and leukemia. It is useful in aiding all chronic diseases. Thought to be an especially effective leukemia treatment. Eliminates pain. Multiplies red corpuscles. Effective against Candida yeast growth. Anti-tumor. Prevents the spread and growth of secondary cancer tumors especially in the blood and skin. Start with small doses to prevent an over-reaction of the immune system and a worsening of symptoms.

Patchouli (*Pogostemon cablin, P. heyneanus*)

It is a bushy herb of the mint family, with erect stems, reaching two or three feet in height and bearing small, pale pink-white flowers.

It is antidepressant, antiphlogistic, antiseptic, aphrodisiac, astringent, cicatrisant, cytophylactic, deodorant, diuretic, febrifuge, fungicide, insecticide, sedative and tonic substance. Patchouli is used to treat dysentery, diarrhoea, colds without fevers, vomiting, and nausea.

In folk medicine, it is considered a sensual oil, and it can ward off negativity and evil. It is also burned in incenses to aid divination and clairvoyance.

Pennyroyal (*Mentha pulegium*)

Pennyroyal is a perennial that grows to 1 1/2 feet high. It tolerates most soils, and prefers direct sun. Grow as you would any member of the mint family.

Pennyroyal herb removes gas from the digestive system. It is also used as a tea, taken a few days before menstruation to aid a suppressed flow. It is used in treatments for colds, upset stomach, and to stimulate blood flow to the pelvis area. It's strong minty smell makes its essential oil useful for externally repelling insects such as mosquitoes, fleas, and flies. It should not be taken or used by pregnant women. Large internal doses have been known to cause convulsions and coma. Pennyroyal oil is an effective insect repellent.

Caution: Pennyroyal oil is poisonous in large dosages, should never be taken internally.

In folk medicine, it is added to protection and exorcism incenses. It aids in making favorable business deals. It is given to arguing couples to cease their fighting and restore harmony in the relationship.

Peony (*Paeonia officinalis*)

Peonies are a perennial shrub-like plant, growing 2 - 4 feet high. They prefer rich, humusy, well-drained soils, and full sun.

Peony root treats menstrual cramps and irregularities, gout, and asthma. It is also used in combination with other herbs to ease emotional nervous conditions. It should be used only under the guidance of a professional, as it can be toxic if taken incorrectly. Caution: Do not use the flowers or leaves internally.

In folk medicine, dried peony roots are carved and made into bracelets and necklaces for protection, as well as for breaking spells and curses. Peonies planted outside the home guard against storm damage and demons. A chain of beads cut from the dried root was worn as a protection against illness and injury, and to cure insanity.

Peppermint (*Mentha piperata*)

Peppermint is a perennial grown in full sun, is tolerant of most soil types, and grows to 3 feet tall.

Peppermint cleans and strengthens the body. It acts as a sedative on the stomach and strengthens the bowels. It is also mild enough to give to children as needed for chills and colds. Used with bitter herbs to improve their taste.

In folk medicine, it is used in charms to heal the sick, as well as in incenses in the sickroom of the patient. It is burned to cleanse the home, and is used in sleep pillows to aid in getting to sleep. Placed beneath the pillow, it can bring dreams that give a glimpse into the future. The essential oil is used in spells to create a positive change in one's life.

Periwinkle (*Vinca major, Vinca minor*)

Periwinkle is a perennial plant that spreads by putting out runners, mostly used for a ground cover in partial to full shade. It prefers moist, well-drained soils. Periwinkle is used made into a tea or salve for external use to treat skin problems such as dermatitis, eczema, and acne.

In folk medicine, it can help restore memory when it is gazed at or carried. It is also hung on a door to protect all within, and to prevent a witch from entering a home.

Plantain (*Plantago lanceolata*)

It is a perennial herb growing to 0.5 m (1ft 8in) by 0.2 m (0ft 8in) Plantains are common weeds, some varieties being annual and some perennial. They are found in all soil types, and prefer full sun.

Plantain is used to clear mucous from the body, and to neutralize poisons. As a mild tea it is used to treat lung problems in children, and as a stronger tea is used to treat stomach ulcers. It is also used for diarrhoea, bladder infections, and for treating wounds.

In folk medicine, it is hung in the car to guard against evil spirits.

Milkwort (*Polygala myrtifolia*)

Milkwort is an upright shrub usually growing 1-2.5 m tall of coastal environs, open woodlands, grasslands and watercourses (*i.e.* riparian areas) in the temperate regions of Australia.

Actions: tonic.

Indications: stress, insomnia, low will power, quitting smoking or other addictions and compulsive behaviour.

Polygala is one of the truly extraordinary tonic herbs in the entire Asian tonic herbal system. This herb first attained wide use in Daoist circles because it was believed to have powerful mind and spirit-developing power. In fact, Polygala root was believed to be an empowering substance in the class with wild Ginseng and Ganoderma in this regard. This root is traditionally used as a Shen tonic to relax the mind, calm the emotions and to aid in the sleep process. However, it does not relax the mind in entirely the same manner as many Shen stabilizing herbs. Instead, many people claim that it enhances dreaming and aids in creative thinking. And it not only aids creative thinking, but the ability to manifest our ideas. In fact, the ancient name for this herb is the "Will Strengthener." The herb is believed to have the ability to strengthen that part of the psyche which we call the "Will." These days, Polygala may be used to strengthen the Will of the spiritual seeker, or it may be used to strengthen the Will of the more earth-bound. It is used in formulations to build enough will power to overcome obstacles and to achieve greater heights. For example, it can be used as the main ingredient in formulas to help stop smoking, or even to break other habits, such as drug abuse, overeating, or compulsive behaviour of any sort. Polygala has the unique power to provide the energy (the "power") to our will, so that we can overcome the obstacles that block us from becoming all that we can become. And it not only helps us break old, bad habits. It also helps strengthen our will to do new things, to achieve new heights. It strengthens our will so that we can start to work on a new project, to start exercising, to start and continue whatever we need to do to grow, to become a great human being. That is the magic of Polygala, the "will strengthener."

Polygala has another unique quality that sets it apart from most other herbs, including the tonic herbs. It has the ability to connect the Kidney (sexual) energy with the Heart (love) energy. It does this by opening the energy flow between the Heart and Kidney known as the penetrating vessel. The penetrating vessel is one of the energy channels that regulate the functions of the body-mind. It is called a "psychic channel" by the Taoists. Commonly, this vessel is blocked, resulting in a de-linking of our sexual energy and our emotional feelings. It is essential for our true health and well being that feelings of love and the functions of sex are united. Consuming polygala for a period of time will have this result. The will strengthener thus has a unique power to deepen our experience and our feelings, and to bring new levels of happiness into our lives.

Pleurisy Root (*Asclepius tuberos*)

It is a fleshy rooted, perennial plant, prefers good peat soil,growing 1 to 1 1/2foot high and bearing corymbs of deep yellow and orange flowers in September.

Actions: diaphoretic, expectorant, febrifuge, anti-inflammatory, antispasmodic and mild-laxative.

Indications: pleurisy, lung disorders, pain when breathing, pneumonia, colds, flus, bronchitis, chest congestion and mucus.

Decreases thickness and increases fluidity of mucus in bronchial tubes, irritates mucuos membranes, stimulates and irritates gastrointestinal tract.

Possible effects: potential mild laxative to cause watery bowel movements, may increase perspiration, may help treat pleurisy. Promotes sweating and opening of lung capillaries, allowing more moisture to enter the lungs to thin the mucus, which can then be expelled more easily.

Poppy (*Papaver somniferum*)

Poppies are perennials that like poor to average soils that tend toward dryness. There are varieties that will grow most anywhere in North America. Their foliage tends to die off by July, after a spectacular showing of flowers in the spring, but the foliage begins rejuvenation around September, which waits until spring to begin growing again. There are many annual varieties, but the perennial ones are most commonly used for healing.

Poppy is used for pain, insomnia, nervousness, and chronic coughs.

In folk medicine, it's seed pods are used in prosperity charms. The seeds are added to food to aid in getting pregnant. To find the answer to a question, write it in blue ink on a piece of white paper. Place the paper inside a poppy seed pod and put it beneath your pillow. The answer will come to you in a dream.

Queen Anne's Lace (*Daucus carota*)

Queen Anne's Lace is found throughout most of North America. It is a wildflower, distinguished by the one red flower in the center of a cluster of many tiny white flowers. It is a biennial that grows to 3 feet tall.

Queen Anne's Lace is used for treating gallstones and kidney stones, as well as water retention and strains and sprains. It is also called Wild Carrot.

Raspberry (*Rubus idaeus*)

Red Raspberry is a biennial or perennial, depending on the variety, growing 3 - 6 feet tall. They need a cold winter and a long cool spring, so they do not do well in the South. They aren't too picky about soil, so long as they get plenty of water. Red Raspberry is one of the most proven female herbs. It strengthens the uterine wall during preganancy, reduces the pain of childbirth, and helps to reduce false labor pains. After childbirth it is used to decrease uterine swelling and cut down on post-partum bleeding. It is used to ease menstrual cramps and to regulate the flow during menstruation. It is also good for vomiting in small children, and dysentery and diarrhea in infants.

In folk medicine, it is served as a love-inducing food. The brambles are hung at the entrance to the home to prevent unwanted spirits from entering.

Rauwolfia or Sarpagandha (*Rauvolfia serpentina*)

It is an evergreen plant, has been in use since 4000 years in Indian medicine. An erect under-shrub, grows well in highly acidic and neutral soil.

Actions: sedative.

Indications: insomnia, nervousness, anxiety, abnormal muscle tension, and twitching. In folk medicine, It has been used for fever, diarrhoea, general weakness, constipation, intestinal diseases, liver problems, rheumatism, water retention, and mental disorders, giddiness, high blood pressure, slow and painful urination, and wounds.

In India, extracts of the plant are used as antidotes for poisonous snake bites. Roots contain several alkaloids, the more important being two chemical classes know as the ajmaline and the serpentine group. The quantity of the total alkaloids has been estimated to be fairly high in the dried roots, the roots also contain a lot of resin and starch and when incinerated, leave an ash consisting mainly of potassium carbonate, phosphate, silicate and traces of iron and manganese. The herb is an effective drug in lowering blood pressure. It is also used to reduce fever during delivery, it is said to stimulate uterine contractions and promote the expulsion of the fetus. This however is not corroborated and may be regarded as folklore. The plant is effective in treating insanity. The herb is effective in treating insomnia because of its sedative properties. Rauwolfia is the best remedy for high blood pressure and it has been adapted by medical fraternity in most countries especially American countries. Rauwlofia is also very effective in treating hysteria. Rauwolfia relives itching in urticaria.

Red Clover (*Trifolium pratense*)

Red Clover is an herb, short-term perennial plant, which is capable of growing to a maximum height ranging from 8 inches to 32 inches (20 cm to 80 cm).

The flowers of the Red Clover are dark pink in color, with a paler bottom, with the length between 3/64 inches and 1/16 inches (12 mm and 15 mm), produced in a thick inflorescence.

Actions: alterative, diuretic, expectorant, antisposmodic, nutritive and supportive tonic.

Indications: cough, bronchitis, skin eruptions, infections, cancer, toxins and spasms.

It is used to pull out and eliminate nitrogen-based toxins, due to its containing small amounts of Molybdenum, which may also aid the body to produce hemoglobin more rapidly out of any iron ingested, and Molybdenum and iron can form specific antibodies to venomous bites. Mild blood purifier suitable for longterm usage. Pleasant taste, mildly strengthening. Can be used with children, the elderly or in conditions of debility. Thought to be an anti-cancer herb, especially good for breast and prostate cancer. Tonic for nerves. Phytoestrogen. Anti-tumor. Mild blood-thinner, contains coumarins. Skin complaints, eruptions, psoriasis and eczema. Useful as a preventative against the recurrence of breast cancer. Good herb to use with nervous exhaustion or wasting diseases. Good to use with any cleansing program. It has also been long used to promote milk production in nursing mothers.

Red Root (*Ceanothus* spp.)

An upright summer annual, generally growing between 1 and 6 feet in height. The common name 'redroot' refers to the pinkish-red color at the base of the stem (sometimes the whole stem) and the taproot. Redroot pigweed seed germination is favored by warm temperatures and high light conditions.

Actions: astringent, expectorant, sedative, stimulant and antispasmodic.

Indications: sore throats, enlarged lymph nodes, menstrual haemorrhage, nose bleeds, haemorrhoids, old ulcers, sprains, bruises, infection, congestion, varicose veins, arthritis, allergies and indigestion.

Used to strengthen the structure of the lymph system, improve the blood's ability to flow, and generally stimulate the cleansing cycle needed to heal. Red root is an excellent lymphatic remedy, stimulating lymph and inter tissue fluid circulation, and for shrinking non fibrous cysts.

Redroot is also an effective astringent reducing blood flow in cases of uterine haemorrhage and nosebleeds, very useful in the bleeding, pain, tenderness and swelling of ovarian cysts and its mild antispasmodic action.

It is an excellent home remedy for bleeding piles, and capillary ruptures from vomiting or coughing. Improves the action of herbs that work on the immune system, by removing the dead infecting cells that the antibiotics have killed.

Reishi (Ling Chi) Mushroom (*Ganoderma lucidum*)

The ling zhi or reishi is a relatively slow growing mushroom that prefers a temperature around 75-85 degrees Fahrenheit, and a high humidity (75-85 per cent relative humidity). It requires some shaded light to develop properly. Reishi requires fresh air, but drying drafts must be avoided. It likes a high humidity, but does not like to be continuously damp.

Actions: adaptogen, deep immune activator, analgesic, anti-inflammatory, hypotensive, antibacterial, antiviral, cardiac tonic, expectorant, antineoplastic, antioxidant.

Indications: cancer, high altitude stress, high cholesterol and hyperlipidemia, high blood pressure, chronic (post-viral) fatigue syndrome, AIDS, weakness of the lung, wasting syndromes, difficulty concentrating, poor digestion, insomnia, poorly regulated immune response, allergies, poisoning.

Reishi is in the most highly rated category of herbs, in terms of multiple benefits and lack of side effects, in Traditional Chinese Medicine. Inhibits tumors and helps with cancer, side effects of cancer treatments including radiation, chemo-therapy and surgery.

The polysaccharides and ergosterols probably work together to stimulate natural immune functions that tend to be suppressed by cancers and immune disorders. Ganoderic acids are responsible for the anti-allergy effects and improved oxygen utilization. Reishi greatly reduced the symptoms of headaches, nausea, vomiting, insomnia, heart palpitations and extreme fatigue of oxygen deprivation among Chinese workers who traveled to the high plateau of Tibet, climbing 15,000 feet in 3 days in the process. Reishi is also effective in reducing the symptoms of cardiovascular blockage and disease, including angina, palpitation, fullness in the chest, dizziness and headache, shortness of breath, insomnia and weariness, and loss of memory in 65 per cent or more of the patients in various studies. Reishi is a true adaptogen, enhancing health and normal functions of the body. For example, while it increases some components of immune response for cancer patients, it also inhibits pathological immune functions in auto-immune diseases such as myasthenia gravis. It has been reported to reduce the histamine release associated with allergic reactions, and help prevent anaphylactic reaction. It also increased immunoglobulin-A levels in 2,000 chronic bronchitis patients. Considered an adaptogen, that is, an herb which balances and modulates the body's actions and responses in whatever direction is needed. Some of the systems and functions to which this normalizing effect is applied are blood pressure and circulation, the immune system, blood sugar levels, heart function, the nervous system, allergic response, digestion, and all major internal organs. Believed by ancient Chinese to confer not only longevity, but immortality as well. It sharpens the memory and mental capacity, awakens the spirit, calms the mind, opens the breath, strengthens the heart, protects the liver, increases stamina and endurance, and increases resistance to disease.

Rose (*Rosa* spp., *Rosa damascena*, *Rosa gallica*, *Rosa moschata*)

Roses of all varieties are adaptable to most soils as long as they have adequate water, and are occasionally fed through the growing season. There are varieties that will grow throughout North America. Plant them where you can enjoy their beauty and fragrance.

Rose hips are very nourishing to the skin, as well as containing vitamin C. It is used as a blood purifier, and for treatment of infections, colds, and flue.

In folk medicine, rose water is used in gourmet dishes and in love potions. Petals are used in healing incense and sachets, and burned to provide a restful night's sleep. The essential oil is used in ritual baths to provide peace, love, and harmony within the self. The hips are strung like beads and worn to attract love. Rose petals sprinkled around the home will calm personal stress and upheavals in the home.

Rosemary (*Rosmarinus officinalis*)

Rosemary is a perennial that prefers mild climates, so it needs to be grown indoors where the winters are harsh, or very heavily mulched. It reaches 2-4 feet in height, and is tolerable of poor soils. Cut back after flowering to keep it from becoming leggy.

Actions: diaphoretic, carminative, stimulant, emmenagogue, tonic, astringent, nervine, stomachic, antispasmodic, anti-septic, anti-bacterial and anti-oxidant.

Indications: headache, memory loss, premature baldness, colic, colds, nervous diseases and depression.

Improves memory function, recent studies have found more than a dozen anti-oxidant compounds, most of which collect in the tissue of the brain, preventing inflammation, stimulating circulation, and generally benefiting the mental functions. Rosemary may even help slow the breakdown of acetylcholine, a degeneration related to Alzheimer's disease. Stimulates hair folicles. The warming and stimulating effects of rosemary help clear phlegm from the head and chest which is why it has traditionally been used as a remedy for upper respiratory ailments such as acute bronchitis, catarrh and colds. Stimulates digestion and is cleansing to the whole body. People with bad circulation can benefit greatly from it.

Rosemary is a stimulant of the circulatory system. It is used to treat bites and stings externally. Internally it is used to treat migraines, bad breath, and to stimulate the sexual organs. It is also used to treat nervous disorders, upset stomachs, and is used to regulate the menstrual cycle and to ease cramps. Mix the crushed leaves generously into meats, fish, potato salads, *etc.* at your next picnic to prevent food poisoning. The essential oil is used in aromatherapy as an inhalant and decongestant, and to enhance memory and clear concentration. It is also used in lotions to ease arthritis and muscle pain.

Often used in Spagyrics as spiritualized essence for energetic balance, it is very stimulating to the mind and thought, also used in Spagyrics for balancing the soul level, it is more cheering, uplifting, and warming.

Rosemary in all of its forms is used in folk medicine for protection and banishment. Rosemary leaves under your pillow do away with evil spirits and bad dreams. It is hung on porches and doors to keep thieves out. Rosemary is grown to attract elves.

Rue (*Ruta graveolens*)

This is a hardy, evergreen shrub of up to one metre tall, it prefers rocky, well-

drained soils and it resists dry weather. with a characteristic greyish color and a sharp unpleasant odour. The leaves are small, oblong, deeply divided, pinnate, glandular dotted. The stems are ramified. The flowers are small (13 mm), yellow and in clusters during Spring and Summer. The fruits are round, small and lobulated. The taste is slightly stinging with strong bitter odour.

Rue is used in small amounts to expel poisons from the system, such as those from snake bites, scorpion, spider, or jellyfish bites. It should not be taken with meals, and it should never be used by pregnant women. Juices from the fresh plant can cause the skin to blister. It is used internally and externally as a remedy for tendonitis.

In folk medicine, it is used in sachets and amulets to ward off illness. The smell of the fresh, crushed herb will chase away thoughts or envy, egotism, and love gone wrong. Rue leaves placed on the forehead will chase away headaches. Added to baths, rue drives away spells and hexes placed on you.

Saffron (*Crocus sativus*)

Saffron grows from a bulb commonly known as a corm. It is a perennial. Plant in the fall, 3 inches deep in light, well-drained soil where it will receive plenty of sun. The three-pronged stigmas that remain after flowering is the part to harvest for healing use.

Actions: alterative, emmenagogue, aphrodisiac, rejuvenative, stimulant, carminative and antispasmodic.

Indications: menstual pain and irregularity, menopause, impotence, infertility, anemia, enlarged liver, hysteria, depression, neuralgia, lumbago, rheumatism, cough, asthma and chronic diarrhea.

As costly as gold, Saffron has been highly valued since ancient times as one of the most sublime medicine/food substances in the world. In Ayurveda, Saffron is considered pure Sattva, the element of spirit and etheric energy, and it is held in high esteem as an aid to meditation, love, devotion, compassion, and spiritual development. It is also considered the best purifying herb, clearing the subtle channels of the etheric body, as well as eliminating toxic buildup in the physical body. It is used as a rejuvenative of the blood, metabolism, circulation, and entire body, and also acts as a catalyst to any other course of therapy with which it is combined. It is said to promote the assimilation of food into deep tissues, and also to stimulate new tissue growth, especially in the female reproductive system. One of the best anti-Pitta herbs and spleen-liver regulators. Best stimulant and aphrodisiac, especially for women. Catalyzes the tonic action of other herbs and promotes tissue growth in the reproductive organs and in the entire body. Modern research has shown that saffron helps the body to regulate lactic and uric acids, and keeps these acids from building up because of muscle strain or injury, or some types of arthritis and gout. Additionally, it is considered one of the best anti-inflammatories, thus helping with symptomatic relief in these cases as well. It is also a natural source of hydrochloric acid, and so is tonic to the digestion. It can be added to other herbs to facilitate their function, or used as a spice to promote assimilation of food into deeper tissues. Saffron is used as a preventative for heart disease, as it prevents the build-up of cholesterol. It is also used to soothe the membranes of the stomach and colon. It is not to be taken in large doses, nor should it be taken by pregnant women.

In Spagyrics used as a spiritualized essence for energetic balance, it is very useful as a mental and spiritual purifier, stimulant to the flow of Qi, and general vitality tonic.

In folk medicine, it is used to clean the hands before rituals. It is used in healing mixtures. The essential oil is used to induce clairvoyance.

Salvia (*Salvia divinorum*)

This plant is a psychoactive plant which can induce visions and other altered and spiritual experiences. Its native habitat is in cloud forest in the isolated Sierra Mazateca of Oaxaca, Mexico, where it grows in shady and moist locations. The plant grows to over a meter high, has hollow square stems, oval leaves and blue-purple flowers, occasional white flowers with violet calyxes and is about 1 metre tall.

Used as a sacred medicine by indigenous shamans and healers to travel into the supernatural world to discover the true cause of a patients trouble, used in a ceremonial manner to induce a visionary trance state within which it is possible to determine the underlying cause of disease and what steps can be taken to remedy it. It is also used to determine the where-abouts of a lost or stolen object. reduce extraneous mental activity and calm the mind; many people find that low doses are quite valuable as an aid to meditation and introspection. Mild doses also increase sensual awareness and can have aphrodisiac effects. A moderate dose produces a gentle expansion of consciousness that is useful for inner-journeying and self-exploration; this can also have an extraordinary memory enhancing effect that can provide access to incredibly detailed memories stretching back into early childhood (clearly this has therapeutic potential). Strong doses alter the fabric of consciousness profoundly, producing rich inner-visions and inspiring insight. Very strong doses can totally immerse one in a visionary world and can propel consciousness into the far reaches of transdimensional/transpersonal space. It has etheogen effects- meaning it gives one contact with the divne within oneself. It can be mood elevating and create a sense of wellbeing and a fresh outlook on life for the user.

Sage (*Salvia officinalis*)

Sage is an evergreen perennial, growing to 2 feet tall. It does best in sandy, limey soil in full sun.

Actions: diaphoretic, expectorant, nervine, astringent, alterative, diuretic, carminative, antispasmodic, antiseptic, tonic, bactericide, antioxidant, antimicrobial, anti-inflammatory and anti-biotic.

Indications: colds, flus, sore throat, laryngitis, swollen lymph glands, night sweats, spermatorrhoea, hair loss, nervous dysfunction, hot flashes, excess perspiration and to cease lactation, diarrhoea, enteritis, venereal disease, snake bites, toothaches, cancer, coughs, digestion, fevers, sweating, mouth sores, profuse excretions, colic, nausea, diarrhoea, colitis, gas, asthma, depression, digestive problems, muscle cramps and spasms, pre menstrual syndrome and flatulence.

It reduces excess secretions, stops sweating, dries up excess mucus from the nose and lungs and salivation, dries up sores and ulcers and stops bleeding. Reduces high kapha. For the brain and nervous system and promoting growth of hair, it combines well with Gotu Kola. Has a special power to clear emotional obstuctions from the mind and promote calmness and clarity. Helps reduce excessive desires and passions. It is specific for calming the heart. May help improve memory and concentration. Used in mouthwashes and gargles. Thought to prolong life. Used for mouth disease.

Sage is one of the most popular herbs in the United States. It has a fragrant aroma and a binding but warm flavor. Greeks and Romans commonly used Sage to cure snakebites and to stimulate the mind and body. It is used to treat colds, fevers, liver trouble, and epilepsy.

Sage is used to relieve excess mucous buildup. It is beneficial to the mind by easing mental exhaustion and by strengthening the concentrating abilities. In a lotion or salve, it is useful for treating sores and skin eruptions, and for stopping bleeding in all cuts. Chewing the fresh leaves soothes mouth sores and sore throats, as will sage tea. It is good for all stomach troubles, diarrhoea, gas, flu and colds. As a hair rinse, it removes dandruff. Sage combined with peppermint, rosemary, and wood betony provides an excellent headache remedy. It is used to regulate the menstrual cycle, to decrease milk flow in lactating women, aids in treating hot flashes, and is used as a deodorant.

In folk medicine, it is used in healing amulets, incenses, and sachets, and is also used in the same manner for bringing prosperity. Sage burned at the altar or in sacred space consecrates the area. Burned in the home, it removes impurities and banishes evil, as well as providing protection. It has euphoric properties creating positivity and happiness for its user. It will help change problems into insights. If a drop is placed on the third eye before sleep it can induce prophetic dreams

Salvia (*Salvia splendens*)

It is a tender herbaceous perennial native to Brazil, growing at 2,000 to 3,000 m elevation where it is warm year-round and with high humidity.

It creates a deep sense of relaxation and well being while enhancing vision and relieving depression. Often used together with its companion herb, Salvia Divinorum. S. splendens, or Crimson Sage, makes the deeply meditative and insightful states of S. divinorum more accessible, and activates the centered and inward focus of the meditation. Used together, they produce a much more productive time of contemplation, which is often followed by great insights and breakthroughs. Sage's pronounced pine-like aroma capitalizes on our most memory-evoking sense, smell.

It is widely used in Indian traditional medicine for the control of diabetes mellitus. Study suggested that the hypoglycemic activity of aqueous and methanolic extracts of S. splendens might be due to the presence of high amounts of anthocyanins and phenolic and flavonoid compounds. In aromatherapy it s prescribed for women for it's effectiveness in pre-menstrual stress, depression, temper tantrums, irritability, emotional hyperactivity and mood swings.

Siberian Ginseng (*Eleutherococcus senticosus*)

It is small, woody shrub of the family Araliaceae native to Northeastern Asia.

Actions: adaptogen, circulatory stimulant, tonic, vasodilator and nervine.

Indications: debility, depression, fatigue and nervous breakdowns, insomnia, irritability, tension and low blood pressure.

This herb may safely be used to increase stamina when under stress. After nearly a thousand studies, Siberian Ginseng has been shown to increase energy and stamina and to help the body resist viral infections, environmental toxins, radiation and chemotherapy. In Chinese Medicine it has been used to prevent bronchial, respiratory and viral infections, provide energy and vitality, increase resistance to stress, treat rheumatic and heart ailments, improve cardiovascular and neurovascular conditions and help restore memory, concentration and cognitive abilities which may be impaired from poor blood supply to the brain. It has a growing reputation for increasing all kinds of body resistance. Also used to regulate low blood pressure and increase circulation. Excellent as a general tonic. Siberian Ginseng increases the levels of neurotransmitters in the brain, and induces a feeling of well-being, helpful in treating both the causes and symptoms of chronic emotional stress. It has stronger nervine and calming properties than Panax, and helps lessen irritability, tension, and other nervous disorders, and greatly helps with insomnia.

St. John's Wort (*Hypericum perforatum*)

St. Johnswort is a perennial reaching 32 inches tall. It is grown throughout much of North America. It prefers rich to moderately rich soils, and full sun. It is not long-lived, so requires replantation every few years. The leaves and flower tops are collected as they bloom.

Actions: antispasmodic, expectorant, astringent, anti-depressant, anti-inflammatory, antibiotic and antibacterial.

Indications: depression, wounds, burns, injured nerves, inflammation, ulcers, anxiety, HIV, damaged skin, burns, nerve injury, fibromyalgia, uterine disorders, chronic fatigue and hypertension.

The relaxing effects of St. John's Wort seem to have a strong intelligence, traveling to the stressed tissues in the body, soothing the membranes and increasing the blood flow and tone of the area, while at the same time slightly stimulating the heart. In addition to being helpful in both depression and hypertension, these actions are often effective for treating uterine disorders and nerve-related problems, such as fibromyalgia or chronic fatigue. Also used to soothe and heal damaged skin. The plant has also generated intense scientific interest because of its potential as an antiviral agent. Some experiments have shown it to have inhibitory effects on the binding of HIV cells to human cells, as well as rendering AIDS non-infective in vitro, and inhibiting the spread of AIDS through cells in patients. Avoid excessive exposure to the sun when using. St. John's wort is useful for bronchitis, internal bleeding, healing wounds, and for dirty, septic wounds. It is used to ease depression, headaches, hysteria, neuralgia, shingles, as well as symptoms that occur during menopause. It is useful in swellings, abcesses, and bad insect stings. Studies are showing that it may be effective in combatting AIDS by increasing the immune functions of the body.

Caution: Do not go into the sun without protection while using this herb, as it causes blistering sunburns, especially in fair-skinned people.

In folk medicine, it is hung around the neck to prevent fevers. Wearing the herb aids you in war and other battles, including those of the will and indecision. Burnt it will banish evil and negativity. Hung in the home or carried, it will prevent spells of others from entering, and it is used in exorcisms. If you pick the plant on the night of St. John and hang it on your bedroom wall, you will dream of your future husband. The red juice of the stems was associated with the blood of John the Baptist, hence the plant's name.

Sandalwood (*Santalum album*)

Sandalwood is a hemi-parasitic plant which is widely scattered in dry deciduous forests of India and Australia.

Sandalwood oil is used to cool the body during fevers and heat stroke. It is also used to aid in the passing of kidney and gall stones, and for infections in the urinary tract.

In folk medicine, it's oil is massaged on the forehead and between the eyes to help center and calm the mind. It is used in healing oils and sachets. It's wood is burned as a purifiying and protective agent.

Sarsaparilla (*Smilax febrifuga, Smilax aspera*)

Sarsaparilla grows in rich, moist woods found in temperate zones, tropics and subtropics worldwide.

Actions: alterative, diuretic, diaphoretic, antispasmodic, antisyphilitic, antirheumatic, antibiotic, anti-microbial, stimulating, antiseptic and general tonic.

Indications: venereal diseases, herpes, skin diseases such as psoriasis and eczema, arthritis, rheumatism, gout, epilepsy, insanity, chronic nervous diseases, abdominal distention, intestinal gas, debility, impotence, turbid urine, infertility, hormonal imbalance, menopause, skin eruptions, wasting diseases, ringworm, syphilis and Multiple Sclerosis.

It purifies the urino-genital tract, dispelling all infection and inflammation. Purifies the blood and increases Agni and dispels accumulated Vata from the intestines. Purifying action extends to the nervous system and it helps cleanse the mind of negative emotions; therefore it is useful in many nervous disorders. Stimulates the production of reproductive hormones and has a tonic action on the sexual organs. As a blood purifier it works well with Burdock root. Increases circulation to rheumatic joints. It contains cortico-steriods, and natural plant sulfur. Sarsaparilla is used for treating psoriasis and other skin diseases, rheumatoid arthritis and kidney disease; for increasing urination to reduce fluid retention; and for increasing sweating. Sarsaparilla is also used along with conventional drugs for treating leprosy and for syphilis. Used in bodybuilding and gland balancing formulas for both males and females. It has endotoxin binding properties. It is also very effective at removing heavy metals and nitrogen-based waste, such as uric acid, which can be factors in some arthritis. Sarsaparilla's beneficial effects collect in the nerve fibers and tissues, and so may be helpful in hardening of these tissues from Multiple Sclerosis.

Shatavari (*Asparagus racemosus*)

Shatavari is a climbing plant which grows in low jungles areas throughout India.

Actions: nutritive tonic, demulcent, emmenogogue and rejuvenative.

Indications: debility of the female organs, sexual debility, infertility, impotence, menopause, diarrhea, dysentary, stomach ulcers, hyperacidity, dehydration, lung abcess, hematemesis, cough, convalecense, cancer, herpes, leucorrhea and chronic fevers.

Main ayurvedic rejuvenative for the female. Effective demulcent for dry and inflamed membranes of the lungs, kidneys, stomach, and sexual organs. Good for ulcers, and with its thirst-relieving and fluid-protecting powers it is good for chronic diarrhea and dysentary. Soothes Vata. Increases milk, semen, and nurtures mucus membranes. Nourishes and cleanses the blood and the female reproductive organs. Supplies many female hormones. Nourishes the ovum and increases fertility, yet its quality is sattvic and aids in love and devotion. Helps one connect with Mother Nature.

Saw Palmetto (*Serenoa repens, Sabal serrulata*)

It is a small palm, growing to a maximum height of around 7–10 ft (2–3 m).

Actions: tonic, rejuvenative, aphrodisiac, expectorant, anti-inflammatory, diuretic and antiseptic.

Indications: wasting diseases, prostate, thyroid disease and dysmenorrhea.

A good remedy for the reproductive glands. Most research centers use to treat benign prostatic hyperplasia. Many prostate problems are connected to levels of dihydrotestosterone, which causes cells in the prostate to over-multiply, resulting in inflammation, blockage, symptoms of frequent and uncomfortable urination, and even rupture. Research indicates that Saw Palmetto inhibits the production of dihydrotestosterone in the body, which can very effectively prevent or reverse prostate problems. Saw Palmetto is also a tonic to the bladder muscles, as well as a tonic to the reproductive organs in either sex, by increasing blood flow to the area. Increases size of mammary glands. Decreases ovarian and uterine irritability. Assists and normalizes thyroid activity. Urinary antiseptic, endocrine agent. It improves digestion and increases absorption of nutrients from food. Used for breasts, digestion, glands, reproductive organs, sex stimulant, weight gain. Note: Needs to be taken with food. Fat-soluble.

It is used for all wasting diseases. Also very useful for all diseases of the reproductive glands and organs. It is also used for the mucous membranes, as well as for treating bronchitis and lung asthma. It's most popular current use is for treating enlargement of the prostate gland.

Schizandra (*Schisandra chinensis, Schisandra glabra*)

It is a climbing perennial vine with deciduous leaves, widely used in Traditional Chinese Medicine. It has been used for thousands of years as a tonic and an anti-aging substance. New studies show it has potent antioxidant activity.

Actions: tonic, adaptogen, aphrodisiac, mild stimulant, deep immune activator, antioxidant, anti-bacterial and anti-depressant.

Indications: indigestion, liver dysfunction, headaches, insomnia, dizziness, palpitations, depression, stress, fatigue and depression.

Schizandra fruits or berries contain various lignans, mainly schizandrins, these lignans have been found to prevent liver damage, stimulate liver repair, and stimulate normal liver functioning. These properties appear to be related to the various antioxidant abilities of the various schizandrins. Schizandra also helps in digestion, regulating gastric acid release. Other studies have shown that extracts of schizandra are stimulating to the central nervous system and are cholinergic. There have been numerous reports on schizandra's ability to quicken reflexes, increase work efficiency, control anger and combat neurasthenias (headaches, insomnia, dizziness, palpitations). Other reports have mentioned increased cognitive function and increased memory. A recent study concluded that schizandra may be a useful herb to reverse depression, particularly that due to adrenergic exhaustion. Recent research studies have indicated that schizandra has numerous biological activities, including: liver protective, anti-depressant, antioxidant, adaptogen and cardiac tonic. In Oriental Medicine, schizandra is recommended for the lungs, liver and kidneys, including as an aphrodisiac (kidney element). Schizandra is also a registered medicine in Russia for vision problems since the herb has been found to prevent eye fatigue and increase acuity. In China, Schizandra is valued by women as a restorative to youth, sexual vitality, and energy, and by both sexes as a protective against all types of stress and fatigue.

Scullcap (*Scutellaria laterifolia*)

Scullcap prefers moist well-drained soils. It is a perennial that reaches to 3 feet in full to partial shade. It is not long-lived, so replant every few years.

It is a food for the nerves. It supports and strengthens as well as giving immediate relief from all chronic and acute diseases that affect the nerves. It is used to regulate sexual desires, and is very useful in remedies for feminine cramps and menstrual troubles. It reduces fevers and aids in easing insomnia and restlessness. It is also used to lessen the affects of epilepsy.

In folk medicine, it is used in spells that bring about peace, tranquility, and relaxation.

Senna (*Cassia senna*)

Senna is a shrub native to North Africa, India, and China, and grows to 3 feet high.

It is used as a very effective laxative, often used as a cleanser during a fast. It is strong, so you should combine it with fennel or ginger. Do not use for prolonged periods to avoid the bowel becoming dependent, and do not use in cases of dehydration.

Solomon's Seal (*Polygonatum multiflorum*)

It is a herbaceous perennial herb that grows from 1 - 3 feet high. It prefers moist woods, thickets, and roadsides. It prefers full to partial sun.

Actions: nutritive tonic, rejuvenative, aphrodisiac, demulcent, expectorant and hemostatic.

Indications: debility, infertility, impotence, chronic bleeding disorders, diabetes, consumption, dry cough, dehydration, malnutrition, burning sensation, broken bones, inflamed mucous membranes. It is used to treat tuberculosis, diabetes, and wasting diseases. It is also used as a kidney tonic, and as a healer of broken bones.

In folk medicine, it is added to protection sachets and incenses. It is also used for exorcisms and cleansing mixtures.

Slippery Elm (*Ulmus fulva*)

It is a deciduous tree that grows 50-80 feet tall. It needs full sun and good soils. It is found from Canada to Florida, west to the Dakotas and Texas.

The inner bark of the slippery elm is the portion used for healing. Slippery Elm is used to neutralize stomach acids. It is used to boost the adrenal glands, draws out impurities and heals all parts of the body. It is most useful for the respiratory system. Externally it is an excellent healer for burns, skin cancers, poison ivy, and wounds.

Shepherd's Purse (*Capsella bursa pastoris*)

Shepherd's Purse is typically found in full sun, mesic to dry conditions, and almost any kind of soil, including cracks in pavement. The size of this plant varies considerably with the fertility of the soil and availability of moisture.

Actions: astringent, hemostatic, alterative, diuretic and anti-inflammatory.

Indications: dysentery, diarrhoea, and inflammation of colon, ulcer, bleeding, haemorrhoids, profuse menstrual flow, high blood pressure, bruises and bleeding.

Used for all types of excessive or unwanted fluid discharge, highly valued for cases of post-birth haemorrhage, as it seems to create an "hourglass effect" in the female organs, closing off any blood flow at the base of the uterus.

Spearmint (*Mentha viridis*)

It is a perennial growing to 3 feet tall and is tolerable of many different growing conditions.

Spearmint is a valuable herb for stopping vomiting during pregnancy. It is gentle enough to use for colic in babies, while aiding in curing colds, flu, and gas.

In folk medicine, it is added to healing incenses and sachets to aid in healing lung diseases and other afflictions. Place some in a sleeping pillow for protection during sleep.

Squawvine (*Mitchella repens*)

Squawvine is a perennial evergreen creeper that grows on forest floors.

It is most beneficial in childbirth. It strengthens the uterus, helps prevent miscarriage, and relieves congestion of the uterus and ovaries. Its antiseptic properties make it valuable for treating vaginal infections, and is a natural nerve sedative. It is most often used in combination with Raspberry.

Suma [*Pfaffia paniculata, Hebanthe paniculata, Acacia suma* (Indian suma)]

It is a considerable, sprawling, shrub-like vine that grows along the ground possessing a complicated, deep and widespread root system. This plant is native to the Amazon basin as well as other tropical regions of southern Brazil, Panama, Ecuador, Peru, Paraguay and Venezuela.

Actions: adaptogen, tonic, aphrodisiac, restorative, astringent, analgesic, anti-inflammatory, antioxidant and anti-tumor.

Indications: wounds, skin rashes, low energy and sexual disinterest, chronic low energy and weakness, recurrent infections, viral syndromes, early onset of various cancers, diabetes, hypertension, rheumatoid arthritis, PMS, benign prostrate enlargement, impotence, rheumatism and peptic ulcers, AIDS and chronic fatigue.

It is being used to improve various ailments including fatigue, anxiety, erectile malaise, and stress. Suma root is an authentic adaptogenic herb, and as such exerts a normalizing influence on your body and can help regulate and enhance your endocrine, nervous, digestive, cardiovascular and immune systems. South American Natives have used Suma for centuries to treat wounds, skin rashes, low energy and sexual disinterest. The overall effect is to give you an increased resistance to stress while having a cell-building and regenerating effect. It also has analgesic and anti-inflammatory properties that can help alleviate chronic and acute pain. Suma root is also quite valuable nutritionally as it contains essential vitamins, minerals, amino acids and trace elements. Researchers have identified 152 chemical constituents in the root, including 19 amino acids, electrolytic and trace minerals such as iron, magnesium, cobalt, silica and zinc, as well as vitamins A, B-1, B-2, E, K and pantothenic acid. It contains high amounts of the trace element germanium, which is a powerful immune stimulator. The germanium may be partly responsible for Suma's powerful ability to bring more oxygen to the cells. Suma root is becoming quite popular because of its high content of several plant hormones. It is considered one of the richest sources of B-Ecdysterone, a plant hormone that can help maintain your youth and strength. B-Ecdysterone can also accelerate wound healing, along with allantoin (comfrey also contains allantoin), a known cellular rebuilder that is present in this plant. Research in Brazil, Japan and the United States has found unique natural substances in Suma called pfaffosides which are believed to reduce tumors and regulate blood sugar levels. Suma helps regulate your blood pressure, cholesterol, hormones (especially estrogen) and acid-base balance. Suma is very nutritionally rich. It contains antioxidants, amino acids, vitamins A, B, E, and K, minerals, and electrolytes. The electrolytes and B vitamins provide its energizing effect. Particularly in men, Suma can be of great help in sexual function, and has been researched for low libido and erectile dysfunction. Even in healthy men, Suma often greatly increases stamina and duration of arousal. Not recommended if pregant or nursing.

Sundew (*Drosera rotundifolia*)

Sundew plant grows in seasonally moist habitats or places that are not continually wet, where the soil has acidic content and the levels of sunlight are high. Accordingly, it grows in the swamps, marshes, fens, bogs and moist steam banks in the wallums of coastal areas, the tepuis (table-top mountains) and the fynbos.

Actions: anti-spasmodic.

Indications: whooping cough, chronic bronchitis and asthma.

Used with advantage in whooping-cough, exerting a peculiar action on the respiratory organs; useful in incipient phthisis, chronic bronchitis, asthma, *etc.*, the juice is said to take away corns and warts. In America it has been advocated as a cure for old age.

Sinicuichi or Sun Opener (*Heimia salicifolia*)

A South American Shaman's herb, traditionally used in divination ceremonies to contact the ancestors. It is called Sun Opener because it adds a yellowish hue to the vision; natives consider the auditory effects created by it to be the voices of the ancestors, who speak to give advice and guidance. It also has an affinity for spiritual work with music or the sound vibrations of etheric energies.

Native in Highlands from Mexico to Northern Argentina, Sinicuichi is an interesting and ancient divination plant. Although many different uses in folk medicine are reported from widely separated parts of its range, only in Mexico, as of yet, has the small shrub been valued as a euphoric/shamanistic herb. Many people report back seeing golden (sun like) aura's or glows around people after drinking sinicuichi brew.

It has mildly intoxicating properties, usually devoid of unpleasant after-effects, with a slight feeling of giddiness followed by euphoria characterized by a darkening of the partaker's surroundings, a great shrinking in size of the world around, auditory hallucinations, an altered sense of time and place, forgetfulness and a removal from a state of reality. Sounds seem to come distorted from a great distance. This plant typifies an hallucinogen of which the hallucinogenic characteristics are mainly auditory, but some visual effects also. The natives believe that sinicuichi has sacred or supernatural qualities, since they hold that it helps them recall events which took place many years earlier as if they had happened yesterday; others assert that they are able, with sinicuichi, to remember pre-natal events.

Syrian rue (*Peganum harmala*)

A woody perennial glabrous shrub with fleshy spikey-looking leaves, growing up to 1 meter tall which grows spontaneously in semi-arid conditions, steppe areas and sandy soils. Its small, brown seeds contains harmine and other harmala alkaloids, native to dry areas from the Middle East to India. This species has segmented leaves, produces small white flowers and reaches 0.5-1 meter in height.

Very large doses have hallucinogenic effects. Mostly used for its monoamine oxidase inhibiting properties. Increases the effects of any hallucinogen if taken shortly after drinking syrian rue extract. Studies suggest a role as an antimicrobial, vasorelaxant, antidepressant, analgesic, or cytotoxic agent, there have been many reports of intoxications following ingestion of specific amounts of P. harmala seeds, care should be taken regarding usage of this plant for therapeutic purposes.

Tarragon (*Artemisia dracunculus*)

Tarragon is a perennial shrub growing to 2 feet high that likes dry areas, full sun, and is found in comparable climates in many places in the world. The leaves of this herb is high in vitamins A and C and the leaves are believed to help stimulate the appetite.

A mild infusion is used to treat insomnia and hyperactivity. It stimulates that appetite and aids in digestion.

Thyme (*Thymus vulgaris*)

Thyme is a perennial that loves warm, sunny fields, and is found throughout North America. It grows to 15 inches tall, and makes an excellent ground cover on dry slopes. Trim it back after flowering to prevent it from becoming woody.

It is a powerful antiseptic. It is used in cases of anaemia, bronchial ailments, and intestinal disturbances. It is used as an antiseptic against tooth decay, and destroys fungal infections as in athlete's foot and skin parasites such as crabs and lice. It is good for colic, flatulence, and colds.

In folk medicine, it is burnt to purge and fumigate magical rooms and spaces, as well as to bring good health. Thyme in a sleeping pillow repels nightmares.

Tibetan or Indian Rhodiola (*Rhodiola sacra*)

Actions: anti-aging and anti-tumor.

This herb grows in the slopes of the Himalayas, known to be the supreme mountain range on the earth. This herb is indigenous to the arctic region, it grows in an unpolluted and cold endurance environment.

Indications: dysentery, depression, back pain, lung inflammation, painful and irregular menstruation, leucorrhoea, epidemic diseases, edema of the limbs, traumatic injury, and to heal burns and eliminate poisonous substances from the body.

Rhodiola has been used by Tibetans as a traditional herbal, anti-stress remedy and adaptogen for more than 1000 years. An adaptogen is defined as a substance that has no toxicity or side effects at normal dosages and that non-specifically increases the body's resistance to disease and to physical and chemical stresses. Put another way, use of Rhodiola as an adaptogenic herb will safely assist the body to combat stress by maintaining its homeostatic balance and in recovering from the effects of adverse weather, emotions, and disease influences. Furthermore, studies conducted at the Institute of Medical and Biological Problems at the Russian Federation Ministry of Health show that Rhodiola can help increase protein synthesis, remove ammonia from the blood and have a detoxifying effect while increasing blood supply to the muscles and the brain, and improving athletic performance. Dating back to 760 AD, Rhodiola was first used by Tibetans to preserve body health and treat disease. It is recorded and published in Tibetan manuscripts that Rhodiola can support vital energy, help the body to resist stress, enrich the blood, strengthen the brain, improve intelligence, and provide nourishment to preserve health. Rhodiola preparations have proved effective to reinforce physical strength, enhance body endurance, compensate hypoxia, relieve tiredness and weakness, improve efficiency of physical and mental work, treat cardiac and pulmonary diseases and supplement the chemotherapy and radiotherapy of cancers. Based on the indications for use, Rhodiola appears to be cooling and detoxifying, and vitalizes blood circulation. Also aids adaptation to high altitudes, thus, as a preventive and treatment for mountain sickness. Rhodiola has even been linked to increased creatine phosphate and ATP (energy) levels. All of this leads scientists to believe that supplemental Rhodiola may increase strength and lean body mass. Researchers speculate that Rhodiola helps reduce the stress that occurs secondary to exercise by stimulating the parasympathetic nervous system, which causes relaxation (as opposed to the sympathetic nervous system which causes stimulation).

Anti-aging Effect: Crenulatin, a chemical component of Rhodiola, has been found to inhibit the production of lipid peroxide in liver cells and save those cells from death. It can also enhance the activity of serum SOD (a natural anti-oxidant in the body and a key enzyme in the metabolism of oxygen), clear free radicals, and improve viral activity of cells. Crenulatin could retard the growth and proliferation of tumors from human lung tissue and reduce mortality rate. Radiation-resistance and Anti-tumor Effect: Rhodiola extract has a central stimulant action that increases the body's tolerance to microwave irradiation by protecting the hematopoietic system, which regulates the growth and formation of tumors. Rhodiola can help regulate brain function, leukocyte count, blood glucose, and promotes protein hydrolysis all the while enhancing the functions of the thyroid gland, adrenals, and ovaries.

Adjustment on Metabolism: The administration of a Rhodiola preparation could prolong the endurance of muscles and promote catabolism of proteins. Both Muscle proteins and glutamic acid levels have been shown to be enhanced when supplementing with Rhodiola. Rhodiola can prolong the loading time of muscles and the timely restoration of normal metabolism after extreme exercise.

Evidence has shown that Tibetan Rhodiola is, in many ways, even more powerful than Ginseng. It is a superb herb for people who work very hard, either physically or mentally. Tibetan Rhodiola has been shown to improve endurance and mental capacity, including memory enhancement. It has been shown to be a superb tonic and energizer for those into physical fitness, and a superior tonic for those stuck behind a desk or at a computer terminal for long periods of time. Consumption of even small amounts of Rhodiola extract significantly improves a person's capacity to absorb and utilize oxygen.

Turmeric (*Curcuma longa*)

It is an upright, relatively short and stout plant that rarely grows more than about 1 meter in height. Its leaves are elongated, dark green, and pointed, often curling slightly along the margins. The rhizomes grow to about 5-8 cm x 1.5 - 2.5cm. When bruised they omit a spicy scent. Turmeric propagates by means of rhizome segments. Originated in India, it is now common throughout Southeast Asia, China and southern Australia and it has been naturalized in all wet tropical regions of the world.

Actions: stimulant, carminative, alterative, vulnerary and antibacterial.

Indications: indigestion, poor circulation, cough, amenorrhea, pharyngitis, skin disorders, diabetes, arthritis, wounds and bruises.

Excellent antibiotic, while at the same time it strengthens digestion and improves intestinal flora. Not only purifies the blood, but also warms and stimulates formation of new blood tissue. Gives the energy of the Divine Mother and grants prosperity. It is effective for cleansing the chakras, and purifying the channels of the subtle body. It helps stretch the ligaments and is good for the practice of hatha yoga. Promotes proper metabolism in the body, correcting both excesses and deficiencies. Aids in the digestion of protein. Tonic to the skin.

Usnea (*Usnea lapponica*)

Usnea is a type of lichen that grows on trees in a combination of fungus and algae that grow together. Lichens grow in colorful, flat patches. Usnea may be whitish, reddish, or black.

Actions: Antibiotic, antibacterial, antimicrobial, expectorant and immune strengthener.

Indications: strep throat, tuberculosis and parasites.

Usnea or old man's beard as it is sometimes referred contains usnic acid, a natural antibiotic proven effective against gram positive bacteria, such as Streptococcus (strep throat), Staphylococcus (impetigo, *etc.*), Mycobacterium tuberculosis, and other fast growing species. Scientists think that usnic acid works by disrupting cellular metabolism, either by preventing the formation of ATP (the cells' energy source), or by the uncoupling of oxidative phosphorylization. Usnea may also be superior to the drug Flagyl (metronidazole). Usnea is effective against Trichomonas, a parasite which can cause among other things, a serious infection of the uterine cervix. Flagyl is widely prescribed for Trichomonas infection, but it is now believed to also cause cancer.

Uva Ursi (*Arctostaphylos uva-ursi*)

Uva Ursi rarely grows more than a few inches tall. It is best propogated from cuttings. It takes an unusually long time to root, so consider instead buying small plants from nurseries. It does poorly in rich soil, as it prefers poor soils in full sun. Once established, it spreads and becomes an attractive, hardy ground cover, surviving temperatures of -50.

Actions: diuretic, astringent and antiseptic.

Indications: bladder, kidney, and urinary tract infections, Bright's Disease, cystitis, diabetes, nephritis. Strengthens and tones urinary tract. Used for bladder and kidney infections. Urinary antiseptic properties. Urinary anesthetic. Urinary antibiotic. Crude extract possess some anti-cancer activity. For cystitis, used with apple juice. Uva Ursi strengthens and tones the urinary tract. It is especially useful for kidney infections, bladder infections, and inflammatory disease of the urinary tract. It is used as a diabetes remedy for excessive sugar in the blood. It is used for postpartum women to return the womb to its natural size, as well as to prevent infection of the womb after childbirth. It should not be used by pregnant women.

In folk medicine, it is added to sachets to increase psychic powers.

Valerian (*Valeriana officinalis*)

An upright perennial plant which can grow up to 3 feet tall. It prefers full sun, and average to rich well-drained soil. Root cuttings are best for propogation, and once the plants are established, they self-sow and spread by root runners. Valerian has a similar effect on cats as catnip, so you may need to protect your patch with chicken wire. Harvest roots for medicinal use in the fall of their second year,grows in both dry grassland and damp grassland habitats, but it does not tolerate permanently waterlogged conditions.

Actions: nervine, antispasmodic, sedative and carminative.

Indications: insomnia, hysteria, delirium, neuralgia, convulsions, epilepsy, vertigo, nervous cough, dysmenorrhea, palpitations, migrane, chronic skin diseases, flatulence, colic and menstrual cramps.

Exellent for vata disorders. Cleanses ama from the colon, blood, joints, and nerves. Grounding, calms muscle spasms, relieves mentrual cramps, stops fermentation in GI tract. Excessive use can dull the mind. Combines well with calamus.

Valerian is a relaxer, and is very effective for insomnia. It is often used as a tranquilizer, but it leaves no sluggish effects on the user. It is used for nervous tension, pain relieving, strengthening the heart, lowering blood pressure, IBS, diverticulosis, menstrual cramps, and for muscle spasms. It should not be taken over a long period of time, as it can cause mental depression in some people after long-term steady use. It is not habit forming.

In folk medicine, it is used to get fighting couples back together, in spells of love, and in purification baths.

Vervain (*Verbena officinalis*)

Vervain is a perennial herb that grows 1-2 feet tall. It prefers full sun, average to rich soils, and is grown throughout temperate North America. It is rather short-lived, but self-sows. Harvest leaves and flower tops as the plants bloom.

It is used to treat the liver and diseases related to the liver, exhaustion, fatigue, fever, insomnia, asthma, post-natal depression, as well as painful or irregular menses. It will also help increase the flow of a mother's milk. The Chinese use it to treat malaria, dysentery, and congestion. It is also a pain reliever and to reduce inflammation. Do not use during pregnancy.

In folk medicine, it is used for cleansing incenses and baths. Buried in a field, it will make your crops abundant. It is burned to attract wealth, and hung above a bed to prevent nightmares, and above a baby's crib to offer protection for the little one, and will enable the child to grow up with a love of learning and a happy outlook. Hung in the home it offers protection from negative spells, and is used as a pledge of mutual faith when given to a friend.

Violet (*Viola odorata*)

Violets are a perennial, prefer partial shade, average to rich well-drained soil, and grow to 8 inches tall.

Violet is effective in healing internal ulcers. It is used both internally and externally for pimples, abscesses, tumors, and swollen glands. It is useful in treating malignant growths as well. Native Americans soaked corn seed in an infusion of yellow violet to prevent insects from eating the seeds.

In folk medicine, it is keptin a pillow will help ease headaches away. Carrying the flowers brings a change in luck, and mixed with lavendar makes a powerful love sachet.

Vitex (*Vitex agnus-castus*)

Vitex is a small shrub or tree native to Asia and Europe. It has been introduced in the United States, and has naturalized throughout the southeast, as far north as Maryland. It typically grows 9 to 16 feet tall, but may grow larger in the warm areas of the deep south.

Vitex is a normalizing herb for the reproductive system. It is most commonly used in treating infertility, pre menstrual syndrome, menopausal problems, and hormonal imbalances. It is most effective when taken over a period of time, and in conjunction with other herbs used for the same purposes. It is also known as chaste tree fruit, or chaste berry.

In folk medicine, it's blossoms and branches were strewn in temples during festivals to honor Demeter. Hera is said to have been born under a Vitex tree. Vesta carried twigs of Vitex as symbols of purity.

Walnut [*Juglans nigra* (Black Walnut)]

Walnuts are trees that grow to 60 feet tall. They prefer full sun, deep and well-drained soil, and regular water. They grow well in areas such as the eastern and mid western United States.

Walnut bark is used to treat dysentery and skin diseases. The nut is used to promote strength and weight gain. The ground hull of the nut is used to treat skin diseases, herpes, head and body lice, and internal parasites. Walnut leaf is used to treat eczema, hives, and boils. Diluted walnut oil is used to treat dandruff. A strong decoction of walnut leaves, painted around doorways and woodwork, will repel ants.

In folk medicine, the nut still in its shell is carried to promote fertility.

Wild Dagga

The wild dagga is a robust shrub which grows up to 2-3m tall and 1.5m wide. It is common and widespread throughout South Africa and grows amongst rocks in grassland. Although this plant is not related to Cannabis, nor does it contain any of the same chemicals, the Hottentots of Africa consider it a wild form of that herb, and it acts in many similar ways. Bringing a calm, open state of lightness, Wild Dagga relaxes the self and brings levity and joyousness. This state allows a deeper connection and merging with astral energies, without the weight of fear or over-seriousness.

It is interesting that Wilde Dagga's medicinal uses are quite similar to Marijuana's.

In folk medicine, it is used to treat fevers, headaches, coughs, dysentery and many other conditions. It is also used as a remedy for snake bite and as a charm to keep snakes away.

Wild Yam (*Dioscorea villosa*)

This is a herbaceous vine found in moist woods, swamps, thickets and hedges, about 5-30' long that branches occasionally. Its slender stems have the capacity to twine around adjacent vegetation and fences, climbing upward and outward with glabrous stems of light green, pale yellow, reddish green, or dark red color. The distinctly veined leafs are decorated with blades tend to hang downward from their petioles.

Actions: nutritive tonic, aphrodisiac, rejuvenative, diuretic, antispasmodic and analgesic.

Indications: impotency, senility, hormonal deficiency, infertility, colic, nervous excitability, hysteria, abdominal pain and cramps, nausea, vomiting and morning sickness.

The most common use of wild yam is for the hormonal effects, as it contains DHEA, a hormone precursor. This constituent has usefulness for both men and women, as it can be made by the body into many other hormones, but it does have a particular slant towards progesterone, and so is especially useful for women. These progesterone compounds are also used for exhausted adrenal conditions, and it is comparable to licorice in that effect and its effect on balancing blood sugar levels. Also used for digestive system problems of a bilious or gaseous nature. Its alkaloid constituents act as strong relaxants to the stomach muscles and nerves, and so can be very useful in nausea, cramping, and vomiting. Stong rejuvenative powers. Increase semen, milk and other horomal secretions, and increases body weight, tonic for female reproductive system. Soothing and harmonizing effect on digestive organs.

Yarrow (*Achillea millefolium*)

Yarrow is a perennial, and its various varieties range from 8 inches to 3 feet tall. It prefers full sun, and average to poor dry soils.

Actions: diaphoretic, astringent, hemostatic, vulnerary and antispasmodic.

Indications: colds, flu, fever, infectious diseases, fever, gastritis, enteritis, mealses, menorrahagia, nosebleed, stomach ulcers, abcesses, hemoptysis, anorexia, indigestion, liver disorders, childbirth, tuberculosis, headaches and haemorrhoids.

It stops bleeding-internally and externally. Reduces excessive mentrual bleeding and relieves cramps. Reduces excess pitta, bile and inflammation in GI tract, strenghtens mucous membanes. It has calmative, nervine action and promotes clarity and perception, induces sweating.

Yarrow is used to stimulate and regulate the liver. It acts as a blood purifier and heals the glandular system. It has been used as a contraceptive, and as a part of diabetes treatment, as well as treating gum ailments and toothache. Also is used in formulas for treating colds, flus, and fevers. It arrests internal and external bleeding during childbirth. It is used to stop the bleeding of external wounds. Pregnant women should avoid this herb.

It is believed that Yarrow has the ability to keep a couple together for 7 years, so in folk medicine, it is used in love sachets as well as a gift to give to newlyweds. When worn it wards off negativity, and if held in your hand it repels fear. Yarrow added to the bath protects from harm.

Yerba Mansa (*Anemopsis californica*)

Perennial herb grows commonly in wet, especially sub alkaline places, especially somewhat alkaline or saline marshy places, below 2000 metres through cismontane and desert areas; with creeping rootstocks. It can grow in semi-shade (light woodland) or no shade. It prefers wet soil and can grow in water.

Actions: anti-inflammatory, anti-bacterial, anti-fungal, stimulant, tonic, astringent and cooling.

Indications: infections of the mouth, intestinal problems, urinary tract infections, arthritis, skin infections, malarial fevers, sore throats and sinus problems.

Has been used for centuries for poorly healing infections of the mouth, such as gum, mouth and throat sores; intestinal problems such as stomach and duodendal ulcers; urinary tract infections; and is useful for arthritis because it stimulates the excretion of uric acid and has an anti-inflammatory effect; when injury has caused inflammation and tissue congestion. Yerba Mansa is used to both directly shrink inflamed tissues and help transport the fluid buildup out of them, it also has aspirin-like effects. This spectrum of actions, as well as its benefit in removing built-up nitrogen-based wastes from the deep tissues and joints, make it a popular remedy for arthritis as well. Its antibacterial effects can also be helpful to long-standing or difficult intestinal infections.

Yerba Mate (*Ilex paraguariensis*)

Yerba mate grows as a shrub and then matures into a tree growing up to 15 meters tall in its native rainforest habitat.

Actions: stimulant, diuretic and tonic.

Indications: mental and physical fatigue, promotes weight loss, headaches, fatigue, nervous depression, rheumatic pains and hangovers.

Based on its caffeine content, Mat» is used as a centrally acting stimulant and as a diuretic. Since Mat» tea is greatly appreciated throughout much of South America as a "national drink", the people ascribe many other effects to it. It is not surprising then that in Europe as well Mat» is praised as "the green gold of the Indios", as a "natural remedy and magic drink" and especially as "the ideal slimming remedy" which facilitates losing weight in a natural way and stills the distressing feeling of hunger and thirst. Its richness in natural mineral salts - calcium, magnesium, iron, manganese, silica, phosphates - as well as vitamins, adds greatly to its usefulness.

Yerba Santa (*Eriodictyon californicum*)

It grows below 1600m elevation on dry rocky hillsides and ridges. It can be found in patches on chaparral slopes, forests, canyons, and along riverbanks.

Actions: expectorant, antispasmodic, carminative, decongestant, expectorant and bitter tonic.

Indications: coughs, congestion, asthma, allergies and hemorrhoids.

Yerba Santa is an expectorant used to treat coughs and congestion, as well as aiding in loosening and expelling phlegm. It works to dilate bronchial tubes, so it is useful to drink a cup of the tea to ease asthma and allergy attacks. Likewise advised for haemorrhoids and chronic catarrh of the bladder. Much used in California as a bitter tonic and a stimulating balsamic expectorant and is a most useful vehicle to disguise the unpleasant taste of quinine.

Yellow Dock (*Rumex crispus*)

It is common perennial weed found on waste ground, grassland and shingle beaches besides in agricultural lands. It is found on nearly all soil types, except highly acid soils.

Actions: alterative, astringent, laxative and antipyretic.

Indications: toxic conditions of the blood, skin eruptions, swollen glands, glandular tumors, venereal diseases, haemorrhoids, stomach acidity, psoriasis, herpes, eruptions, eczema and acne, anaemia, liver congestion and skin diseases.

General cleanser of the blood and lymph, good for most toxic conditions of the circulatory system. Reduces high Pitta, relieves toxic heat, and clears infections, thereby reducing pain and inflammation. Contains large amounts of iron, and can help to build the blood, but is mainly for a Pitta kind of anaemia (where bile thins the blood) in a Vata kind of anaemia it will only cause further weakness. Has value in clearing and promoting cerebral circulation. Treats liver and gallbladder disorders. Helps regulate bowel movement. Stimulates flow of bile. High thiamine content. Useful in treating the spleen, liver and lymphatic system. Used topically in salves and ointments for various skin disorders. Yellow Dock is a powerful blood purifier and astringent. It is used in treating all diseases of the blood and skin. It is very high in iron, making it useful for treating anaemia. It nourishes the spleen and liver, detoxifies the liver, and cleanses and enriches the blood.

Yohimbe (*Pausinystalia johimbe*)

It prefers rich soils in a protected part sun to shady position. Mostly found in central and west Africa where it grows upto 100 ft in height.

Actions: aphrodisiac and impotency tonic.

Indications: impotence, Parkinson's disease and lack of libido.

It has traditionally been used in Africa as an aphrodisiac, which contains hormone precursors, especially testosterone. It is also used for building muscle and endurance and it works to lower blood pressure and dilate the capillaries at the skin's surface, especially in the sex organs. Yohimbe is a rich source of alkaloids which can block the uptake of neuro-transmitters at Alpha-2 receptor sites and has also been used in dieting and weight management programs.

Yucca (*Yucca* spp.)

Yucca loves hot, dry, sunny locations. It is a perennial, and there are many varieties available to the home gardener.

Yucca is used to treat joint pain caused by arthritis, and to reduce inflammation in the joints. Shampoo made from the root is used to treat dandruff and other scalp conditions.

In folk medicine, it is used to protect home from evil influences. It can be used to make a soap and/or shampoo that is used to cleanse the body prior to rituals.

Wild Mint (*Mentha arvensis*)

Wild Mint is an erect or ascending native perennial forb of common mint family, grows in moist meadows and moist areas around marshes and streams. It grows from rhizomes and spreads by rhizome growth, forming colonies. Full sun is preferred, partial shade tolerated. Actions: stimulant, diaphoretic, carminative, nervine, analgesic

Indications: colds, fever, sore throat, laryngitis, earache, digestive upset, nervous agitation, headache and dysmenorrhoeal.

Mild soothing effect on the digestion and nerves, helps relax the body and clear the mind and senses. Contains large amounts of the element ether, whose action is soothing, cooling, clarifying, and expanding. Through this ethereal nature it helps relieve mental and emotional tension and congestion.

Wild Red Asparagus or Tian Men Dong (*Asparagus officinalis*)

Asparagus is a herbaceous, perennial plant growing to 100–150 cm tall in salt marshes, estuary banks and low-lying fields and meadows near the sea.

Actions: tonic.

Indications: dry skin, depression, sexual weakness, impotence and frigidity.

Wild Asparagus root is a marvelous Shen tonic and Yin tonic. It was credited by Chinese wise men as being able to open the heart, prolong life, and also to tonify the sexual functions of both men and women. It is traditionally used in Chinese herbalism as a major lung tonic. Prolonged consumption will make the skin soft, supple and smooth. In the art of radiant health, this kind of skin is a sure sign of attainment. Beautiful skin is the result of pure blood and healthy lungs. Asparagus root is useful for those who are experiencing dry skin due to a dry environment or due to internal dryness.

Wild Asparagus is said to open up the Heart Center, allowing Shen to flourish, manifesting as feelings of love, good will, patience and peace of mind. Regular consumption of good quality Asparagus root seems to lift a person's spirits in a way that is consistent with the Daoist philosophy of attaining happiness. If you take it for some time, you tend to see things from a broader view, indeed almost an unlimited view. Daoists mention that by consuming wild Asparagus root a person gains the ability to "fly." This flying is really the ability to rise above things that are limited and mundane, even if they seem very important at the time. This spiritual ability to "fly" is in fact the freedom of spirit one experiences when one has attained harmony with Dao and is guided by universal love. Those who are seeking spiritual attainment should consume good wild Asparagus whenever it is available.

Wild Asparagus root is said to strike a balance into the internal functions of the body. It promotes the production of Kidney Yin, and prolonged use is beneficial for sexual weakness. Even though its greatest value is in its "love tonic" attributes, wild Asparagus root is often used in tonics designed to overcome impotence or frigidity on the physical level.

The Three Great Traditional Healing Systems and Modern Western Medicine

	Unani	Ayurveda	Chinese	Modern Western
Origins	Persia, ca 980 AD	India, ca 2000 BCE	China, ca 2700 BCE	Europe, United States, late 19th century
Primary dynamic elements	ROh (spirit force)	Prana (life breath)	Chi (life energy)	Brain and heart
Disease correlates	Humors	Tridoshas	Yin-yang; chi	Named pathology
Disease causes	Imbalance of humoral temperament	Ama is the "harbinger of misery," the cause of disease	Systemic imbalances; no over-riding emphasis on one	Bacteria, viruses, fungi, parasites, metabolic disturbances, trauma
Basis of diagnosis	4 humors: blood, phlegm, yellow bile, black bile	Tridosha (3 humors): vata, pitta, kapha	4 methods of diagnosis of TCM; *i.e*, observation, auscultation, and olfaction; interrogation; pulse taking; and palpation	Based on patient's history, physical examination, laboratory testing
Diagnostic models	Restore balance to humors and organ systems	Concept of Shiva-Shakti; balance in the tridosha or 3 humors system	Achieve balance of yin (passive) and yang (active) physiological functions	Specifically named pathology
Chief diagnostic modality	Differential, mizāj or temperament assessed for each of the 4 humors	Differential, states of consciousness aligned with each of the 3 humors	Differential; questioning, observation, palpation, and listening; zang fu organ syndromes	Differential, named diseases
Diagnostic tests	Observation, life-style, pulse, urine, stool, and palpation	Tongue, pulse, urine, and palpation	Tongue, pulse, and palpation	Urinalysis, radiography, standard blood tests, sampling organ tissues, diagnostic X-rays, angiography (Note: In the United States, medical practitioners performed about $16 billion in laboratory diagnostic tests in 2011.6

Contd...

	Unani	Ayurveda	Chinese	Modern Western
Pulse diagnosis	Reveals humoral imbalance in organ system; the physician takes with 3 fingers at radial pulse of wrist; evaluates more than 1000 potential factors in seconds	Correlates pulse to the tridosha or 3 humors; the physician takes with the index finger; describes qualities of the pulse in terms of several animals	Direct manifestation of the circulatory energy of the body; classical 5-phase pulse correspondences; the physician takes on the wrist; about 40 per cent reliable as a sole diagnostic method by most TCM practitioners	Speed: fast pulse vs slow pulse
Elements of nature	4: fire, air, water, earth	5: fire, earth, water, air, ether	5: fire, earth, metal, water, wood	22 basic elements of chemistry
Main dietary influences	Nonalcoholic; regular fasting; nonporcine	Vegetarian	Rice and vegetables	Refined sugars, alcohol, fats, drugs
Patient's participation and will	Empower patient to make changes in diet and lifestyle	High objectivity	Personal determination	Not significant
Deity of system	Abrahamic Monotheism, primarily God (Allah) of Islam	Polytheistic, pantheistic, non-dualistic, gods of Hinduism	Nontheistic Confucianism, Taoism, Buddhism	Secular atheism, agnosticism, modern evolutionary nihilism
Primary treatment modalities	Diet, herbs, fasting, cupping, purgation, baths, attars (essential oils or medicinal scented perfumes)	Panchakarma (detoxification), 7 herbs, diet, emetic therapies	Acupuncture, herbs, cupping, moxibustion, diet	Chemotherapy, radiation therapy, pharmaceutical drugs, surgery, rehabilitative physical therapy
Primary treatment objective	Mizan: restore to balance, provoke the "healing crisis"	Clear the entire gastrointestinal tract, regulate the bowels, improve digestion	Tonification of energy	Symptom suppression, killing germs and bacteria, palliative end-of-life management
Some instruments used	Glass cups	Glass cups	Glass cups, acupuncture needles	Ophthalmoscopes, laryngo-scopes, radiographs, sphyg-momanometers, electrocardiograms, chemical tests of body fluids and tissues
Side effects	Overdose of herbal substances; very rare	Overdose of herbal substances; very rare	Potential for acute symptoms from improper needling techniques or over-dose of herbal substances; all very rare	106 000 die annually from improper medications and from severe and frequent drug reactions; very common

Contd...

Contd...

	Unani	Ayurveda	Chinese	Modern Western
Chief complaints	Nonregulation of practitioners; lack of clinics	Nonregulation of practitioners; lack of clinics	Obtuse language	Adverse reactions; patients' dissatisfaction; skyrocketing medical costs
Direction of development	Training practitioners in powers of observation; building schools; sources for formulations	Training practitioners; building schools; developing formulations	Integration with Western hospital medicine	Higher costs; more complex diagnostics; genetic medicine; new health-care reform laws
Typical cost of treatment, USD	$15-200	$150-200	$45-300	$200-4000

Abbreviations: TCM, traditional Chinese medicine.

Glossary of English Terms

Alterative- tending to restore normal health, cleanses and purifies the blood; alters existing nutritive and excretory processes, gradually restoring normal body processes.

Amenorrhea-absence or suppression of menstration.

Anabolic- contructive phase of metabolism; building up (growth and repair) of body substance

Analgesic- relieves pain

Anthelmintic- helps destroy and dispel parasites including worms, bacteria, fungus and yeast

Antibiotic- inhibits growth of or destroys micro-organisms

Antipyretic-dispels heat, fire, and fever

Antispasmodic-relieves spasms of voluntary and involuntary muscles

Aperiant-mild laxative

Aphrodisiac-reinvigorates the body by reinvigorating the sexual organs

Aromatic- herbs which contain volatile essential oils which aid digestion and relieve gas.

Astringent- firms tissues and organs; reduces discharges and secretions

Bitter tonic- bitter herbs, which in small amounts stimulate digestion and otherwise help regulate fire in the body

Carminative-relieves intestinal gas, pain and distention; promotes peristalsis

Catabolic-destructive phase of metabolism

Cathartic-strong laxative which causes rapid evacuation

Demulcent- soothes, protects and nurtures internal membranes

Deobstruent- removes body obstructions

Diaphoretic-causes perspiration and increased elimination through the skin

Diuretic- promotes activity of kidney and bladder and increases urination

Dysmenorrhea-painful or difficult menstruation

Dyspnea-difficulty in breathing

Emetic- induces vomiting

Emmenagogue- helps promte and regulate menstraution

Emollient- soothes, softens and protects the skin

Enteritis- inflammation of the small intestine

Epistaxis-nose bleed

Expectorant-promotes discharge of phlegm and mucus from the lungs and throat

Febrifuge-reduces fever

Gastritis-inflammation of the stomach

Hemostatic-stops the flow of blood, type of astringent that stops internal bleeding or hemmorhaging

Laxative-promotes bowel movements

Lithoptriptic-substance that dissolves and discharges gall bladder and urinary stones and gravel

Malabsorption- inadequate absorption of nutrients from the intestinal tract

Menorrhagia- excessive bleeding during mestruation

Nephritis-inflammation of the kidney

Nervine-strengthens functional activity of the nervous system; may be stimulants and sedatives

Neurasthenia-severe nerve weakness; nervous exhaustion

Nutritive tonic- promotes weight and density and nourishes the body

Refrigerant-reduces body temperature and relieves thirst

Rejuvenative-prevents decay, postpones aging, revitalizes the organs

Salt/Essence level- Salt is the lowest and most dense level, the level of solidity, the body and the physical being. In Alchemical healing, the Salt level corresponds to the physical body of a person, different for each of us, and reflecting in a material way the more ethereal aspects of our selves. This is not always the level of the root causes of illness, but it is often the level at which symptoms manifest. The planetary rulership of the Salt level gives insights into the physical qualities, weaknesses, and disease tendencies that you have.

Sedative-calms or tranquilizes by lowering functional activity of organ or body part

Stimulant- increases internal heat, dispels internal chill, and strengthens metabolism and circulation

Stomachic-strengthens stomach function

Urticaria-skin condition characterized by itching, welts or hives

Vasodilator-causes relaxation of the blood vessels

Vermicidal-kills parasites in the intestines

Vulnerary-assists in healing of wounds by protecting against infection and stimulating cell growth

Glossary of Sanskrit Terms

Agni-biological fire governing metobolism; cosmic force of transformation

Ama- toxins; undigested food or unelimimated waste materials

Kapha (K)-the bodily water humour; earth and water elements in the body

Pitta (P)-the bodily fire humour; fire and water elements in the body

Sattva-principle of light, perception, intelligence and harmony

Vata (V)- the bodily air humour; air and ether in the body

Yoga- a methodology of the practical and coordinated application of knowledge; spiritually, the science of self-realization

Bibliography

1. American Folk Medicine by Clarence Meyer, Meyerbooks publisher, Glenwood, Illinois, 1973.

2. Back to Eden, by Jethro Kloss; Back to Eden Publishing Co., Loma Linda, CA, 1994.

3. How Indians Use Wild Plants for Food, Medicine and Crafts, by Frances Densmore, Dover Publications, Inc., NY,1974.

4. Chinese Medicinal Herbs, by Shih-Chen Li, Georgetown Press, San Francisco, CA, 1973.

5. Culpeper's Complete Herbal and English Physician by Nicholas Culpeper, Meyerbooks, publisher, Glenwood, Illinois 1990.

6. Eastern/Central Medicinal Plants, by Steven Foster and James A. Duke., Houghton Mifflin Company, Park Avenue South, NY.

7. The Herb Book, by John Lust, Bantam Books, NY, 1974.

8. Indian Herbalogy of North America, by Alma R. Hutchens, Shambala Publications, Inc., Boston, Massachusetts, 1973.

9. Planetary Herbology, by Michael Tierra, C.A., N.D., O.M.D., Lotus Press, Twin Lakes. WI, 1992.

10. The Herbalist Almanac, by Clarence Meyer, Meyerbooks, publisher, Illinois, 1994.

11. The Rodale Herb Book: How to Use, Grow, and Buy Nature's Miracle Plants (An Organic gardening and farming book), edited by William H. Hylton, Rodale Press, Emmaus, PA, 1974.

12. A Useful Guide to Herbal Health Care, HCBL, Naples, FL,1996.

13. The Yoga of Herbs: An Ayurvedic Guide to Herbal Medicine, by Dr. David Frawley and Dr. Vasant Lad, Lotus Press, 1988.

14. 'Varahamihira's – Brihat Samhita', by M. Ramakrishna Bhat; Publishers : Motilal Banarsidass Publishers Pvt. Ltd, Delhi, 1997.

15. 'Brihat Parasara Hora Sastra' by R. Santhanam; Publishers: Ranjan Publications, Delhi, 1995.

16. Sri Sarvartha Chintamani by Sri Venkatesa Daivagna Trans by B. Suryanarain Rao; Publishers: by Motilal Banarsidass Pvt. Ltd; Delhi, 1997.

17. 'Saravali' by Kalyana Varma,Trans and Commentary by R. Santhanam; Publishers: Ranjan Publication, Delhi, 1996.

18. 'Phala Deepika' by Sri Mantreswar; Trans by G.S Kapoor; Publishers: Ranjan Publications, Delhi, 1996.

19. 'Kala Prakasika' by N.P. Subramaniya Iyer; Publishers: Asian Educational services, New Delhi, 1991.

20. 'Esoteric Principles of Vedic Astrology' by Bepin Behari; Publishers: Ranjan Publications, New Delhi, 1996.

21. 'Muhurta Chintamani' of Daivagnya Acharya Shri Ram; Trans by G.C. Sharma; Publishers: Sagar Publications, New Delhi, 1996.

22. 'Light on Life- An Introduction to the Astrology of India' by Hart de Fouw, and Robert Svoboda; Publishers: Penguin Books, New Delhi, 1996.

23. 'How to Judge a Nativity' by Alan Leo; Publishers: Sagar Publications, New Delhi, 1970.

24. 'The Art of Synthesis' by Alan Leo; Publishers: Sagar Publications; New Delhi, 2000.

25. 'The Healing Power of Gemstones' by Harish Johari; Publishers: Destiny Books, Vermont, 1996.

26. 'Navagraha Temples of Tamil Nadu (Kaveri Delta)' by Padma Raghavan and Savitri Narayan; English Edition Publishers and Distributors India, Pvt. Ltd, Mumbai, 2004.

27. 'Astrology of the Seers- A guide to Vedic/Hindu Astrology' by Dr David Frawley (Pandit Vamadeva Shastri); Publishers: Motilal Banarsidass Pvt. Ltd, Delhi, 2004.

28. 'Ayurvedic Astrology' by Dr David Frawley; Publishers: Motilal Banarsidass Publishers Pvt. Ltd, Delhi, (2007).

29. 'The Oracle of Rama' by Goswami Tulsidas, Commentary by Dr David Frawley; Publishers: Motilal Banarsidass Pvt. Ltd, Delhi, 1999.

30. 'The Nakshatras' by Dennis M. Harness; Publishers: Motilal Banarsidass Pvt. Ltd, Delhi, 2000.

31. 'The Seven Human Temperaments' by Geoffrey Hodson; Publishers: The Theosophical Publishing House; Adyar, Chennai, 2001.

32. 'Ayurvedic Healing'- A Comprehensive Guide, by Dr David Frawley; Motilal Benarsidass Pvt. Ltd, Delhi, 1995.

33. 'Ancient Hindu Astrology for the Modern Western Astrologer' by James.T. Braha; Hermetician Press, Miami, 1993.

34. 'The Art and Practice of Ancient Hindu Astrology' by James.T. Braha; Hermetician Press, Miami, 2001.

35. 'Systems Approach for Interpreting Horoscopes' by Prof. V.K. Choudhry and Rajesh Chaudhary; Sagar Publications, Pvt. Ltd, New Delhi, 2006.

36. Healthy People 2000: National Health Promotion and Disease Prevention Objectives. U.S. Department of Health and Human Services, Public Health Service. Washington, DC, U.S. Government Printing Office.

37. Gelb M. Body Learning: An Introduction to the Alexander Technique. New York, 1994, Henry Holt and Company.

38. Eisenberg DM, Davis RB, Ettner SL, *et al.,* Trends in alternative medicine use in the United States, 1990-1997. JAMA. 1998;280:1569-1575.

39. Bartner S, Knapp C, Vineyard M. Defining the Technique and our profession. ASAT News. 1999;44:7-9.

40. De Alcantara P. Indirect Procedures: A Musician's Guide to the Alexander Technique. Oxford: Clarendon Press;1997.

41. Tinbergen N. Ethology of Stress Diseases.,Nobel Laureate Lecture, England Science 185:20-27, 1974.

42. Wasley D, Taylor A. Changes in psycho-physiological response to a musical performance after 16 weeks of aerobic exercise and Alexander technique training. *J Sports Sci* 21:292-293, 2003.

43. Bagchi, K. (Ed.) Music, Mind and Mental Health New Delhi: Society for Gerontological Research.

44. Crandall, J. 1986 Self-transformation through Music, New Age, New Delhi.

45. Predictive Astrology of the Hindus, by Pandit Gopesh Kumar Ojha, Taraporevala Sons and Co, Bombay.

46. An Overview of Ayurveda, Yoga and Naturopathy, Unani, Siddha, and Homoeopathy (AYUSH) in India, (New Delhi: Dept. of Ayurveda, Yoga and Naturopathy, Unani, Siddha, and Homeopathy, Ministry of Health and Family Welfare, Government of India, 2003).

47. Materia Medica and Drug Pathogenesy by Debasish Kundu, IBPS, New Delhi, 1993.

48. Diseases of Women By Shefalika Kundu, Debasish Kundu, IBPS, New Delhi, 1993.

49. Aids and Homeopathy – by Debasish Kundu, S.Azizur Rahman, IBPS, New Delhi, 1995.

50. Pharmacodynamics in Electrohomoepathy by Debasish Kundu, B Jain, New Delhi, 1992.

51. Materia Medica for the New Age Man, Debasish Kundu, A.Jayasuriya, M Ahmed, Originals, New Delhi, 2015.

52. Essentials of Acupuncture, Debasish Kundu, Dr.Choo Led Sin, Originals, New Delhi, 2016.

Index

N

O

P

www.ingramcontent.com/pod-product-compliance
Lightning Source LLC
Chambersburg PA
CBHW050505190326
41458CB00005B/1443